Plant Biotechnology and Sustainable Agriculture: Bridging the Gap for Global Food Security

CABI BIOTECHNOLOGY SERIES

Biotechnology, in particular the use of transgenic organisms, has a wide range of applications including agriculture, forestry, food, and health. There is evidence that it could make a major impact in producing plants and animals that are able to resist stresses and diseases, thereby increasing food security. There is also potential to produce pharmaceuticals in plants through biotechnology, and provide foods that are nutritionally enhanced. Genetically modified organisms can also be used in cleaning up pollution and contamination. However, the application of biotechnology has raised concerns about biosafety, and it is vital to ensure that genetically modified organisms do not pose new risks to the environment or health. To understand the full potential of biotechnology and the issues that relate to it, scientists need access to information that not only provides an overview of and background to the field, but also keeps them up to date with the latest research findings.

This series, which extends the scope of CABI's successful "Biotechnology in Agriculture" series, addresses all topics relating to biotechnology including transgenic organisms, molecular analysis techniques, molecular pharming, in vitro culture, public opinion, economics, development, and biosafety. Aimed at researchers, upper-level students, and policy makers, titles in the series provide international coverage of topics related to biotechnology, including both a synthesis of facts and discussions of future research perspectives and possible solutions.

Titles Available

1. Animal Nutrition with Transgenic Plants
   Edited by G. Flachowsky

2. Plant-derived Pharmaceuticals: Principles and Applications for Developing Countries
   Edited by K.L. Hefferon

3. Transgenic Insects: Techniques and Applications
   Edited by M.Q. Benedict

4. Bt Resistance: Characterization and Strategies for GM Crops Producing *Bacillus thuringiensis* Toxins
   Edited by Mario Soberón, Yulin Gao and Alejandra Bravo

5. Plant Gene Silencing: Mechanisms and Applications
   Edited by Tamas Dalmay

6. Ethical Tensions from New Technology: The Case of Agricultural Biotechnology
   Edited by Harvey James

7. GM Food Systems and their Economic Impact
   Tatjana Brankov and Koviljko Lovre

8. Endophyte Biotechnology: Potential for Agriculture and Pharmacology
   Edited by Alexander Schouten

9. Forest Genomics and Biotechnology
   Edited by Richard Meilan and Matias Kirst

10. Transgenic Insects: Techniques and Applications, 2nd Edition
    Edited by M.Q. Benedict and Maxwell J. Scott

11. Plant Omics: Advances in Big Data Biology
Edited by Hajime Ohyanagi, Eiji Yamamoto, Ai Kitazumi and Kentaro Yano

12. Next-generation Sequencing and Agriculture
Edited by Phillip Bayer and Dave Edwards

13. Aquaculture and Fisheries Biotechnology: Genetic Approaches
Edited by Rex A. Dunham

14. Algal Biotechnology
Edited by Qiang Wang

15. Genetic Engineering of Vegetable Crops
Edited by Pritam Kalia

16. Plant Breeding Technology: Future Trends and Challenges
Edited by Zeba Khan

17. Stress-resilient Crops: Coordinated Omics-CRISPR-Nanotechnology Strategies
Edited by Jen-Tsung Chen

18. High-throughput Plant Metabolomics
Edited by Jen-Tsung Chen

19. Plant Biotechnology and Sustainable Agriculture: Bridging the Gap for Global Food Security
Edited by Tariq Aftab

Plant Biotechnology and Sustainable Agriculture: Bridging the Gap for Global Food Security

Edited by

Tariq Aftab

CABI

CABI is a trading name of CAB International

CABI
Nosworthy Way
Wallingford
Oxfordshire OX10 8DE
UK

Tel: +44 (0)1491 832111
E-mail: info@cabi.org
Website: www.cabi.org

CABI
200 Portland Street
Boston
MA 02114
USA

T: +1 (617)682-9015
E-mail: cabi-nao@cabi.org

A catalogue record for this book is available from the British Library, London, UK.

ISBN-13: 9781800627635 (hardback)
9781800627642 (ePDF)
9781800627659 (ePub)

DOI: 10.1079/9781800627659.0000

Commissioning Editor: Rebecca Stubbs
Editorial Assistant: Emma McCann
Production Editor: James Bishop

Typeset by Exeter Premedia Services Pvt Ltd, Chennai, India
Printed and bound in the UK by CPI Group (UK) Ltd, Croydon, CR0 4YY

Contents

Contributors ix

About the Editor xv

Preface xvii

1 **Plant Biotechnology and the Quest for Sustainable Agriculture: An Introduction** 1
Kaiser Iqbal Wani, Abdullah M. Naeem and Tariq Aftab

2 **Harnessing Genetic Engineering in Crop Improvement: Towards Resilient Agriculture** 8
Iqra Noor, Hamza Sohail and Muhammad Tanveer Akhtar

3 **Biotechnological Tools to Improve Soil Health and Fertility** 28
Silvina M.Y. Lopez, Virginia Martínez Alcántara, Juan M. Reparaz and Pedro A. Balatti

4 **Harnessing Plant–Microbial Interaction for Sustainable Crop Production Using Biotechnology** 43
Iqra Owais, Tahleel and Muhammad Khashi u Rahman

5 **Metabolic Engineering of Tea for Enhanced Health Benefits** 54
Misbah Naz, Muhammad Rahil Afzal, Adalberto Benavides-Mendoza, Kashif Akhtar, Nazeer Ahmed and Shi Yuanzhi

6 **Molecular Hydrogen Treatments in Food Supplies: Enhanced Growth, Better Yields, and Prolonged Postharvest Storage?** 67
John T. Hancock and Jennifer May

7 **Application of Biotechnologies for Crop Improvement and Food Production: Case Studies From the BIOALI Network** 83
Sandra Sharry, Lorenzo Burgos, Daniela Dalzotto, Patricia Boeri, Lucrecia Piñuel and Iselen Trujillo

8 **Development of Doubled Haploid Wheat Genotypes through Intergeneric Hybridization with Maize** 96
Md. Yeakub Khan, Sabrina Parvin Mita and Akbar Hossain

9 **Bioeconomy and Biotechnology: Innovative Strategies in the Utilization of Resources in the Argentine North Patagonia** 113
Daniela Dalzotto, Miguel José Morón Rivera, Luciana Di Sario, Agustina Cabrera, Mariana Dalponte, Sandra Sharry and Patricia Boeri

10 **Biofortification: Improving Nutritional Content for Human Health and Food Security** 126
Rakesh Kumar Bairwa, Om Prakash Gupta, Arun Gupta, Antim Kundu, Anamika Pandey, Dharmendra Kumar and Mohd. Kamran Khan

11 **Sustainable Crop Management: Agroecology of the Future** 151
Braulio Edgar Herrera-Cabrera, Rafael Salgado Garciglia, Alejandra García Hernández, Adriana Delgado Alvarado, Luis Germán López Valdez, Francisco Javier Verduzco Miramón and Hebert Jair Barrales-Cureño

12 **Bio-Inoculants as Biological Tools to Mitigate Water Deficit in Crops** 165
Urley A. Pérez-Moncada, Pablo Cornejo, Christian Santander, Cledir Santos, Humberto Aponte, Sivagnanam Silambarasa and Gladys Vidal

13 **Genetic Engineering of Plants for Remediation of Toxic Metals and Metalloids** 185
Syeda Gul Nisa, Maria Hameed, Ishmal Eman, Noor Fatima, Shahab Sadiq, Sawaira Rafique, Zeeshan Khan and Tariq Shah

14 **Oxidative Stress Generation and Antioxidant Defense Response in Plants During Abiotic Stress** 205
Faamiya Shajar, Azha Ufaq Nabi, Aamina Manzoor, Seerat Saleem, Naveed Ul Mushtaq and Reiaz Ul Rehman

15 **Plant Defense Mechanisms: Innate and Induced Immunity in Plants** 221
Sucharita Mohapatra, Lellapalli Rithesh, Rahul Kumar Tiwari, Laxmipriya Behera Milan Kumar Lal and Ravinder Kumar

16 **Alley Cropping: A Sustainable Agroforestry System for Enhancing Productivity, Soil Health and Biodiversity** 235
Sagar Maitra, Masina Sairam, Lalichetti Sagar, Sumit Ray, D.T. Santosh, Dinkar Jagannath Gaikwad, Anjana J. Atapattu and Harun I. Gitari

17 **Regulatory and Ethical Issues for Plant Biotechnology and Genetic Engineering** 249
Ankur Singh and Aryadeep Roychoudhury

Index 257

Contributors

Abdullah, Department of Botany, Aligarh Muslim University, Aligarh, India

Tariq Aftab, Department of Botany, Aligarh Muslim University, Aligarh, India

Muhammad Rahil Afzal, Key Laboratory of Green Pesticide and Agricultural Bioengineering, Ministry of Education, Guizhou University, Guiyang, China

Nazeer Ahmed, State Key Laboratory of Green Pesticide, Guiyang, China

Kashif Akhtar, State Key Laboratory for Conservation and Utilization of Subtropical Agro-bioresources, Guangxi Key Laboratory of Sugarcane Biology, College of Life Science and Technology, Guangxi University, Nanning, China

Muhammad Tanveer Akhtar, School of Horticulture and Landscape Architecture, Yangzhou University, Yangzhou, China

Virginia Martínez Alcántara, Centro de Investigaciones de Fitopatología (CIDEFI CIC-UNLP), Córdoba, Argentina

Adriana Delgado Alvarado, Colegio de Postgraduados, Campus Puebla, Puebla, México

Humberto Aponte, Departamento de Ciencias Químicas y Recursos Naturales, Universidad de La Frontera, Temuco, Chile

Anjana J. Attapattu, Agronomy Division, Coconut Research Institute of Sri Lanka, Lunuwila, Sri Lanka

Rakesh Kumar Bairwa, ICAR-Indian Institute of Wheat and Barley Research, Karnal, India

Pedro A. Balatti, Comisión de Investigaciones Científicas de la Provincia de Buenos Aires (CICPBA), La Plata, Argentina

Hebert Jair Barrales-Cureño, Tecnológico Nacional de México, Zamora, México

Laxmipriya Behera, ICAR-National Rice Research Institute, Cuttack, India

Adalberto Benavides-Mendoza, Department of Horticulture, Antonio Narro Autonomous Agrarian University, Saltillo, Mexico

Patricia Boeri, Universidad Nacional de Río Negro, Viedma, Argentina

Lorenzo Burgos, Fruit Tree Biotechnology Group – Plant Breeding Department, CEBAS-CSIC, El Puntal, Spain

Agustina Cabrera, CIT-RIO NEGRO, Sede Viedma, Universidad Nacional de Río Negro (UNRN-CONICET), Viedma, Argentina

Pablo Cornejo, Núcleo Científico Tecnológico en Biorecursos (BIOREN-UFRO), Universidad de La Frontera, Temuco 4811230, Chile and Centro de Recursos Hídricos para la Agricultura y la Minería (CRHIAM), Centro ANID Fondap, Victoria 1295, Concepción, Chile

Mariana Dalponte, CIT-RIO NEGRO, Sede Viedma, Universidad Nacional de Río Negro (UNRN-CONICET), Viedma, Argentina

Daniela Dalzotto, Universidad Nacional de Río Negro, Viedma, Argentina

Luciana Di Sario, CIT-RIO NEGRO, Sede Viedma, Universidad Nacional de Río Negro (UNRN-CONICET), Viedma, Argentina

Ishmal Eman, Department of Plant Biotechnology, Atta-ur-Rahman School of Applied Biosciences, National University of Sciences and Technology, Islamabad, Pakistan

Noor Fatima, Department of Plant Biotechnology, Atta-ur-Rahman School of Applied Biosciences, National University of Sciences and Technology, Islamabad, Pakistan

Dinkar Jagannath Gaikwad, Research Centre for Smart Agriculture, Centurion University of Technology and Management, Paralakhemundi, India

Rafael Salgada Garciglia, Instituto de Investigaciones Químico Biológicas, Universidad Michoacana de San Nicolás de Hidalgo, Morelia, México

Harun I. Gitari, Department of Agricultural Science and Technology, School of Agriculture and Environmental Sciences, Kenyatta University, Nairobi, Kenya

Arun Gupta, ICAR-Indian Institute of Wheat and Barley Research, Karnal, India

Om Prakash Gupta, ICAR-Indian Institute of Wheat and Barley Research, Karnal, India

Maria Hameed, Department of Botany, Faculty of Sciences, PMAS Arid Agriculture University, Rawalpindi, Pakistan

John T. Hancock, School of Applied Sciences, University of the West of England, Bristol, UK

Alejandra García Hernández, Instituto de Investigaciones Químico Biológicas, Universidad Michoacana de San Nicolás de Hidalgo, Morelia, México

Braulio Edgar Herrera-Cabrera, Colegio de Postgraduados, Campus Puebla, Puebla, México

Akbar Hossain, Soil Science Division, Bangladesh Wheat and Maize Research Institute, Dinajpur, Bangladesh

Md. Yeakub Khan, Biotechnology Division, Bangladesh Wheat and Maize Research Institute, Dinajpur, Bangladesh

Mohd. Kamran Khan, Department of Soil Science and Plant Nutrition, Faculty of Agriculture, Selcuk University, Turkiye

Zeeshan Khan, Department of Plant Biotechnology, Atta-ur-Rahman School of Applied Biosciences, National University of Sciences and Technology, Islamabad, Pakistan

Dharmendra Kumar, ICAR-Central Potato Research Institute, Shimla, India

Ravinder Kumar, ICAR-Indian Agricultural Research Institute, New Delhi, India

Antim Kundu, ICAR-Central Soil Salinity Research Institute, Karnal, India

Milan Kumar Lal, ICAR-National Rice Research Institute, Cuttack, India

Silvina M.Y. Lopez, Consejo Nacional de Investigaciones Científicas y Técnicas (CONICET), Godoy Cruz, Argentina

Sagar Maitra, Department of Agronomy and Agroforestry, M. S. Swaminathan School of Agriculture, Centurion University of Technology and Management, Paralakhemundi, India

Aamina Manzoor, Department of Bioresources, School of Biological Sciences, University of Kashmir, Srinagar, India

Jennifer May, School of Applied Sciences, University of the West of England, Bristol, UK

Francisco Javier Verduzco Miramón, Tecnológico Nacional de México, Zamora, México

Sabrina Parvin Mita, Biotechnology Division, Bangladesh Wheat and Maize Research Institute, Dinajpur, Bangladesh

Sucharita Mohapatra, Department of Plant Pathology, College of Agriculture, Odisha University of Agriculture and Technology, Bhubaneswar, India

Naveed Ul Mushtaq, Department of Bioresources, School of Biological Sciences, University of Kashmir, Srinagar, India

Ahza Ufaq Nabi, Department of Bioresources, School of Biological Sciences, University of Kashmir, Srinagar, India

M. Naeem, Department of Botany, Aligarh Muslim University, Aligarh, India

Misbah Naz, State Key Laboratory of Green Pesticide, Guiyang, China

Syeda Gul Nisa, Department of Plant Biotechnology, Atta-ur-Rahman School of Applied Biosciences, National University of Sciences and Technology, Islamabad, Pakistan

Iqra Noor, School of Horticulture and Landscape Architecture, Yangzhou University, Yangzhou, China

Iqra Owais, Department of Botany, Emerson University, Multan, Pakistan

Anamika Pandey, Department of Soil Science and Plant Nutrition, Faculty of Agriculture, Selcuk University, Turkiye

Urley A. Pérez-Moncada, Doctorado en Ciencias de Recursos Naturales, Universidad de La Frontera, Temuco, Chile

Lucrecia Piñuel, CIT-RIO NEGRO, Universidad Nacional de Río Negro, Viedma, Argentina

Sawaira Rafique, Department of Plant Biotechnology, Atta-ur-Rahman School of Applied Biosciences, National University of Sciences and Technology, Islamabad, Pakistan

Muhammad Khashi u Rahman, Institute for Agribiotechnology Research (CIALE) and Microbiology and Genetics Department, University of Salamanca, Salamanca, Spain

Sumit Ray, Department of Agronomy and Agroforestry, M. S. Swaminathan School of Agriculture, Centurion University of Technology and Management, Paralakhemundi, India

Reiaz Ul Rehman, Department of Bioresources, School of Biological Sciences, University of Kashmir, Srinagar, India

Juan M. Reparaz, Cátedra de Microbiología Agrícola – Facultad de Ciencias Agrarias y Forestales – Universidad Nacional de La Plata, La Plata, Argentina

Lellipalli Rithesh, Department of Plant Pathology, College of Agriculture, Kerala Agricultural University, Thiruvananthapuram, India

Miguel José Morón Rivera, Universidad Nacional de Río Negro (UNRN), Sede Atlántica. Viedma, Argentina

Aryadeep Roychoudhury, Discipline of Life Sciences, School of Sciences, Indira Gandhi National Open University, New Delhi, India

Shahab Sadiq, Department of Plant Biotechnology, Atta-ur-Rahman School of Applied Biosciences, National University of Sciences and Technology, Islamabad, Pakistan

Lalichetti Sagar, Department of Agronomy and Agroforestry, M. S. Swaminathan School of Agriculture, Centurion University of Technology and Management, Paralakhemundi, India

Masina Sairam, Department of Agronomy and Agroforestry, M. S. Swaminathan School of Agriculture, Centurion University of Technology and Management, Paralakhemundi, India

Seerat Saleem, Department of Bioresources, School of Biological Sciences, University of Kashmir, Srinagar, India

Christian Santander, Departamento de Ciencias Químicas y Recursos Naturales, Universidad de La Frontera, Temuco, Chile

Cledir Santos, Departamento de Ciencias Químicas y Recursos Naturales, Universidad de La Frontera, Temuco, Chile

D.T. Santosh, Research Centre for Smart Agriculture, Centurion University of Technology and Management, Paralakhemundi, India

Tariq Shah, Department of Agronomy, Faculty of Crop Production Sciences, University of Agriculture, Peshawar, Pakistan

Faamiya Shajar, Department of Bioresources, School of Biological Sciences, University of Kashmir, Srinagar, India

Sandra Sharry, National Research System (SNI), Panama and Universidad Nacional de Río Negro, Viedma, Argentina

Shi Yuanzhi, Key Laboratory of Tea Biology and Resource Utilization of Tea, Tea Research Institute, Chinese Academy of Agricultural Sciences, Hangzhou, China

Sivagnanam Silambarasan, Laboratorio de Ecología Microbiana del Suelo y Biogeoquímica, Instituto de Ciencias Agroalimentarias, Pecuarias y Ambientales (ICA3), Universidad de O'Higgins, San Fernando, Chile

Ankur Singh, KIIT-Technology Business Incubator (KIIT-TBI), KIIT University, Bhubaneswar, India

Hamza Sohail, School of Horticulture and Landscape Architecture, Yangzhou University, Yangzhou, China

Tahleel, Department of Botany, Emerson University, Multan, Pakistan

Rahul Kumar Tiwari, ICAR-Indian Institute of Sugarcane Research, Lucknow, India

Iselen Trujillo, Centro de Estudios para el Desarrollo Agroecológico Tropical (CEDAT), Instituto de Estudios Científicos y Tecnológicos (IDECYT), Universidad Nacional Experimental Simón Rodríguez (UNESR), Caracas, Venezuela

Luis Germán López Valdez, Laboratorio de Productos Naturales, Texcoco, México

Gladys Vidal, Grupo de Ingeniería y Biotecnología Ambiental (GIBA-UDEC), Facultad Ciencias Ambientales, Universidad de Concepción, Concepción, Chile and Centro de Recursos Hídricos para la Agricultura y la Minería (CRHIAM), Centro ANID Fondap, Victoria 1295, Concepción, Chile

Kaiser Iqbal Wani, Department of Botany, Aligarh Muslim University, Aligarh, India

About the Editor

Tariq Aftab received his PhD in the Department of Botany at Aligarh Muslim University, India, and is currently an Assistant Professor there. He is the recipient of a prestigious Leibniz-DAAD fellowship from Germany, Raman Fellowship from the Government of India, and Young Scientist Awards from the State Government of Uttar Pradesh (India) and Government of India. After completing his doctorate, he worked as a Research Fellow at the National Bureau of Plant Genetic Resources, New Delhi, and as Postdoctorate Fellow at Jamia Hamdard, New Delhi, India. Dr Aftab also worked as Visiting Scientist at the Leibniz Institute of Plant Genetics and Crop Plant Research (IPK), Gatersleben, Germany, and in the Department of Plant Biology, Michigan State University, USA. He is a member of various scientific associations from India and abroad.

He has edited 20 books with international publishers, including Elsevier Inc., Springer Nature, and CRC Press (Taylor & Francis Group), co-authored several book chapters, and published over 100 research papers in peer-reviewed international journals. His research interests include physiological, proteomic, and molecular studies on medicinal and crop plants.

Preface

In an era when the global population is rapidly increasing and challenges to food security are becoming ever more daunting, the need for innovative and sustainable agricultural practices has never been more critical. Traditional methods of agriculture, while foundational, are insufficient to meet the demands of the present and future. This is where plant biotechnology steps in as a beacon of hope, offering advanced solutions to enhance crop production, improve resilience to environmental stresses, and ensure sustainable food security for the global community.

This edited volume brings together leading experts from around the world to explore the multifaceted applications of biotechnology in agriculture. The chapters within this book delve into various biotechnological advancements and their potential to revolutionize agriculture, thus addressing the pressing issue of global food security.

The book begins with an introductory chapter by Kaiser Iqbal Wani, M. Naeem, and Tariq Aftab, setting the scene for the journey through the landscape of sustainable agriculture driven by plant biotechnology. From harnessing genetic engineering for crop improvement (Chapter 2 by Iqra Noor *et al.*) to exploring biotechnological tools for enhancing soil health and fertility, each chapter offers a deep dive into specific areas of research and application.

Notably, the book covers a wide array of topics including the role of biotechnological tools in improving soil health (Chapter 3 by Silvina M.Y. Lopez *et al.*), the harnessing of plant–microbial interactions for sustainable crop production (Chapter 4 by Iqra Owais and colleagues), and the metabolic engineering of tea for enhanced health benefits (Chapter 5 by Misbah Naz *et al.*). It also addresses innovative treatments like molecular hydrogen for enhanced growth and prolonged post-harvest storage (Chapter 6 by John T. Hancock and Jennifer May).

Further, the volume explores practical applications and case studies, such as the development of doubled haploid wheat genotypes through intergeneric hybridization with maize (Chapter 8 by Md. Yeakub Khan *et al.*), and the implementation of biofortification to improve nutritional content for human health and food security (Chapter 10 by Rakesh Kumar Bairwa and co-authors). The insights into sustainable crop management through agroecology (Chapter 11 by Braulio Edgar Herrera-Cabrera *et al.*) and the use of bio-inoculants to mitigate water deficit in crops (Chapter 12 by Urley A. Pérez-Moncada and team) highlight the book's commitment to presenting practical, research-backed solutions.

Moreover, the volume addresses the critical areas of plant defense mechanisms, genetic engineering for environmental remediation, and the ethical and regulatory issues surrounding plant biotechnology. Chapters such as "Plant Defense Mechanisms: Innate and Induced Immunity in

Plants" (Chapter 15 by Sucharita Mohapatra *et al.*), "Genetic Engineering of Plants for Remediation of Toxic Metals and Metalloids" (Chapter 13 by Syeda Gul Nisa and colleagues), and "Regulatory and Ethical Issues for Plant Biotechnology and Genetic Engineering" (Chapter 17 by Ankur Singh and Aryadeep Roychoudhury) provide a comprehensive overview of these essential aspects.

This book is not just a collection of academic contributions but a call to action for researchers, policymakers, and practitioners to embrace biotechnological innovations in agriculture. It serves as a vital resource for understanding the current state and future potential of plant biotechnology in fostering a sustainable and food-secure world.

The chapter authors have provided a valuable contribution to the field of agricultural biotechnology, offering insights and practical solutions that bridge the gap between scientific research and real-world agricultural practices. It is our hope that this volume will inspire further research and implementation of biotechnological advancements in agriculture, ultimately contributing to the global effort to achieve food security and sustainable agricultural practices.

Tariq Aftab
Department of Botany, Aligarh Muslim University, Aligarh, India

1 Plant Biotechnology and the Quest for Sustainable Agriculture: An Introduction

Kaiser Iqbal Wani, Abdullah, M. Naeem and Tariq Aftab*
Department of Botany, Aligarh Muslim University, Aligarh, India

Abstract
In the face of escalating global challenges such as climate change, population growth, and dwindling natural resources, the quest for sustainable agriculture has never been more urgent. Traditional agricultural practices are increasingly proving insufficient to meet the demands of a growing population while preserving environmental integrity. Consequently, plant biotechnology emerges as a pivotal tool in addressing these challenges, offering innovative solutions to enhance crop productivity, resilience, and sustainability. By integrating biotechnological innovations into agricultural practices, a more sustainable and resilient food production system can be achieved. This integration not only addresses the immediate need for increased food production but also promotes long-term ecological balance and resource conservation. Plant biotechnology therefore holds the promise of transforming agriculture into a more adaptive and sustainable enterprise, capable of meeting the nutritional needs of a growing population while safeguarding the planet's environmental health for future generations.

Keywords: food security, global climate change, crop productivity, environmental resilience, sustainability

1.1 Introduction

The global agricultural sector is at a pivotal point as it faces increasing demands for food production amidst challenges such as climate change, dwindling natural resources, and a growing population (Smith, 2020). These challenges not only threaten food security but also result in environmental degradation, pushing the boundaries of conventional agricultural practices. Plant biotechnology has emerged as a critical tool in addressing these challenges, offering innovative solutions for sustainable agriculture. This chapter provides an overview of the current state of plant biotechnology and its potential to transform agriculture into a more sustainable practice. Moreover, it highlights key innovations that are driving this transformation, paving the way for future food systems that are resilient, resource efficient, and environmentally sound.

Plant biotechnology encompasses a range of techniques and tools used to modify plants' genetic makeup to achieve desirable traits such as increased yield, pest resistance, and tolerance to abiotic stresses like drought and salinity (Fedoroff, 2015). Recent advances in gene-editing technologies, such as CRISPR-Cas9, have further accelerated progress, allowing for more precise modifications that can be made in a shorter time frame (Ricroch *et al.*, 2017). These advancements can potentially revolutionize agricultural practices by reducing reliance on chemical inputs, conserving water, and improving crop resilience to adverse

*Corresponding author: tarik.alig@gmail.com

DOI: 10.1079/9781800627659.0001

environmental conditions (Smith, 2020). In addition, plant biotechnology offers opportunities to enhance the nutritional quality of crops, addressing issues of malnutrition and micronutrient deficiencies in regions where food security is a pressing concern (Aftab and Roychoudhury, 2021). By integrating biotechnology into agricultural systems, the goal of achieving higher productivity while minimizing environmental impact is becoming increasingly attainable.

Genetically modified organisms (GMOs) have been at the forefront of plant biotechnology. By introducing specific genes into crops, scientists have developed varieties that exhibit superior traits compared to their conventional counterparts. For instance, Bt cotton and Bt maize, engineered to express *Bacillus thuringiensis* (Bt) toxin, have significantly reduced the need for chemical pesticides, thereby minimizing environmental contamination and promoting a healthier ecosystem (Qaim and Zilberman, 2003). Beyond pest resistance, researchers are now exploring GMOs that can improve tolerance to climate-induced stresses such as heat and water scarcity, potentially providing vital solutions for regions vulnerable to climate change (Raza, 2022). As the global climate continues to shift, these innovations will be crucial in ensuring the stability and sustainability of crop production in diverse environments.

1.2 The Role of Plant Biotechnology in Agriculture

Plant biotechnology encompasses a range of techniques including genetic modification, marker-assisted breeding, and genome editing. These technologies enable the development of crops that are more resilient to environmental stresses, have higher nutritional value, and can be cultivated with reduced inputs, thereby promoting sustainability and improving resource efficiency (Pérez-Massot *et al.*, 2013). As the world faces increasing challenges related to climate change, water scarcity, and soil degradation, plant biotechnology offers viable solutions to support global food security while minimizing the environmental impact of agricultural practices. By fostering innovation in crop development, biotechnology is helping to build a more sustainable agricultural future.

1.2.1 Genetic modification and crop improvement

Genetic modification (GM) involves the insertion of genes into a plant's genome to confer desirable traits such as pest resistance, herbicide tolerance, and improved nutritional content. For example, Bt cotton and Bt maize, which contain genes from the bacterium *Bacillus thuringiensis*, are resistant to certain insect pests, reducing the need for chemical pesticides (James, 2013). This not only lowers production costs for farmers but also decreases the environmental impact associated with pesticide use. Additionally, GM crops have been developed to enhance food quality and nutritional value. Biofortified crops like Golden Rice, engineered to contain higher levels of vitamin A, have the potential to address micronutrient deficiencies in populations where malnutrition is prevalent (Wu *et al.*, 2021). Golden Rice involved the addition of two genes, *phytoene synthase* (*Psy1*) from *Zea mays* and *carotene desaturase* (*CrtI*) gene from the soil bacterium *Erwinia uredovora*, into a *japonica* rice variety. This completed the carotenoid pathway in the grain and resulted in the accumulation of β-carotene (pro-vitamin A) into (carotenoid-free) rice endosperm (Paine *et al.*, 2005).

As research continues, the scope of GM technology is expanding to include traits such as enhanced drought tolerance and improved nitrogen use efficiency, offering solutions to some of the most pressing challenges in agriculture today.

1.2.2 Marker-assisted breeding

Marker-assisted breeding uses molecular markers linked to desirable traits to select plants for breeding more efficiently. This method accelerates the breeding process and increases the precision of developing new crop varieties with specific traits such as drought tolerance and disease resistance (Varshney *et al.*, 2005). By focusing on specific genetic markers, breeders can identify and select plants that carry the most advantageous traits early in the breeding cycle, significantly shortening the time it takes to develop new crop varieties. For example, marker-assisted selection has been widely used to develop rice varieties that are more resistant to flooding and drought (Dixit *et al.*, 2017), which are increasingly important traits as climate variability intensifies. In addition, this approach enables breeders to combine multiple beneficial traits into a single variety, improving both crop resilience and productivity in a more targeted manner.

1.2.3 Genome editing

Technologies such as CRISPR-Cas9 have revolutionized plant biotechnology by allowing precise edits to be made to the plant genome. This can result in crops with enhanced traits such as increased yield, improved stress tolerance, and reduced need for fertilizers and water (Zhang *et al.*, 2018). The ability to edit genes with such accuracy enables scientists to address specific agricultural challenges more effectively and quickly than with traditional breeding or genetic modification methods. Moreover, genome editing holds great promise in tackling emerging threats like new pests and diseases, as well as helping crops adapt to changing environmental conditions (Sanz Juste *et al.*, 2023). As these technologies continue to advance, they are also being used to explore novel applications, such as enhancing carbon sequestration in crops to mitigate the effects of climate change (Ma *et al.*, 2022). The flexibility and efficiency of genome editing tools like CRISPR-Cas9 make them invaluable in the ongoing effort to create sustainable agricultural systems that can meet future demands.

1.3 Benefits of Plant Biotechnology for Sustainable Agriculture

Plant biotechnology offers benefits that are critical for promoting agricultural sustainability (Fig. 1.1). By enhancing crop resilience, reducing environmental impact, and improving food security, biotechnology plays a pivotal role in addressing the global challenges posed by population growth, climate change, and resource depletion.

1.3.1 Environmental sustainability

Plant biotechnology can significantly reduce the environmental impact of agriculture. GM crops with pest resistance lessen the need for chemical pesticides, which can harm beneficial insects and contaminate water supplies (Brookes and Barfoot, 2018). By decreasing pesticide use, biotechnology promotes biodiversity and safeguards ecosystem health. Additionally, crops engineered for herbicide tolerance allow for more efficient weed management, reducing soil erosion and preserving soil health through no-till farming practices (Duke and Powles, 2009). No-till farming not only minimizes soil disturbance but also enhances carbon sequestration, hydraulic conductivity, and water retention capacity of soil (Feng *et al.*, 2020). These practices demonstrate how biotechnology can be integrated into sustainable farming systems that protect natural resources while maintaining productivity.

1.3.2 Economic benefits

Biotechnology can enhance the economic viability of farming by increasing crop yields and reducing input costs. For instance, GM crops have been shown to provide higher yields and lower production costs, which can increase profitability for farmers, especially in developing countries (Qaim and Kouser, 2013). In addition to direct financial benefits, biotechnology helps reduce crop losses caused by pests and diseases, improving overall farm productivity. This is particularly important for smallholder farmers who often face significant challenges accessing agricultural inputs. By lowering the barriers to adopting improved crop varieties, biotechnology can contribute to poverty alleviation and rural development.

1.3.3 Nutritional enhancement

Biotechnology also holds promise for addressing nutritional deficiencies. Biofortified crops, such as Golden Rice, engineered to produce provitamin A, can help alleviate vitamin A deficiency in populations that rely heavily on rice as a staple

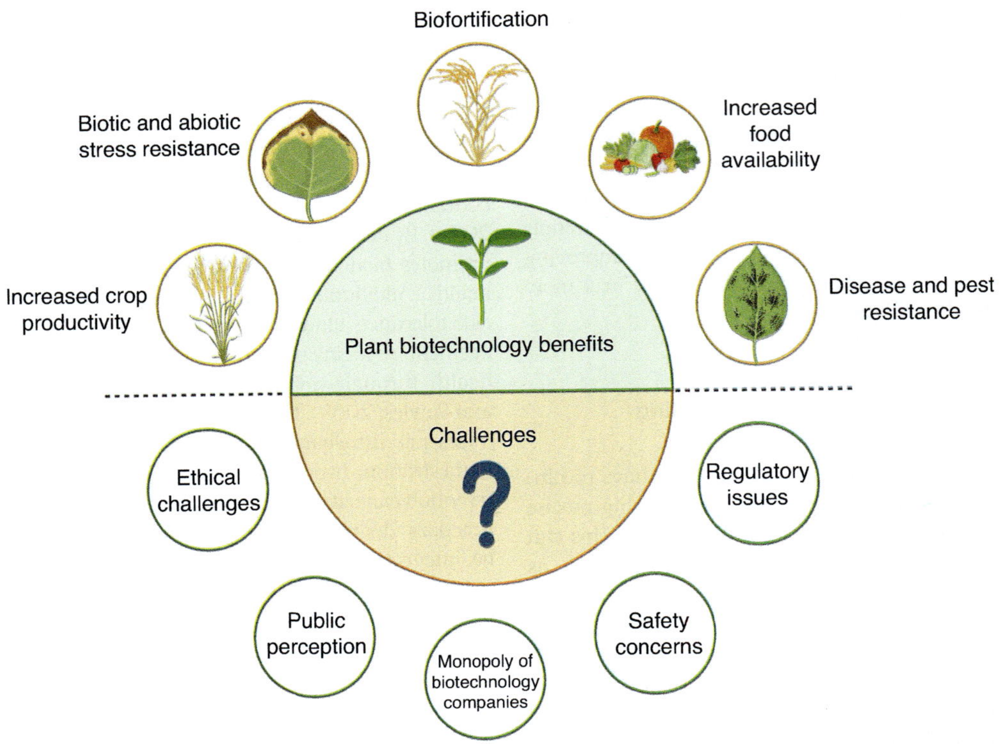

Fig. 1.1. Potential benefits and challenges of plant biotechnology in agriculture.

food (Tang *et al.*, 2009). This targeted approach to nutrition allows for the development of crops that directly address public health issues, offering a cost-effective solution to malnutrition. Furthermore, advancements in biotechnology are being applied to develop crops with enhanced levels of essential nutrients, such as iron and zinc (Stangoulis and Knez, 2022), providing a pathway to improving global health outcomes, particularly in regions where access to diverse, nutrient-rich foods is limited.

1.4 Addressing Abiotic and Biotic Stresses

One of the critical areas where plant biotechnology has shown promise is in addressing abiotic stresses, which are nonliving factors such as drought, salinity, and extreme temperatures that negatively impact crop productivity (Aftab and Roychoudhury, 2021; Raza *et al.*, 2023). Advances in biotechnology have enabled the development of drought-tolerant crops through the introduction of genes responsible for water use efficiency and osmotic stress tolerance. For example, transgenic rice expressing the *DREB1A* gene, which is involved in drought response, has demonstrated enhanced tolerance to water scarcity (Agarwal *et al.*, 2006). As climate change intensifies, the ability to engineer crops that can thrive under suboptimal growing conditions becomes increasingly important.

Similarly, biotic stresses, which include damage caused by living organisms like pests, diseases, and weeds, pose significant threats to agricultural productivity. Biotechnological interventions, such as the development of disease-resistant crops through gene editing technologies like CRISPR-Cas9, offer promising solutions (Fedoroff, 2015). These technologies allow for precise modifications in the plant genome, conferring resistance to pathogens while maintaining the crop's overall health and productivity (Zaidi *et al.*, 2020). As new pests and diseases are emerging due to global trade and environmental changes, biotechnological approaches will be key to maintaining food security.

1.5 Enhancing Nutritional Value and Food Security

Beyond improving crop resilience and yield, plant biotechnology also plays a crucial role in enhancing the nutritional value of food crops. Biofortification, the process of increasing the nutrient content of crops through biotechnological means, addresses micronutrient deficiencies that affect millions of people worldwide. Golden Rice, engineered to produce β-carotene, is a notable example of how biotechnology can combat malnutrition and improve public health (Paine *et al.*, 2005). Additionally, crops biofortified with iron, zinc, and other essential nutrients have the potential to alleviate widespread nutritional deficiencies, particularly in developing countries (Szerement *et al.*, 2022). By improving the nutritional profile of staple crops, biotechnology offers a long-term, sustainable approach to addressing hidden hunger.

Moreover, plant biotechnology contributes to food security by developing crops that can withstand the pressures of climate change and other environmental challenges. The ability to cultivate crops in marginal lands, previously deemed unsuitable for agriculture, represents a significant step toward ensuring a stable food supply for future generations (Fedoroff, 2015). As the global population continues to grow, securing a reliable food supply will depend on innovative biotechnological solutions that enhance both the quantity and quality of food produced.

1.6 Challenges, Ethical Considerations, and Future Directions

Despite the numerous benefits, the adoption of plant biotechnology is not without controversy. Ethical considerations, regulatory challenges, and public perception play critical roles in the acceptance and implementation of biotechnological innovations. Concerns about the safety of GMOs, potential environmental impacts, and the dominance of large biotechnology firms in the agricultural sector necessitate a balanced and informed approach to biotechnology adoption (Nicolia *et al.*, 2014). Addressing these concerns through transparent communication, rigorous safety assessments, and equitable access to biotechnological advancements is essential for ensuring that the benefits of biotechnology are widely shared.

The regulatory landscape for GM crops varies widely across the world, with some regions adopting strict regulations that can delay the deployment of biotech solutions. Public perception of GM crops is often negative, influenced by concerns about safety, environmental impact, and corporate control of the food supply (Marris, 2001). The emergence of resistance in pests and weeds to GM traits necessitates the development of integrated pest and weed management strategies. This includes rotating crops, using multiple modes of action, and incorporating non-GM methods to manage resistance (Gassmann *et al.*, 2011). Such integrated approaches will help ensure the long-term sustainability of biotechnological solutions.

Public engagement and transparent communication about the benefits and risks of plant biotechnology are essential to garnering support and addressing misconceptions (Woźniak-Gientka *et al.*, 2022). Rigorous scientific assessments and regulatory frameworks ensure that biotechnological advancements are safe and beneficial for society and the environment. These assessments include comprehensive risk evaluations, covering potential ecological impacts, food safety, and long-term sustainability (Entine *et al.*, 2021). Regulatory frameworks are designed to monitor the entire life cycle of biotech crops, from development and field testing to commercialization and postmarket surveillance, ensuring that no unintended consequences arise (Wani and Aftab, 2022). Furthermore, international collaborations between regulatory bodies, scientists, and policymakers facilitate the sharing of best practices and the harmonization of safety standards, fostering global confidence in the safety of biotechnological innovations. Public transparency and stakeholder involvement in these processes are also crucial in building trust and addressing societal concerns (Carson *et al.*, 2021; Harfouche *et al.*, 2021). Continued public discourse will help build trust and ensure that innovations in biotechnology are aligned with societal values and priorities.

The future of plant biotechnology lies in the continued innovation and integration of new technologies. Advances in synthetic biology, systems biology, and computational modeling hold promise for the development of crops with even greater precision and efficiency (Liu and Stewart, 2015). Additionally, collaboration between public and private sectors, along with effective communication and engagement with the public, will be crucial for the sustainable advancement of plant biotechnology. Expanding access to these technologies across different regions and agricultural systems will be key to realizing their full potential in promoting global food security.

1.7 Conclusion

Plant biotechnology is a powerful tool in the quest for sustainable agriculture, offering solutions that can enhance food security, reduce environmental impact, and improve economic outcomes for farmers. Through continued research, ethical considerations, and public engagement, plant biotechnology can pave the way for sustainable and resilient agriculture. While challenges remain, continued research, innovation, and collaborative efforts will be key to realizing the full potential of plant biotechnology in creating a sustainable agricultural future.

References

Aftab, T. and Roychoudhury, A. (eds) (2021) *Plant Perspectives to Global Climate Changes: Developing Climate-Resilient Plants*. Academic Press, Oxford.

Agarwal, P.K., Agarwal, P., Reddy, M.K. and Sopory, S.K. (2006) Role of DREB transcription factors in abiotic and biotic stress tolerance in plants. *Plant Cell Reports* 25(12), 1263–1274.

Brookes, G. and Barfoot, P. (2018) Environmental impacts of genetically modified (GM) crop use 1996–2016: Impacts on pesticide use and carbon emissions. *GM Crops & Food* 9(3), 109–139.

Carson, S.G., Myskja, B.K. and Myhr, A.I. (2021) Public engagement in biotechnology innovation – the need for research and the role of ethics. In: Schübel, H. and Wallimann-Helmer, I. (eds) *Justice and Food Security in a Changing Climate*. Wageningen Academic Publishers, Wageningen, pp. 325–347.

Dixit, S., Singh, A., Sandhu, N., Bhandari, A., Vikram, P. *et al.* (2017) Combining drought and submergence tolerance in rice: Marker-assisted breeding and QTL combination effects. *Molecular Breeding* 37(12), 143.

Duke, S.O. and Powles, S.B. (2009) Glyphosate-resistant crops and weeds: Now and in the future. *AgBioForum* 12(3 and 4), 346–357.

Entine, J., Felipe, M.S.S., Groenewald, J.-H., Kershen, D.L., Lema, M. *et al.* (2021) Regulatory approaches for genome edited agricultural plants in select countries and jurisdictions around the world. *Transgenic Research* 30(4), 551–584.

Fedoroff, N.V. (2015) Food in a future of 10 billion. *Agriculture & Food Security* 4(1), 1–10.

Feng, Q., An, C., Chen, Z. and Wang, Z. (2020) Can deep tillage enhance carbon sequestration in soils? A meta-analysis towards GHG mitigation and sustainable agricultural management. *Renewable and Sustainable Energy Reviews* 133, 110293.

Gassmann, A.J., Petzold-Maxwell, J.L., Keweshan, R.S. and Dunbar, M.W. (2011) Field-evolved resistance to Bt maize by western corn rootworm. *PLOS One* 6(7), e22629.

Harfouche, A.L., Petousi, V., Meilan, R., Sweet, J., Twardowski, T. *et al.* (2021) Promoting ethically responsible use of agricultural biotechnology. *Trends in Plant Science* 26(6), 546–559.

James, C. (2013) *Global status of commercialized biotech/GM crops: 2013* ISAAA Brief No. 46. ISAAA, New York, p. 46.

Liu, W. and Stewart, C.N. (2015) Plant synthetic biology. *Trends in Plant Science* 20(5), 309–317.

Ma, Z., Cheah, W.Y., Ng, I.S., Chang, J.S., Zhao, M. *et al.* (2022) Microalgae-based biotechnological sequestration of carbon dioxide for net zero emissions. *Trends in Biotechnology* 40(12), 1439–1453.

Marris, C. (2001) Public views on GOMs: Deconstructing the myths. *EMBO Reports* 2(7), 545–548.

Nicolia, A., Manzo, A., Veronesi, F. and Rosellini, D. (2014) An overview of the last 10 years of genetically engineered crop safety research. *Critical Reviews in Biotechnology* 34(1), 77–88.

Paine, J.A., Shipton, C.A., Chaggar, S., Howells, R.M., Kennedy, M.J. *et al.* (2005) Improving the nutritional value of golden rice through increased pro-vitamin A content. *Nature Biotechnology* 23(4), 482–487.

Pérez-Massot, E., Banakar, R., Gómez-Galera, S., Zorrilla-López, U., Sanahuja, G. *et al.* (2013) The contribution of transgenic plants to better health through improved nutrition: Opportunities and constraints. *Genes & Nutrition* 8(1), 29–41.

Qaim, M. and Zilberman, D. (2003) Yield effects of genetically modified crops in developing countries. *Science* 299(5608), 900–902.

Qaim, M. and Kouser, S. (2013) Genetically modified crops and food security. *PLOS One* 8(6), e64879.

Raza, A. (2022) Metabolomics: A systems biology approach for enhancing heat stress tolerance in plants. *Plant Cell Reports* 41(3), 741–763.

Raza, A., Tabassum, J., Fakhar, A.Z., Sharif, R., Chen, H. *et al.* (2023) Smart reprograming of plants against salinity stress using modern biotechnological tools. *Critical Reviews in Biotechnology* 43(7), 1035–1062.

Ricroch, A., Clairand, P. and Harwood, W. (2017) Use of CRISPR systems in plant genome editing: Toward new opportunities in agriculture. *Emerging Topics in Life Sciences* 1(2), 169–182.

Sanz Juste, S., Okamoto, E.M., Nguyen, C., Feng, X. and López Del Amo, V. (2023) Next-generation CRISPR gene-drive systems using Cas12a nuclease. *Nature Communications* 14(1), 6388.

Smith, J. (2020) Advances in plant biotechnology for sustainable agriculture. *Journal of Agricultural and Food Chemistry* 68(38), 10224–10235.

Stangoulis, J.C.R. and Knez, M. (2022) Biofortification of major crop plants with iron and zinc – achievements and future directions. *Plant and Soil* 474(1–2), 57–76.

Szerement, J., Szatanik-Kloc, A., Mokrzycki, J. and Mierzwa-Hersztek, M. (2022) Agronomic biofortification with Se, Zn, and Fe: An effective strategy to enhance crop nutritional quality and stress defense—A review. *Journal of Soil Science and Plant Nutrition* 22(1), 1129–1159.

Tang, G., Qin, J., Dolnikowski, G.G., Russell, R.M. and Grusak, M.A. (2009) Golden rice is an effective source of vitamin A. *American Journal of Clinical Nutrition* 89(6), 1776–1783.

Varshney, R.K., Graner, A. and Sorrells, M.E. (2005) Genomics-assisted breeding for crop improvement. *Trends in Plant Science* 10(12), 621–630.

Wani, K.I. and Aftab, T. (2022) Limitations, biosafety, ethics, regulatory issues in molecular farming in plants. In: *Plant Molecular Farming: Applications and New Directions*. Springer International Publishing, Cham, pp. 61–74.

Woźniak-Gientka, E., Agata, T., Milica, P., Anna, B., Dennis, E. *et al.* (2022) Public perception of plant gene technologies worldwide in the light of food security. *GM Crops & Food* 13(1), 218–241.

Wu, F., Wesseler, J., Zilberman, D., Russell, R.M., Chen, C. *et al.* (2021) Allow golden rice to save lives. *Proceedings of the National Academy of Sciences* 118(51), e2120901118.

Zaidi, S.S., Mahfouz, M.M., Mansoor, S. and Mitter, N. (2020) CRISPR-Cas9: A powerful tool for crop genome editing to accelerate climate resilience, disease resistance, and yield. *Frontiers in Plant Science* 11, 565456.

Zhang, H., Zhang, J., Wei, P., Zhang, B., Gou, F. *et al.* (2018) The CRISPR/Cas9 system: Versatile tools for genome engineering and plant breeding. *Biotechnology Advances* 36(2), 302–309.

2 Harnessing Genetic Engineering in Crop Improvement: Towards Resilient Agriculture

Iqra Noor*, Hamza Sohail and Muhammad Tanveer Akhtar
School of Horticulture and Landscape Architecture, Yangzhou University, Yangzhou, PR, China

Abstract
Ensuring sufficient nutrition for the growing population is a major challenge, especially in rapidly changing climate conditions. Genome editing has the potential to revolutionize plant breeding and contribute to the global food supply. In a short amount of time, it has significantly impacted crop development, outperforming conventional breeding in terms of precision and efficiency in genetic alteration. In order to increase crop productivity, genetic engineering has been essential in both enhancing already existing features and adding new, desired traits. This strategy has also enhanced hybrid breeding processes, simplifying the removal of undesirable traits or addition of desired features, enabling precise adjustment of crop traits even within a single generation. Here, we review the development and application of genome editing tools in plants. This study outlines the fundamentals of various genome editing technologies, with an emphasis on genetic engineering methods, i.e., resilient agriculture, uses of genetic engineering, and research endeavors to enhance genetic transformation for higher crop yield and quality.

Keywords: genome editing, plant breeding, genetic engineering, crop productivity, climate resilience

2.1 Introduction

Many populations are facing threats to food and nutritional quality around the world due to climate change. Plant breeders are facing more challenges in developing resistant varieties because of droughts, changes in temperature, floods, increases in CO_2 levels in the atmosphere, salinity, plant diseases, insects, and pests (Van Oort and Zwart, 2018). The main reasons for food insecurity include increasing population, urbanization, climatic conditions, growing scarcity of water, dependence on inputs from the carbon-based economy, and deterioration of cultivated land. These factors will ultimately influence crop productivity (Tian *et al.*, 2021). Therefore, the challenge of meeting the crop yield and quality requirements of 10 billion people deserves extensive consideration. Due largely to the Green Revolution and advances in plant breeding techniques, current crop yields can provide sufficient food for a majority of the human population. With diminishing resources and limited arable land, sustainable production to support food and nutritional demands has been a daunting task. Just like human health,

*Corresponding author: iqranoor@yzu.edu.cn

DOI: 10.1079/9781800627659.0002

plant health must also be protected and secured by adopting preventive measures (Lamichhane *et al.*, 2016).

Improved agriculture is based on genetic variation. The use of these genetic variations is the main goal of plant breeding. Herrera-Estrella *et al.* (1983) initiated the plant genetic engineering era by producing herbicide-resistant tobacco plants through genetic engineering, and Somssich (2019) later reviewed this work. Horsch *et al.* (1985) followed this with a report on leaf disc procedures for successful *Agrobacterium*-mediated plant transformation and regeneration in tomato, tobacco, and petunia. In 1996, Mackelprang and Lemaux (2020) successfully commercialized the first batch of genetically engineered crops. As a result, numerous advanced techniques have emerged, including *Agrobacterium tumefaciens* transformation, biolistic method, ovary injection, ultrasound-mediated transformation, polyethylene glycol induction, pollen-tube pathway method, and liposome-mediated transfection (Luo *et al.*, 2014).

To increase crop production, genetic engineering has been very important for both altering traits that were already there and adding new ones. Genetic engineering changes the genetic makeup of recipient cells to achieve the required results by changing genomic DNA at the molecular level using modern biological tools (Lanigan *et al.*, 2020). For example, herbicide resistance, pest/pathogen tolerance, and improved nutritional value are the main advantageous traits offered by genetically modified crops. Currently, scientists have approved up to 525 different transgenic events across 32 crops for cultivation (Kumar *et al.*, 2020). However, the challenges associated with plant transformation continue to be costly, time-consuming, and difficult, particularly for transformation-recalcitrant plant species and crop genotypes.

This review summarizes the principles of several genome editing technologies, focusing on genetic engineering techniques, applications of genetic engineering for resilient agriculture, and researchers' efforts to improve genetic transformation to increase crop yield and quality. We also discuss the various techniques used for the development of genetically modified crops.

2.2 Historical Perspective of Genetic Engineering

Traditional breeding techniques utilized genetic variations originating from varietal germplasms for intergeneric or interspecific hybridizations, causing mutations at the cell or tissue level to enhance the plant's resilience to environmental obstacles. However, these techniques have many limitations. In the past 20 years, genomics and next-generation sequencing (NGS) technologies have changed the way we think about the molecular basis of economically important traits. They have also made it easier to use modern breeding methods to create climate-resilient crop varieties (Varshney *et al.*, 2021). Experts believe that both conventional and novel breeding techniques, along with advanced transformation technologies, play a crucial role in maintaining global food security (Landry and Mitter, 2019).

Pathogens caused historical food famines like the Irish potato famine of 1845–1849 and the Bengal famine of 1943, but the Green Revolution in the 1960s and 1970s gave farmers higher yielding and more resilient semi-dwarf crop varieties. The Green Revolution had a significant impact on food security but its effects on human nutrition, agricultural stress tolerance, and the environment were not evenly distributed (Pingali, 2012). Plant and crop transformation is a crucial component of the second Green Revolution that is about to begin, which will deliver new crop traits and digital technologies to farmers and agriculture (Shepherd *et al.*, 2020). Therefore, crop production needs to lessen its significant environmental impact and adapt to more varied environmental conditions. To meet the demands of the current population, productivity must also rise far faster than it has in the past. Plant genomics is largely responsible for crop development, including the identification of genetic diversity that contributes to improved performance and increased plant breeding efficiency (Fig. 2.1).

Plant genomics plays a central role in crop improvement, including the discovery of genetic variation that underlies improved performance and increased plant breeding efficiency. Transgenic breeding, a key breakthrough in plant breeding, introduces genes or traits from other organisms into crop plants,

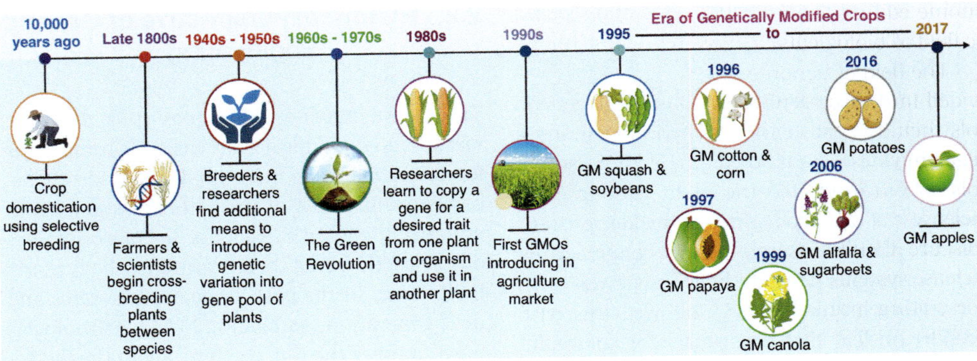

Fig. 2.1. Historical events of genetic engineering and genetically modified crops.

resulting in higher yields, reduced pesticide use, and improved nutrition. Nevertheless, only a few transgenic crops have been utilized thus far (Raman, 2017), as this technique randomly integrates foreign DNA into plant genomes, and these genetically modified organisms are subject to strict government regulations. This is not too soon because to ensure food security for 10 billion people by 2050, we need to increase crop productivity while lowering the environmental impact of agriculture, reducing crop losses due to diseases and unfavorable weather patterns, and improving the nutritional value of crops for the benefit of human health. Additionally, we must achieve a wider acceptance of genetically modified crops through genome-scale breeding, precisely calibrated genetic engineering, and more accurate agronomic management techniques. Future agricultural production will depend increasingly on tools and technology to improve sustainability; more nutritious crops will reduce our reliance on animal protein and introduce novel fiber and plant-based medicines to the market.

2.3 Genetic Engineering Techniques for Crop Improvement

2.3.1 Gene editing technologies

Genome editing, also known as targeted mutation, is a technique that can produce desired alterations in the genome. Gene editing may be used to alter specific DNA sequences by adding, removing, or even substituting one or more genetic bases. By precisely altering certain regions of the genome, genome editing can enable genome modification. Additionally, the chemicals required for this process can be introduced into the cell without introducing DNA into the genome (Ma *et al.*, 2020). Another remarkable feature of this technology is that it creates inheritable mutations in the genome with a low probability of generating off-targets.

Genome editing techniques have emerged as a novel breeding technology with the ability to address these issues. They have been successfully utilized to modify crop genomes in a wide range of plants without introducing foreign genes. They have significantly impacted crop development in a short time, outperforming conventional breeding in terms of precision and efficiency in genetic alteration (Chen *et al.*, 2019). CRISPR-Cas9 DNA, RNA (*in vitro* transcripts of Cas9 and single guide RNA; sgRNA), and RNP (ribonucleoprotein; a combination of Cas9 protein and *in vitro*-transcribed sgRNA) are three popular types of genomes editing reagents (Ran *et al.*, 2017). RNA and RNP can only be introduced into plant cells through particle bombardment, but DNA can be introduced into plant cells through both *Agrobacterium*-mediated transformation and particle bombardment (Liang *et al.*, 2017).

Sequence-specific nucleases that can be programmed are used to modify the plant genome. The precise modifications in the genomes of plants and animals achieved by genome editing techniques have revolutionized the field of biological sciences.

The field of genome editing can be broadly divided into three generations: First-generation tools include zinc finger nucleases (ZFNs) and mega nucleases (MegaN); second-generation tools are transcription activator-like effector nucleases (TALENs); and third-generation tools are CRISPR-associated protein 9 (Cas9) nuclease systems. Third-generation GEs, such as base editing, prime editing, CRISPR/Cas9, and CRISPR-CRISPR from *Prevotella* and *Francisella* 1 (Cpf1), have been demonstrated to be effective tools for the precise modification of genome sequence (Lin *et al.*, 2020). The strategy of gene editing has also improved hybrid breeding methods; undesirable traits can be easily removed or desired traits can be added to elite types, making it possible to accurately modify crop attributes even within a single generation. In order to overcome the challenges encountered by traditional breeding programs, it is imperative that these technologies be used for the production of novel and better agricultural varieties.

Zinc finger nucleases

Zinc finger nucleases are created by combining the nonspecific cutting activity of Fok1 endonuclease with the specific binding activity of chemically modified zinc finger protein (ZFP) (Carroll, 2011). A number of amino acid residues on ZFP's α helix can pair with bases next to the target site and are directly involved in its recognition.

Since their initial development through the use of chemically modified nucleases, ZFNs have emerged as one of the most promising methods for genome editing that specifically target double-strand breaks (Fig. 2.2). This technique led to the discovery of the functional Cys2-His2 zinc finger domain (Gaj *et al.*, 2013). ZFNs are composed of two domains that are fundamental to their structure: (1) The DNA-binding domain, which has 300–600 zinc finger repeats (Carlson *et al.*, 2012), and (2) the DNA cleavage domain, sometimes referred to as the nonspecific cleavage domain of the type II restriction endonuclease Fok1, which functions as the DNA cleavage domain in zinc finger repeats (ZFNs), and each zinc finger repeat can monitor and read between 9 and 18 base pairs (BP) (Carroll *et al.*, 2006). Plants have been manipulated to exhibit stable and site-specific mutations through the discovery of ZFNs and modified ZFs (Carroll, 2011). Combining multiple ZFs to create an array of DNA-binding ZFs may improve the affinity and specificity for recognizing the target DNA sequence (Segal *et al.*, 2003).

Although ZFNs are effective genome editing elements, their widespread adoption has been hindered by the complicated nature of their interaction with DNA. According to Sander *et al.* (2011), other restrictions consist of the intrinsic difficulty of engineering, interest-dependent specificity, and the challenge of authenticating such proteins for a specific DNA region of context. ZFN has been effectively used for genome editing in model plants, including tobacco (Townsend *et al.*, 2009), maize (Shukla *et al.*, 2009), soybeans (Curtin *et al.*, 2011), and *Arabidopsis* (Osakabe *et al.*, 2010). However, their use has been limited by the availability of targeting sites in the genome, construction complexity, high but variable off-target rate, and high cost and skill needed for analysis.

Transcription activator-like effector nuclease (TALEN)

Transcription activator-like effector nuclease was created by combining the DNA-binding domain of the transcription activator-like effector (TALE) protein with the FokI cleavage domain (see Fig. 2.2). The fundamental domain of the proteins involved in the TALEN system is responsible for nuclear localization sequence and DNA binding (Schornack *et al.*, 2006). This genome editing technique is regarded as the precise genome editing method, as researchers can easily design TALENs because there is a one-to-one recognition convention among protein repeats and nucleotide sequences; hence, it can target nearly any sequence of interest presently by assembling the arrays (Luo *et al.*, 2019).

The ZFPs and TALENs vary structurally in that ZFPs lack the TALE binding domain used by TALENs, whose binding function is mostly dependent on highly variable amino acids located at positions 12 and 13 on the TALE. The first application of TALEN was done to improve rice by disrupting *OsSWEET14*, a

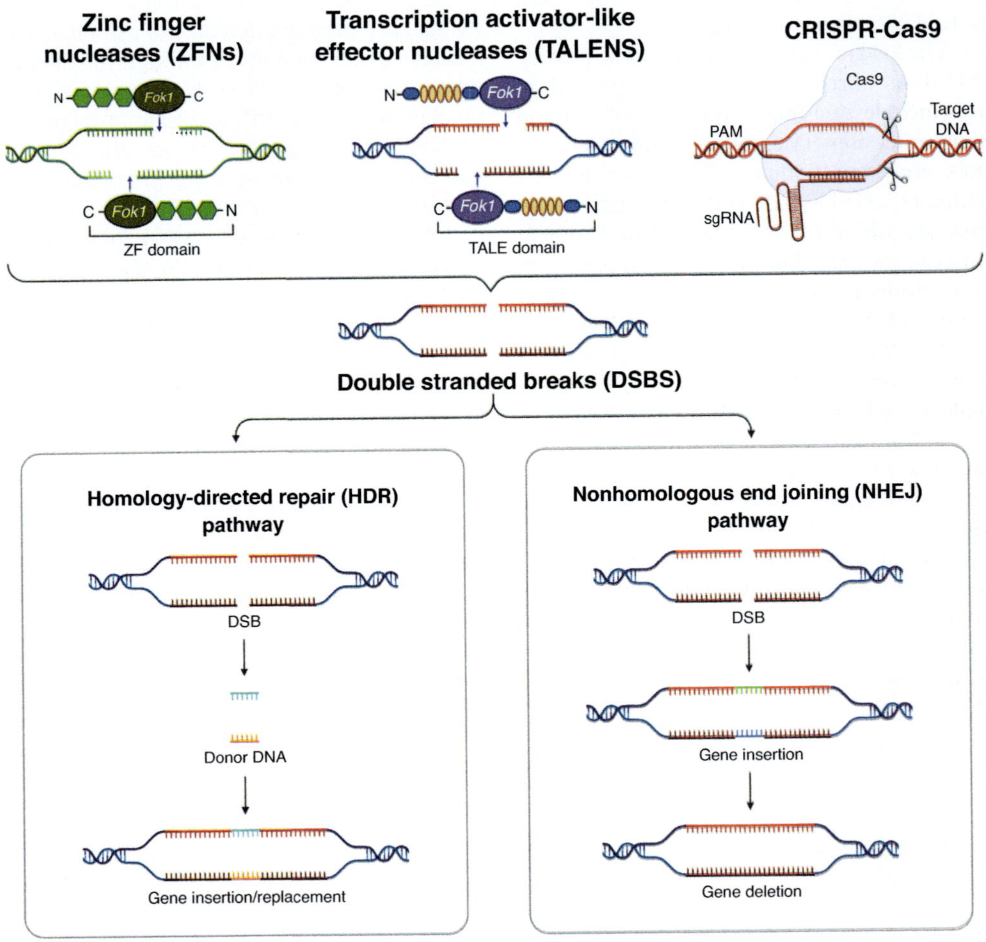

Fig. 2.2. Schematic representation of ZFNs, TALENs, and CRISPR-Cas9.

bacterial blight susceptibility gene (Li *et al.*, 2012). One example is the production of soybeans through the disruption of fatty acid desaturase genes, which results in a reduction in linoleic acid and an increase in oleic acid content. As a result, the heat stability and shelf-life of soybean oil were enhanced (Demorest *et al.*, 2016). Because TALENs have greater target binding selectivity and produce fewer off-targets than ZFNs, they have been extensively used in crop genome editing (Liu *et al.*, 2021). Although TALEN genome editing is more affordable and efficient due to their ease of construction and validation, their widespread acceptance in genome editing applications has been hampered by the challenges associated with protein design, synthesis, and validation.

CRISPR-Cas9

The central principle of gene targeting with CRISPR-Cas9, or other-directed DNA endonucleases, is that a double-stranded DNA break is generated in the cell of interest. The two main outcomes after a chromosomal break are homology-directed repair (HDR) and nonhomologous end-joining (NHEJ) repair (Chang *et al.*, 2017). NHEJ typically results in a small insertion or deletion of DNA sequence at the break (indel), which causes frameshifts in mRNA transcripts and premature termination codons, which cause nonsense-mediated mRNA decay and loss of protein expression when the break is directed to a coding exon in a gene (Popp and Maquat, 2016). During DNA repair, the HDR pathway duplicates a template, allowing the insertion of altered genetic sequences in the form of DNA donors. A new era in crop improvement has emerged as a result of the development of clustered regularly interspaced short palindromic repeats (CRISPR)-associated protein (Cas) technology as a genome-editing tool (Zhang *et al.*, 2020).

The advent of the CRISPR-Cas9 system (see Fig. 2.2), which is derived from the immune systems of bacteria and archaea, has been instrumental in the current broad application of gene editing technology in crop improvement. It can be used to modify genes in both plants and animals. CRISPR-Cas9 technology is regarded as one of the most significant discoveries of the 20th century (Barrangou and Doudna, 2016). The Cas9 and Cas12a complexes, which are both single-effector proteins that carry out nucleic acid cleavage, are the most often utilized CRISPR-Cas systems (Chen *et al.*, 2019). Plant genome editing has also recently been made possible via the Cas12b mechanism (Ming *et al.*, 2020). Utilizing CRISPR-Cas9, a number of microRNAs (miR319, miR169, miR398, miR169c, miR408, miR398a/b) have been found to be involved in a variety of abiotic stress tolerances (cold, drought, and salinity) to overcome abiotic stress tolerance (Bustamante *et al.*, 2018). AREB1 and *OsSAPK2* in *Arabidopsis* and rice, respectively, are two important genes linked to drought stress that have been studied using CRISPR technology (Lou *et al.*, 2017; Shinwari *et al.*, 2020). Sun *et al.* (2017) made further efforts to enhance the nutritional qualities and fine structure of rice grains.

The adaptability, effectiveness, and broad variety of applications of the CRISPR-Cas9 system have led to its implementation in numerous research domains, including genetic engineering, biotechnology, and fundamental and applied biology.

2.3.2 Transgenic approaches

Transgenic technology is a technique that transfers a DNA piece to the target cell in line with the aim of researchers through physical, chemical, biological, or comprehensive strategies and integrates it into the genome to produce the desired expression (Lanigan *et al.*, 2020). Transgenic technology cannot accomplish real site-specific alteration without the use of a transgenic library. Cells in the library can be modified at each location. The synthesis of nano devices and nanomaterials can lessen the effect of substantial challenges on food and energy production while optimizing the use of scarce resources, such as water or nutrients (Giraldo *et al.*, 2019). Nanobiotechnology procedures have enhanced the accuracy of plant breeding and created innovative, exciting opportunities for gene selection and transition by cutting down on the time needed to eliminate undesirable genes and providing access to crucial genes from large plantations (Pérez-de-Luque, 2017). According to prior statistics, roughly 1.7 million hectares of land were utilized for transgenic crop development in 1996 whereas approximately 191.7 million hectares were employed for transgenic crop growth in 2018. Papaya plants resistant to the papaya ringspot virus have been considered the first effective commercial application of genetic engineering technology in the fruit crop (Gonsalves, 1998).

Bt crops

The introduction of Bt genes, which confer resistance to *Bacillus thuringiensis*, has been studied in commercially significant plants, such as cotton and maize. One of the most frequently designed features is insect resistance, achieved by introducing an insecticidal protein from the *Bacillus thuringiensis* (Bt) bacterium. Since 1938, both conventional and organic growers have utilized the proteins themselves in commercial sprays to manage crop-damaging insects (Sanahuja *et al.*, 2011).

There are many types of Bt toxins, including vegetative insecticidal proteins (Vips) and Cry proteins. Each protein kills a specific group of insect species (Rashid *et al.*, 2018). Genetically modified crops express Cry genes, which encode the Cry protein. According to Labbé *et al.* (2017), the Cry protein causes pores to form in the midgut cell membranes of insects, ultimately resulting in paralysis and death. Pyramiding, or combining many Bt traits into one plant, can slow the development of insect resistance and create smaller refuge regions (Wei *et al.*, 2015). Increased yield and ensuing profitability over nonadopters are further advantages for Bt crop users. Researchers created genetically engineered cotton by adding either Cry1Ac or stacking Cry1Ac and Cry1Ab to protect cotton plants from *Heliothis virescens*, *Pectinophora gossypiella*, *Helicoverpa zea*, and *Helicoverpa armigera* (Romeis *et al.*, 2008). Chakroun *et al.* (2016) successfully inserted the Vip gene, which encodes vegetative insecticidal proteins, for the development of transgenic cotton and maize.

Herbicide-resistant crops

Farmers widely use herbicides to boost crop yield and labor efficiency (Schütte *et al.*, 2017). However, one major risk with herbicide use is that it may drift and run off onto nontarget plants, in addition to damaging target crops. Herbicide-tolerant crops get rid of herbicides in two ways: (1) they change an enzyme or other herbicide target in plants so that it cannot be killed by herbicides; or (2) they use an enzyme or enzyme system to break down the herbicide before it can harm the plant (Batterman and Leemans, 1988). Genetically engineered crops that are resistant to glyphosate herbicides include lucerne, canola, maize, and soybean (Kumar *et al.*, 2020). Some South American nations, notably Brazil and Argentina, have relatively high acceptance rates for genetically modified soybeans, maize, and cotton. In 2017, researchers genetically modified 97% of Brazilian soybeans and 100% of Argentina's soybeans and cotton for herbicide resistance, often paired with insect resistance (James, 2017).

RNA interference (RNAi)

RNA editing, unlike DNA editing, is more prone to recovery and posttranscriptional regulation, making it very safe for the treatment of genetic diseases. However, this makes it challenging to apply to agricultural features. dCas13 may also fuse with fluorescent proteins to localize target RNA in cells, giving researchers a novel tool for studying RNA localization (Abudayyeh *et al.*, 2017). Overall, Cas13 is more accurate than standard silencing approaches like RNAi, although its additional usage in crops needs more investigation. Cas13 is more precise than other gene silencing approaches, such as RNAi or artificial microRNA, and does not rely on plants' natural immune functions. In plants, RNAi, or artificial microRNA, links dsRNA with target RNA and binds to Argonaut and Dicer-like enzymes to form an RNA-induced silencing complex, thereby silencing target genes. Researchers have utilized this approach, which silences a specific gene in the pathogen, to alter papaya with the coat protein gene of the papaya ringspot virus (Gonsalves *et al.*, 2004). To eliminate virus-mediated damage, plant researchers are rapidly developing transgenic plants that are virus resistant. Researchers have also effectively used this approach to generate GM cassava (Odipio *et al.*, 2014), summer squash (Klas *et al.*, 2011), and soybean (Abbas *et al.*, 2021).

This modern technology primarily enhances the nutritional value, quality, and resistance to pests and diseases in various crops. For instance, researchers used the RNAi approach to produce *Lycopersicon esculentum* with elevated levels of carotenoid and flavonoid compounds. In this case, Davuluri *et al.* (2005) used RNAi along with a fruit-specific promoter to turn off DET1, a tomato gene that controls photomorphogenesis. Furthermore, Regina *et al.* (2006) utilized the RNAi method to increase the amylose content of wheat by more than 70% while concurrently reducing the expression of SBEIIa and SBEIIb.

2.4 Applications of Genetic Engineering for Stress Tolerance

Agrobacterium tumefaciens has mediated genetic transformation in plant biotechnology, leading to the discovery and application of several genes to enhance agricultural traits (Table 2.1). However, plants comprise an array of genetic regulatory mechanisms, and a gene that is functional in one crop may not necessarily be functional in another (Biswas *et al.*, 2021).

Table 2.1. Targeted genes involved in improving nutritional quality and stress tolerance in various crops

Crop	Phenotype	Genes	Reference
Cocoa (*Theobroma cacao*)	Disease resistance	*TcNPR3*	Fister *et al*. (2018)
Rice (*Oryza sativa*)	Drought resistance	*OsSAPK2*	Lou *et al*. (2017)
	Enhancing blast resistance	*OsERF922*	Wang *et al*. (2016)
	Drought and salinity tolerance	*OsDST*	Kumar *et al*., 2020)
	Salinity tolerance	*OsNAC041*	Bo *et al*. (2019)
	Herbicide tolerance	ALS	Endo *et al*. (2016)
	Drought tolerance	*OsDREB1F*	Singh *et al*. (2015)
Kiwi (*Actinidia chinensis*)	Growth performance	*AcCEN4, AcCEN*	Varkonyi-Gasic *et al*. (2019)
Rapeseed (*Brassica napus*)	Seed oil quality improvement	*ALCATRAZ (ALC)*	Braatz *et al*. (2017)
	High oil production	*BnTT8*	Zhai *et al*. (2020)
	Resistance to *Fusarium graminearum*	*TaNFXL1*	Brauer *et al*. (2020)
	Improved resistance against powdery mildew	*TaEDR1*	Zhang *et al*. (2017)
Gold of Pleasure (*Camelina sativa*)	Improvement in oleic acid	*FAD2*	Jiang *et al*. (2017)
Soybean (*Glycine max*)	Herbicide resistance	*ALS1*	Li *et al*. (2015)
	Low polyunsaturated fats	*FAD2-1A, FAD2-1B*	Demorest *et al*. (2016)
	Salt and drought tolerance	*Drb2a and Drb2b*	Curtin *et al*. (2011)
	Herbicide tolerance	*ALS*	Yadava *et al*., 2016
	Herbicide resistance	*PAT*	Schornack *et al*. (2006)
Watermelon (*Citrullus lanatus*)	Herbicide resistance	*ALS*	Tian *et al*. (2018)
Lettuce (*Lactuca sativa*)	Yield increase	*LsNCED4*	Bertier *et al*. (2018)
Tobacco (*Nicotiana tabacum*)	Auxin biosynthesis	*NtPIN4*	Xie *et al*. (2017)
Tomato (*Solanum lycopersicum*)	Plant chilling tolerance	*SlCBF1*	Li *et al*. (2018)
	Powdery mildew tolerance	*SlPMR4*	Santillán Martínez *et al*. (2020)
	Parthenocarpy (leading to seedless fruit)	*SlIAA9*	Ueta *et al*. (2017)
	Heat stress	*SlAGL6*	Klap *et al*. (2017)
Chrysanthemum (*Chrysanthemum morifolium*)	Increased peroxidase activity and proline content along with reduced malondialdehyde content	*CgHSP70*	Song *et al*. (2014)

Continued

Table 2.1. Continued

Crop	Phenotype	Genes	Reference
Potato (*Solanum tuberosum*)	Increased amylopectin content	*GBSS*	Andersson *et al*. (2017)
	Herbicide resistance in transgene-free potato	*StALS2*	Veillet *et al*. (2019)
Barley (*Hordeum vulgare*)	Characterization of gene utility for crop enhancement	*HvPM19*	Lawrenson *et al*. (2015)
Cotton (*Gossypium hirsutum*)	Insect pest resistance	*Cry1Ac, Cry2Ac*	Puspito *et al*. (2015)
	Herbicide tolerance	*EPSPS*	D'Halluin *et al*. (2013)
Citrus	Resistance to citrus canker	*CsLOB1*	Jia *et al*. (2017)
Sugarcane (*Saccharum officinarum*)	Herbicide tolerance	*ALS*	Oz *et al*. (2021)
Canola (*Brassica napus*)	Drought tolerance	BnaA6.RGA (protein)	Wu *et al*. (2020)
Petunia	Increased shelf-life	*PhACO1*	Xu *et al*. (2020)
Carrot (*Daucus carota*)	Purple depigmented carrot	*DcPDS, DcMYB113*	Xu *et al*. (2019)

Therefore, it may be possible to overcome this crop improvement obstacle by isolating the diverse alleles from various natural resources. Recent advances in genome editing technologies, such as TALENs, ZFNs, and CRISPR-Cas, have enabled the application of insights from model plants like *Arabidopsis thaliana* to cultivated species (Chen *et al*., 2019). These methods have generated numerous editing events in staple crops, which are crucial for ensuring significant advances in the development of crops that are tolerant to environmental constraints.

2.4.1 Abiotic stress tolerance

Abiotic stresses such as salt, drought, heat, cold, and floods have detrimental impacts on crop growth and development. However, sometimes resistance against abiotic stressors is transient because resistance mechanisms rely on the introduction of a single resistant gene into plants (Steiner *et al*., 2019). In order to produce transgenic crops with improved stress resistance, it is critical to identify the specific genes that contribute to plant tolerance to various abiotic stresses. Once quantitative trait locus (QTL) mapping identifies the genes regulating tolerance, hybridization and genetic engineering with marker-assisted selection can incorporate these genes into the genetic background of any selected cultivar (Gantait and Mondal, 2018). Stressful circumstances trigger an intricate transcriptional module in plants, influenced by the type of stress and specific environmental conditions.

These adaptive responses by plants against abiotic stresses not only mitigate yield losses but also foster superior growth and development (Rejeb *et al*., 2014). Furthermore, Calleja-Cabrera *et al*. (2020) have noted that a well-developed root system is crucial for plant adaptation to diverse soil conditions and fluctuating climates. Ultimately, more tolerant transgenic plants are necessary to feed growing populations and lessen the damage that abiotic stresses do to crop yields (Noman *et al*., 2017). Recent breakthroughs in biotechnology have radically altered the potential of gene discovery, presenting several opportunities for scientists to build stress-resistant crops.

Drought resistance

Drought is a major constraint under global warming scenarios and hinders agricultural productivity worldwide (Mishra *et al.*, 2012). Plants have a variety of adaptation strategies against drought stress, and reactive oxygen species (ROS) play an integral part in this regard. ROS accumulation from extreme climatic conditions severely damages plants. In response, plants make antioxidative enzymes such as ascorbate peroxidase (APX), superoxide dismutase (SOD), peroxidase (POD), and catalase (CAT), all of which remove ROS from the body. Additionally, the high amounts of malondialdehyde (MDA) and electrolyte leakage show that plant cell membranes are breaking down (Noor *et al.*, 2022). A lot of ROS cause the transcription of many genes involved in cells, enzymes, defense responses, photosynthesis, and chlorophylls (Anwar *et al.*, 2018). At lower concentrations, the hydrogen peroxide signaling pathway (H_2O_2) triggers the plant's defense system to mitigate the harmful effects of abiotic stress. However, at higher concentrations, it becomes highly toxic and exerts a severe impact on the plant's defense system (Gill and Tuteja, 2010).

Enhancing crop resilience to drought and high temperatures is crucial in the face of climate change. Plants adjust to drought conditions by altering their physiological and metabolic processes, primarily mediated at the transcriptional level (Gupta *et al.*, 2020). Since roots are the most important component of a plant for sensing and establishing drought responses, the majority of studies on drought have concentrated on roots (Dinneny, 2019). Consequently, researchers have identified a variety of QTLs that modify the morphology of roots in a variety of crops, including wheat and tomatoes, in response to drought stress (Teramoto *et al.*, 2020). Furthermore, Rady (2011) discovered that the overexpression of *OsNAC6* in rice led to alterations in the root structure, an increase in the number and size of roots, and enhanced resilience to drought stress. Singh *et al.* (2015) discovered that the isolated gene betA from *Rhizobium meliloti*, which encodes glycine betaine and choline dehydrogenase, enhances drought resistance. The overexpression of *AtNAC2* demonstrated enhanced resistance to reduced water availability, indicating its potential as a gene for agricultural water stress tolerance. Furthermore, overexpressing the *MuNAC4* gene made it much easier for the horse gram to handle water stress by protecting membrane structures and making it easier to control osmotic and antioxidant enzymes. Other studies observed that *OsNAC9* and *OsNAC14* not only contributed to drought resistance in rice but also increased grain yield (Redillas *et al.*, 2012; Shim *et al.*, 2018). Lee *et al.* (2017) found that *OsNAC6* alters root architecture, further enhancing the drought resistance of rice plants.

Overall, these results show that the NAC family of genes is important for making plants better at handling water stress. They also suggest some interesting genetic targets for making crops more resistant to drought stress.

Salinity tolerance

Salinity poses a significant challenge to crop production worldwide, as the majority of agricultural plants are vulnerable to its destructive impacts (Thitisaksakul *et al.*, 2015). Salinity affects plant performance through several mechanisms, such as changes in metabolic processes, disruptions in cell membranes, abnormal cell division and growth, decreased photosynthetic activity and protein production, and disturbances in enzyme functions (Fig. 2.3). Furthermore, excessive Na^+ and Cl^- ions in plant cells generate a high level of ROS, which is exceedingly harmful and causes oxidative stress (Kumar *et al.*, 2015). High salinity additionally decreases the absorption of water and nutrients by disrupting the osmotic balance, resulting in biochemical and enzymatic alterations (Chappell *et al.*, 2015). The salt overly sensitive (SOS) signaling system, which consists of SOS1, SOS2, and SOS3, has a profound impact on ion homeostasis and salinity tolerance. SOS1 is crucial for Na^+ and H^+ trafficking in the plasma membrane.

Significant progress has been made in rice's ability to survive salt stress (Bhat *et al.*, 2021). *OsCBL4* and *OsCIPK24*, which are CBL-interacting protein kinases in rice, enhance the transport activity of *OsSOS1* to reduce the buildup of Na^+ in the cell and mitigate the adverse effects of salt stress (Martínez-Atienza *et al.*, 2007). High-affinity potassium transporters (HTKs) play a crucial

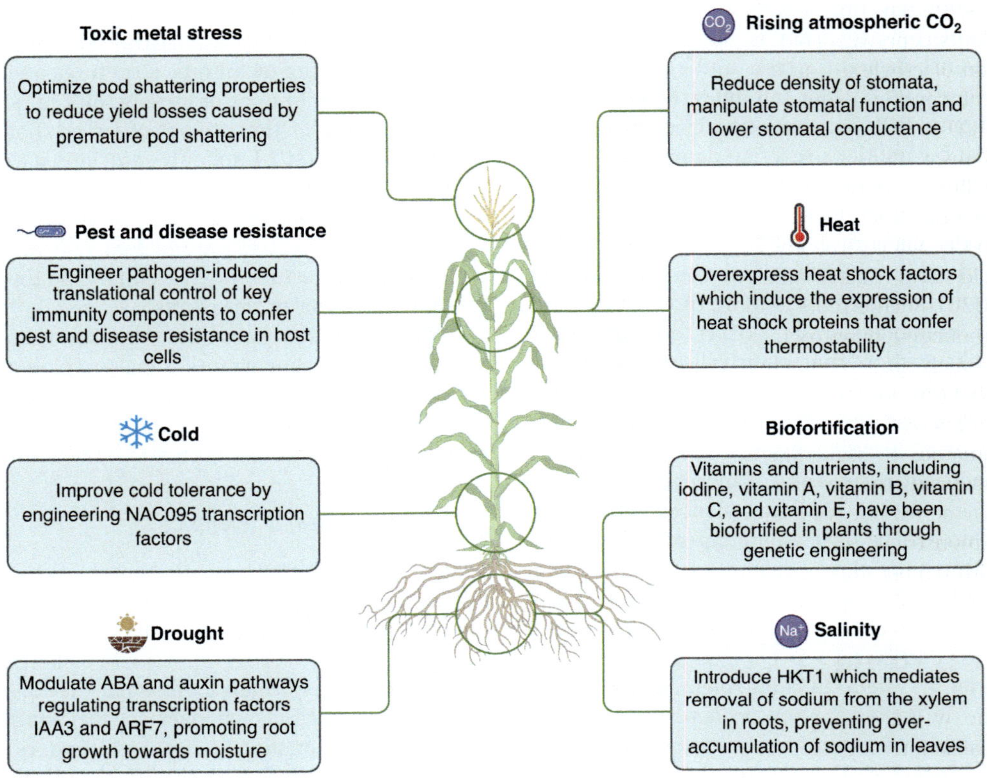

Fig. 2.3. Applications of genetic engineering for improving nutritional quality and stress tolerance.

role in controlling potassium levels in plants and modulating the effects of salt stress. Rice ability to withstand salt stress was enhanced by HKT genes, including *OsHKT1;1*, *OsHKT1;4*, and *OsHKT1;5* (Wang *et al.*, 2016). According to Kobayashi *et al.* (2017), *Oryza sativa* salt tolerance is regulated by the major QTL SKC1 (encode *OsHKT1;5*). The overexpression of *OsHTK1;5* enhanced salt tolerance by inhibiting the accumulation of MDA and H_2O_2 and activating antioxidant enzyme activities.

Thus, we assumed that some crop varieties that are resistant to salinity express salinity-tolerant genes; these genes can be mapped using microsatellite markers to facilitate the breeding of salt-tolerant lines. Moreover, plant biotechnology can further help in the generation of transgenes from the identified genes that may be implicated in salt stress resistance.

Heat and cold tolerance

Temperature is a crucial environmental factor that plays a vital role in the development and ecological dispersion of plant species worldwide. Temperature extremes adversely affect plant growth, reproduction, and output at all stages of development (see Fig. 2.3). Even a slight fluctuation in temperature stress can cause a substantial decrease in crop output, especially while the plants are in the flowering stage (Anwar *et al.*, 2018). The global production of major crops, such as wheat, rice, and maize, decreases by 6%, 3.2%, and 7.4%, respectively, with each 1°C increase in global temperature. Therefore, it is imperative to develop stress-resistant crops to address this threat (Zhao *et al.*, 2017).

A variety of crops have identified genes that encode metabolites and proteins that can withstand elevated temperatures (Chen *et al.*, 2017). According to Zhang *et al.* (2016), new research has shown that miR396b helps *Arabidopsis* plants cope with cold by reducing the production of ethylene and levels of the transcript 1-aminocyclopropane-1-carboxylic acid oxidase (ACO). Similarly, studies have linked a number of microRNAs and transcription factors to the ability to withstand temperature stress. Reports suggest that miR319, miR535, miR396b, and *OsNAC095* can enhance rice resistance to cold stress (Bustamante *et al.*, 2018; Sun *et al.*, 2020). Furthermore, *PIF4* is essential for auxin production, hypocotyl elongation, and the ability to withstand high temperatures under stress (Nadeem *et al.*, 2019). Genetic modification has enhanced the ability of transgenic apples and tomatoes to withstand high temperatures by increasing the expression of the cytosolic ascorbate peroxidase (cAPX) gene. These plants have demonstrated the capacity to resist field settings with temperatures reaching up to 40°C (Wang *et al.*, 2005).

Toxic metal tolerance

The world is now experiencing rapid industrialization and technological growth, which is causing heavy metal toxicity. Heavy metal toxicity is a serious global problem since it not only depletes agricultural soil fertility but also poses a hazard to human health (Sharma *et al.*, 2016). Heavy metals either exchange important metal ions or block functional groups, causing oxidative damage. Redox-active metals like Fe and Cu produce ROS directly through redox processes, whereas other metals, including Pb, Cd, Ni, Al, Mn, and Zn, produce ROS indirectly (Anwar *et al.*, 2018). Opdenakker *et al.* (2012) reported on the various kinetics of mitogen-activated protein kinase (MAPK) cascades in response to metal stress. Researchers have also implicated cation diffusion facilitators, also known as the metal tolerance protein (MTP), in heavy metal tolerance. According to Delhaize *et al.* (2007), the number of MTP genes identified in *Arabidopsis* and rice to date is 12 and 10, respectively. The *AhMTP1* gene is thought to be involved in Zn hypertolerance in *A. halleri*, a Zn hyperaccumulator plant (Shahzad *et al.*, 2010). Unlike the *AtMTP1* gene, *AtMTP3* primarily expresses in the roots and is believed to contribute to Zn homeostasis by tolerating excess Zn (Lan *et al.*, 2013). Recently, silencing of *PpNRAMP5* has improved manganese toxicity tolerance in peach (Noor *et al.*, 2023).

2.4.2 Biotic stress resistance

There are many factors that reduce the yield of food cereals; it is estimated that global annual losses from biotic stressors are approximately 25% (Pandey *et al.*, 2017). Biotic stressors (insects, diseases, and weeds) are the main source of pre- and postharvest losses in the field (Singla and Krattinger, 2016). The global agricultural industry is very concerned about the suddenly falling crop yield and productivity as a result of biotic stresses (see Table 2.1). For a number of years, we have used chemical pesticides to reduce biotic stress damage but a number of unfavorable factors, such as high production costs with multiple applications that most small-scale farmers cannot afford, have discouraged their use in recent times (Sharifzadeh *et al.*, 2018). In the past, crop domestication was essential for livelihood but a burgeoning population and escalating demand in the contemporary world are establishing new objectives for plant breeding (Alseekh *et al.*, 2021).

Pest and disease resistance

Plant diseases and insect pests are the two main biotic stresses in agriculture, causing 11–13% of average global loss (Savary *et al.*, 2019). Interestingly, the global increase in temperatures also influences the extent of plant damage, amplifying its severity (see Table 2.1 and Fig. 2.3). Consequently, insect infestations have a significant impact on global food security, particularly in regions where there is already a food shortage. The rapid evolution of pathogens due to pesticide resistance is diminishing the efficacy of pesticides (Berger *et al.*, 2017). Furthermore, pests can quickly overcome resistance acquired through a single gene, so developing unique and effective strategies for improving crop stress resistance is crucial for increasing yield and quality in the long run (Bailey-Serres *et al.*, 2019).

Since most traits are multigene governed, evolving resistance to biotic stressors (pesticides, diseases, pests, and weeds) means amending multiple genes (Gupta and Pandey, 2014). For example, wheat breeders will need to select blast resistance for wheat while maintaining immunity to stem rust (Islam *et al.*, 2016). Al Atalah *et al.* (2014) successfully expressed a lectin from *Oryza sativa* seedlings in transgenic tobacco, demonstrating insecticidal effects against beetroot armyworm, green peach aphid, and pea aphid. Additionally, Jagdish *et al.* (2020) effectively explored protease inhibitors like trypsin inhibitor and potato protease inhibitor II to confer insect resistance in several crops.

Biofortification

Malnutrition is a global concern that affects an estimated 800 million people, the majority of whom are in underdeveloped nations. Considering this concern, one of the key goals of breeding programs is to enhance crop nutritional quality in order to prevent diseases associated with malnutrition (Sinha *et al.*, 2019). By mainstreaming nutrition as a vital component in the breeding of basic crops, biofortified crop cultivars for different nutrients could be generated to alleviate this problem (see Fig. 2.3). Currently, 30 countries have introduced a total of 262 biofortified cultivars, spanning 12 different crops.

Using the natural variation within the ICTP 8203 population as a case study, researchers introduced the 'Dhanshakti' and 'Chakti' pearl millet varieties, enriched with high levels of Fe and Zn, in India and Africa (Govindaraj *et al.*, 2019). Under the control of GluB-1, the rice seed-storage protein glutelin promoter, researchers developed transgenic rice with soybean ferritin to enhance the iron level in rice 2–3-fold (Lucca *et al.*, 2001). By modifying the genes *AK* (*lysC*) and *DHPS* (*dapA*), which code for important enzymes involved in lysine biosynthesis, rice lysine content has been increased by up to 25 times (Yang *et al.*, 2016). CRISPR-Cas9-mediated editing of *BnTT8* homologs, yellow-seeded mutants in rapeseed, resulted in a 9.47% increase in oil content in the T2 generation (Zhai *et al.*, 2020). Furthermore, many other vitamins and nutrients, including iodine, vitamin A, vitamin B, vitamin C, and vitamin E, have been biofortified in plants.

Biofortification offers numerous possibilities to improve the nutritional value of plants and thereby ameliorate the malnutrition issue.

Yield improvement through enhancements in crop traits

Modern crops have a high level of efficiency in utilizing carbon and nutrients. This allows them to rapidly expand their leaf canopies in order to optimize light interception. However, in the process of photosynthesis, there are situations where the crops exhibit reduced efficiency in converting absorbed light energy into sugars (Zhu *et al.*, 2010). This could be due to the proteins and enzymes involved in photosynthesis having evolved in a marine environment with low light levels, which differs greatly from current atmospheric and agronomic conditions (Dann and Leister, 2017).

We employed genetic engineering to promote the production of enzymes that control photosynthesis and nonphotochemical quenching. This resulted in enhanced photosynthesis and a 14–20% growth in vegetative biomass (Kromdijk *et al.*, 2016). Similarly, overexpression of both the small and large Rubisco subunits, along with an assembly chaperone protein, resulted in an increase in maize photosynthesis and fresh weight (Salesse-Smith *et al.*, 2018). Overexpressing sedoheptulose-1,7-biphosphatase in wheat led to better photosynthesis and more biomass (Driever *et al.*, 2017). The potential of these genetic modifications in important crops provides an opportunity to assess their effectiveness and introduce useful traits to elite cultivars.

2.5 Future Prospects

Developing crops with high-yield potential, resistance to biotic and abiotic stresses, and climate resilience is essential for meeting the food demands of a growing global population. Plant genetic resources play a crucial role in enhancing the sustainability of the ecosystem, ensuring food security, and addressing the root cause of hunger. Bioinformatics, an emerging discipline within the agricultural sciences, is facilitating the integration of several disciplines to enhance our understanding of the complex mechanisms behind diverse crop plant characteristics. Crop production, soil profile, crop irrigation, weeding, and crop monitoring are just a few of the variables where AI-based solutions can assist in controlling and enhancing efficiency. As AI advances and incorporates more sophisticated algorithms, its application in agriculture will continue to grow. We can utilize genome editing, another marvel of this era, to produce crops with desirable traits.

By integrating the capabilities of bioinformatics, harnessing the potential of artificial intelligence, and employing genome engineering techniques, it is possible to create crops that are resilient to stress. These crops would not only thrive in a variety of conditions, but would also significantly increase production to address the global issue of malnutrition.

References

Abbas, M.A., Haroun, S.A. and Mowafy, A.M. (2021) Dual inoculation of Bradyrhizobium and enterobacter alleviates the adverse effect of salinity on glycine max seedling. *Notulae Botanicae Horti Agrobotanici Cluj-Napoca* 49(3), 12461.

Abudayyeh, O.O., Gootenberg, J.S., Essletzbichler, P., Han, S., Joung, J. *et al.* (2017) RNA targeting with CRISPR-Cas13. *Nature* 550(7675), 280–284.

Al Atalah, B., Smagghe, G. and Damme, E.J.M. (2014) Orysata, a jacalin-related lectin from rice, could protect plants against biting-chewing and piercing-sucking insects. *Plant Science* 221, 21–28.

Alseekh, S., Scossa, F., Wen, W., Luo, J., Yan, J. *et al.* (2021) Domestication of crop metabolomes: Desired and unintended consequences. *Trends in Plant Science* 26(6), 650–661.

Andersson, M., Turesson, H., Nicolia, A., Fält, A.S., Samuelsson, M. *et al.* (2017) Efficient targeted multiallelic mutagenesis in tetraploid potato (*Solanum tuberosum*) by transient CRISPR-Cas9 expression in protoplasts. *Plant Cell Reports* 36(1), 117–128.

Anwar, A., Liu, Y., Dong, R., Bai, L., Yu, X. *et al.* (2018) The physiological and molecular mechanism of brassinosteroid in response to stress: A review. *Biological Research* 51(1), 46.

Bailey-Serres, J., Parker, J.E., Ainsworth, E.A., Oldroyd, G.E.D. and Schroeder, J.I. (2019) Genetic strategies for improving crop yields. *Nature* 575(7781), 109–118.

Barrangou, R. and Doudna, J.A. (2016) Applications of CRISPR technologies in research and beyond. *Nature Biotechnology* 34(9), 933–941.

Batterman, J. and Leemans, J. (1988) Engineering herbicide resistance in plants. *Trends in Genetics* 4, 219–222.

Berger, S., El Chazli, Y., Babu, A.F. and Coste, A.T. (2017) Azole resistance in *Aspergillus fumigatus*: A consequence of antifungal use in agriculture? *Frontiers in Microbiology* 8, 1024.

Bertier, L.D., Ron, M., Huo, H., Bradford, K.J., Britt, A.B. *et al.* (2018) High-resolution analysis of the efficiency, heritability, and editing outcomes of CRISPR/Cas9-induced modifications of *NCED4* in lettuce (*Lactuca sativa*). *G3: Genes, Genomes, Genetics* 8(5), 1513–1521.

Bhat, M.A., Mir, R.A., Kumar, V., Shah, A.A., Zargar, S.M. *et al.* (2021) Mechanistic insights of CRISPR/Cas-mediated genome editing towards enhancing abiotic stress tolerance in plants. *Physiologia Plantarum* 172(2), 1255–1268.

Biswas, S., Zhang, D. and Shi, J. (2021) CRISPR/Cas systems: Opportunities and challenges for crop breeding. *Plant Cell Reports* 40, 979–998.

Bo, W., Zhaohui, Z., Huanhuan, Z., Xia, W., Binglin, L. *et al.* (2019) Targeted mutagenesis of NAC transcription factor gene, *OsNAC041*, leading to salt sensitivity in rice. *Rice Science* 26(2), 98–108.

Braatz, J., Harloff, H.J., Mascher, M., Stein, N., Himmelbach, A. *et al.* (2017) CRISPR-Cas9 targeted mutagenesis leads to simultaneous modification of different homoeologous gene copies in polyploid oilseed rape (*Brassica napus*). *Plant Physiology* 174(2), 935–942.

Brauer, E.K., Balcerzak, M., Rocheleau, H., Leung, W., Schernthaner, J. *et al.* (2020) Genome editing of a deoxynivalenol-induced transcription factor confers resistance to *Fusarium graminearum* in wheat. *Molecular Plant-Microbe Interactions* 33(3), 553–560.

Bustamante, A., Marques, M.C., Sanz-Carbonell, A., Mulet, J.M. and Gomez, G. (2018) Alternative processing of its precursor is related to miR319 decreasing in melon plants exposed to cold. *Scientific Reports* 8(1), 15538.

Calleja-Cabrera, J., Boter, M., Oñate-Sánchez, L. and Pernas, M. (2020) Root growth adaptation to climate change in crops. *Frontiers in Plant Science* 11, 544.

Carlson, D.F., Fahrenkrug, S.C. and Hackett, P.B. (2012) Targeting DNA with fingers and TALENs. *Molecular Therapy. Nucleic Acids* 1, e3.

Carroll, D. (2011) Genome engineering with zinc-finger nucleases. *Genetics* 188(4), 773–782.

Carroll, D., Morton, J.J., Beumer, K.J. and Segal, D.J. (2006) Design, construction and *in vitro* testing of zinc finger nucleases. *Nature Protocols* 1(3), 1329–1341.

Chakroun, M., Banyuls, N., Bel, Y., Escriche, B. and Ferré, J. (2016) Bacterial vegetative insecticidal proteins (vip) from entomopathogenic bacteria. *Microbiology and Molecular Biology Reviews* 80(2), 329–350.

Chang, H.H.Y., Pannunzio, N.R., Adachi, N. and Lieber, M.R. (2017) Non-homologous DNA end joining and alternative pathways to double-strand break repair. *Nature Reviews. Molecular Cell Biology* 18(8), 495–506.

Chappell, J., Watters, K.E., Takahashi, M.K. and Lucks, J.B. (2015) A renaissance in RNA synthetic biology: New mechanisms, applications and tools for the future. *Current Opinion in Chemical Biology* 28, 47–56.

Chen, K., Wang, Y., Zhang, R., Zhang, H. and Gao, C. (2019) CRISPR/Cas genome editing and precision plant breeding in agriculture. *Annual Review of Plant Biology* 70, 667–697.

Chen, Z.C., Yamaji, N., Horie, T., Che, J., Li, J. *et al.* (2017) A magnesium transporter OsMGT1 plays a critical role in salt tolerance in rice. *Plant Physiology* 174(3), 1837–1849. DOI: 10.1104/pp.17.00532.

Curtin, S.J., Zhang, F., Sander, J.D., Haun, W.J., Starker, C. *et al.* (2011) Targeted mutagenesis of duplicated genes in soybean with zinc-finger nucleases. *Plant Physiology* 156(2), 466–473.

Dann, M. and Leister, D. (2017) Enhancing (crop) plant photosynthesis by introducing novel genetic diversity. *Philosophical Transactions of the Royal Society B: Biological Sciences* 372(1730), 20160380.

Davuluri, G.R., Tuinen, A., Fraser, P.D., Manfredonia, A., Newman, R. *et al.* (2005) Fruit-specific RNAi-mediated suppression of *DET1* enhances carotenoid and flavonoid content in tomatoes. *Nature Biotechnology* 23(7), 890–895.

Delhaize, E., Gruber, B.D., Pittman, J.K., White, R.G., Leung, H. *et al.* (2007) A role for the AtMTP11 gene of *Arabidopsis* in manganese transport and tolerance. *Plant Journal* 51(2), 198–210.

Demorest, Z.L., Coffman, A., Baltes, N.J., Stoddard, T.J., Clasen, B.M. *et al.* (2016) Direct stacking of sequence-specific nuclease-induced mutations to produce high oleic and low linolenic soybean oil. *BMC Plant Biology* 16(1), 1–8.

D'Halluin, K., Vanderstraeten, C., Hulle, J., Rosolowska, J., Den Brande, I. *et al.* (2013) Targeted molecular trait stacking in cotton through targeted double-strand break induction. *Plant Biotechnology Journal* 11(8), 933–941.

Dinneny, J.R. (2019) Developmental responses to water and salinity in root systems. *Annual Review of Cell and Developmental Biology* 35, 239–257.

Driever, S.M., Simkin, A.J., Alotaibi, S., Fisk, S.J., Madgwick, P.J. *et al.* (2017) Increased sbpase activity improves photosynthesis and grain yield in wheat grown in greenhouse conditions. *Philosophical Transactions of the Royal Society B: Biological Sciences* 372(1730), 20160384.

Endo, M., Mikami, M. and Toki, S. (2016) Biallelic gene targeting in rice. *Plant Physiology* 170(2), 667–677.

Fister, A.S., Landherr, L., Maximova, S.N. and Guiltinan, M.J. (2018) Transient expression of CRISPR/Cas9 machinery targeting *TcNRP3* enhances defense response in *Theobroma cacao*. *Frontiers in Plant Science* 9, 268.

Gaj, T., Liu, J., Anderson, K.E., Sirk, S.J. and Barbas, C.F. (2013) Protein delivery using Cys2-His2 zinc-finger domains. *ACS Chemical Biology* 9, 1662–1667.

Gantait, S. and Mondal, S. (2018) Transgenic approaches for genetic improvement in groundnut (*Arachis hypogaea* L.) against major biotic and abiotic stress factors. *Journal of Genetic Engineering & Biotechnology* 16, 537–544.

Gill, S.S. and Tuteja, N. (2010) Reactive oxygen species and antioxidant machinery in abiotic stress tolerance in crop plants. *Plant Physiology and Biochemistry* 48(12), 909–930.

Giraldo, J.P., Wu, H., Newkirk, G.M. and Kruss, S. (2019) Nanobiotechnology approaches for engineering smart plant sensors. *Nature Nanotechnology* 14(6), 541–553.

Gonsalves, D. (1998) Control of papaya ringspot virus in papaya: A case study. *Annual Review of Phytopathology* 36(1), 415–437.

Gonsalves, D., Gonsalves, C., Ferreira, C. and Fitch, M. (2004) Transgenic virus-resistant papaya: From hope to reality in controlling papaya ringspot virus in Hawaii. Available at: www.apsnet.org/edcenter/apsnetfeatures/Pages/papayaringspot.aspx

Govindaraj, M., Rai, K.N., Cherian, B., Pfeiffer, W.H., Kanatti, A. *et al.* (2019) Breeding biofortified pearl millet varieties and hybrids to enhance millet markets for human nutrition. *Agriculture* 9(5), 106.

Gupta, A., Rico-Medina, A. and Caño-Delgado, A.I. (2020) The physiology of plant responses to drought. *Science* 368(6488), 266–269.

Gupta, S.K. and Pandey, M.K. (2014) Breeding for disease and insect-pest resistance. In: Abrol, D.P. (ed.) *Integrated Pest Management*. Academic Press, San Diego, pp. 401–417.

Herrera-Estrella, L., Depicker, A., Montagu, M. and Schell, J. (1983) Expression of chimaeric genes transferred into plant cells using a Ti-plasmid-derived vector. *Nature* 303(5914), 209–213.

Horsch, R.B., Fry, J.E., Hoffmann, N.L., Wallroth, M., Eichholtz, D. *et al.* (1985) A simple and general method for transferring genes into plants. *Science* 227(4691), 1229–1231.

Islam, M.T., Croll, D., Gladieux, P., Soanes, D.M., Persoons, A. *et al.* (2016) Emergence of wheat blast in Bangladesh was caused by a South American lineage of *Magnaporthe oryzae*. *BMC Biology* 14(1), 84.

Jagdish, T., Morris, J.J., Wade, B.D. and Blount, Z.D. (2020) Probing the deep genetic basis of a novel trait in escherichia coli. In: Morris, J.J. (ed.) *Evolution in Action: Past, Present and Future*. Springer, New York, pp. 107–122.

James, C. (2017) Global status of commercialized biotech/GM crops in 2017: Biotech crop adoption surges as economic benefits accumulate in 22 years. Available at: www.isaaa.org/resources/publications/briefs/53/download/isaaa-brief-53-2017.pdf

Jia, H., Zhang, Y., Orbović, V., Xu, J., White, F.F. *et al.* (2017) Genome editing of the disease susceptibility gene CsLOB1 in citrus confers resistance to citrus canker. *Plant Biotechnology Journal* 15(7), 817–823.

Jiang, W.Z., Henry, I.M., Lynagh, P.G., Comai, L. and et al, A. (2017) Significant enhancement of fatty acid composition in seeds of the allohexaploid, *Camelina sativa*, using CRISPR/Cas9 gene editing. *Plant Biotechnology Journal* 15, 648–657.

Klap, C., Yeshayahou, E., Bolger, A.M., Arazi, T., Gupta, S.K. *et al.* (2017) Tomato facultative parthenocarpy results from SLAGAMOUS-LIKE 6 loss of function. *Plant Biotechnology Journal* 15(5), 634–647.

Klas, E., Fuchs, M. and Gonsalves, D. (2011) Fruit yield of virus-resistant transgenic summer squash in simulated commercial plantings under conditions of high disease pressure. *Journal of Horticulture and Forestry* 3(2), 46–52.

Kobayashi, N.I., Yamaji, N., Yamamoto, H., Okubo, K., Ueno, H. *et al.* (2017) OsHKT1;5 mediates Na^+ exclusion in the vasculature to protect leaf blades and reproductive tissues from salt toxicity in rice. *Plant Journal* 91(4), 657–670.

Kromdijk, J., Głowacka, K., Leonelli, L., Gabilly, S.T., Iwai, M. *et al.* (2016) Improving photosynthesis and crop productivity by accelerating recovery from photoprotection. *Science* 354(6314), 857–861.

Kumar, K., Gambhir, G., Dass, A., Tripathi, A.K., Singh, A. *et al.* (2020) Genetically modified crops: Current status and future prospects. *Planta* 251, 1–27.

Kumar, M., Choi, J.-Y., Kumari, N., Pareek, A. and Kim, S.R. (2015) Molecular breeding in brassica for salt tolerance: Importance of microsatellite (SSR) markers for molecular breeding in Brassica. *Frontiers in Plant Science* 6, 688.

Labbé, P., Jean-Philippe, D.H., Alout, P.M., Djogbenou, L., Pasteur, N. *et al.* (2017) Evolution of resistance to insecticide in disease vectors. In: Tibayrenc, M. (ed.) *Genetics and Evolution of Infectious Diseases*, 2nd edn. Elsevier, New York, pp. 313–339.

Lamichhane, J.R., Dachbrodt-Saaydeh, S., Kudsk, P. and Messéan, A. (2016) Toward a reduced reliance on conventional pesticides in European agriculture. *Plant Disease* 100(1), 10–24.

Lan, H.X., Wang, Z.F., Wang, Q.H., Wang, M.M., Bao, Y.M. *et al.* (2013) Characterization of a vacuolar zinc transporter *OZT1* in rice (*Oryza sativa* L.). *Molecular Biology Reports* 40(2), 1201–1210.

Landry, M.P. and Mitter, N. (2019) How nanocarriers delivering cargos in plants can change the GMO landscape. *Nature Nanotechnology* 14(6), 512–514.

Lanigan, T.M., Kopera, H.C. and Saunders, T.L. (2020) Principles of genetic engineering. *Genes* 11(3), 291.

Lawrenson, T., Shorinola, O., Stacey, N., Li, C., Østergaard, L. *et al.* (2015) Induction of targeted, heritable mutations in barley and *Brassica oleracea* using RNA-guided Cas9 nuclease. *Genome Biology* 16, 258.

Lee, D.K., Chung, P.J., Jeong, J.S., Jang, G., Bang, S.W. *et al.* (2017) The rice OsNAC6 transcription factor orchestrates multiple molecular mechanisms involving root structural adaptions and nicotianamine biosynthesis for drought tolerance. *Plant Biotechnology Journal* 15(6), 754–764.

Li, R., Zhang, L., Wang, L., Chen, L., Zhao, R. *et al.* (2018) Reduction of tomato-plant chilling tolerance by CRISPR-Cas9-mediated SlCBF1 mutagenesis. *Journal of Agricultural and Food Chemistry* 66(34), 9042–9051.

Li, T., Liu, B., Spalding, M.H., Weeks, D.P. and Yang, B. (2012) High-efficiency TALEN-based gene editing produces disease-resistant rice. *Nature Biotechnology* 30(5), 390–392.

Li, Z., Liu, Z.B., Xing, A., Moon, B.P., Koellhoffer, J.P. *et al.* (2015) Cas9-guide RNA directed genome editing in soybean. *Plant Physiology* 169(2), 960–970.

Liang, Z., Chen, K., Li, T., Zhang, Y., Wang, Y. *et al.* (2017) Efficient DNA-free genome editing of bread wheat using CRISPR/Cas9 ribonucleoprotein complexes. *Nature Communications* 8, 14261.

Lin, Q., Zong, Y., Xue, C., Wang, S., Jin, S. *et al.* (2020) Prime genome editing in rice and wheat. *Nature Biotechnology* 38(5), 582–585.

Liu, Q., Yang, F., Zhang, J., Liu, H., Rahman, S. *et al.* (2021) Application of CRISPR/Cas9 in crop quality improvement. *International Journal of Molecular Sciences* 22(8), 4206.

Lou, D., Wang, H., Liang, G. and Yu, D. (2017) OsSAPK2 confers abscisic acid sensitivity and tolerance to drought stress in rice. *Frontiers in Plant Science* 8, 993.

Lucca, P., Hurrell, R. and Potrykus, I. (2001) Genetic engineering approaches to improve the bioavailability and the level of iron in rice grains. *Theoretical and Applied Genetics* 102(2–3), 392–397.

Ma, X., Zhang, X., Liu, H. and Li, Z. (2020) Highly efficient DNA-free plant genome editing using virally delivered CRISPR-cas9. *Nature Plants* 6(7), 773–779.

Martínez-Atienza, J., Jiang, X., Garciadeblas, B., Mendoza, I., Zhu, J.-K. *et al.* (2007) Conservation of the salt overly sensitive pathway in rice. *Plant Physiology* 143(2), 1001–1012.

Ming, M., Ren, Q., Pan, C., He, Y., Zhang, Y. *et al.* (2020) CRISPR-Cas12b enables efficient plant genome engineering. *Nature Plants* 6(3), 202–208.

Mishra, K.B., Iannacone, R., Petrozza, A., Mishra, A., Armentano, N. *et al.* (2012) Engineered drought tolerance in tomato plants is reflected in chlorophyll fluorescence emission. *Plant Science* 182, 79–86.

Luo, M., Li, H., Chakraborty, S., Morbitzer, R., Rinaldo, A. *et al.* (2019) Efficient TALEN-mediated gene editing in wheat. *Plant Biotechnology Journal* 17(11), 2026–2028.

Luo, X., Liu, Z., Zhang, S., Shen, S., Wang, D. *et al.* (2014) The progress of plant transgenic technology. *Chinese Agricultural Science Bulletin* 30, 234–240.

Mackelprang, R. and Lemaux, P.G. (2020) Genetic engineering and editing of plants: an analysis of new and persisting questions. *Annual Review of Plant Biology* 71, 659–687.

Nadeem, M., Li, J., Yahya, M., Wang, M., Ali, A. *et al.* (2019) Grain legumes and fear of salt stress: Focus on mechanisms and management strategies. *International Journal of Molecular Sciences* 20(4), 799.

Noman, A., Aqeel, M., Deng, J., Khalid, N., Sanaullah, T. *et al.* (2017) Biotechnological advancements for improving floral attributes in ornamental plants. *Frontiers in Plant Science* 8, 530.

Noor, I., Sohail, H., Sun, J., Nawaz, M.A., Li, G. *et al.* (2022) Heavy metal and metalloid toxicity in horticultural plants: Tolerance mechanism and remediation strategies. *Chemosphere* 303(Pt 3), 135196.

Noor, I., Sohail, H., Zhang, D., Zhu, K., Shen, W. *et al.* (2023) Silencing of *PpNRAMP5* improves manganese toxicity tolerance in peach (*Prunus persica*) seedlings. *Journal of Hazardous Materials* 454, 131442.

Odipio, J., Ogwok, E., Taylor, N.J., Halsey, M., Bua, A. *et al.* (2014) RNAi-derived field resistance to cassava brown streak disease persists across the vegetative cropping cycle. *GM Crops & Food* 5(1), 16–19.

Opdenakker, K., Remans, T., Vangronsveld, J. and Cuypers, A. (2012) Mitogen-Activated Protein (MAP) kinases in plant metal stress: Regulation and responses in comparison to other biotic and abiotic stresses. *International Journal of Molecular Sciences* 13(6), 7828–7853.

Osakabe, K., Osakabe, Y. and Toki, S. (2010) Site-directed mutagenesis in arabidopsis using custom-designed zinc finger nucleases. *Proceedings of the National Academy of Sciences USA* 107(26), 12034–12039.

Oz, M.T., Altpeter, A., Karan, R., Merotto, A. and Altpeter, F. (2021) CRISPR/Cas9-mediated multi-allelic gene targeting in sugarcane confers herbicide tolerance. *Frontiers in Genome Editing* 3, 673566.

Pandey, P., Irulappan, V., Bagavathiannan, M.V. and Senthil-Kumar, M. (2017) Impact of combined abiotic and biotic stresses on plant growth and avenues for crop improvement by exploiting physio-morphological traits. *Frontiers in Plant Science* 8, 537.

Pérez-de-Luque, A. (2017) Interaction of nanomaterials with plants: What do we need for real applications in agriculture. *Frontiers in Environmental Science* 5, 12.

Pingali, P.L. (2012) Green Revolution: Impacts, limits, and the path ahead. *Proceedings of the National Academy of Sciences USA* 109(31), 12302–12308.

Popp, M.W. and Maquat, L.E. (2016) Leveraging rules of nonsense-mediated mRNA decay for genome engineering and personalized medicine. *Cell* 165(6), 1319–1322.

Puspito, A.N., Rao, A.Q., Hafeez, M.N., Iqbal, M.S., Bajwa, K.S. *et al.* (2015) Transformation and evaluation of Cry1Ac+Cry2A and *GTGene* in *Gossypium hirsutum* L. *Frontiers in Plant Science* 6, 943.

Rady, M.M. (2011) Effect of 24-epibrassinolide on growth, yield, antioxidant system and cadmium content of bean (*Phaseolus vulgaris* L.) plants under salinity and cadmium stress. Scientia Horticulturae 129(2), 232–237.

Raman, R. (2017) The impact of Genetically Modified (GM) crops in modern agriculture: A review. *GM Crops & Food* 8(4), 195–208.

Ran, Y., Liang, Z. and Gao, C. (2017) Current and future editing reagent delivery systems for plant genome editing. *Science China Life Sciences* 60, 490–505.

Rashid, M., Hasan, M. and Matin, M. (2018) Socio-economic performance of Bt eggplant cultivation in Bangladesh. *Bangladesh Journal of Agricultural Research* 43(2), 187–203.

Redillas, M.C.F.R., Jeong, J.S., Kim, Y.S., Jung, H., Bang, S.W. *et al.* (2012) The overexpression of *OsNAC9* alters the root architecture of rice plants enhancing drought resistance and grain yield under field conditions. *Plant Biotechnology Journal* 10(7), 792–805.

Regina, A., Bird, A., Topping, D., Bowden, S., Freeman, J. *et al.* (2006) High-amylose wheat generated by RNA interference improves indices of large-bowel health in rats. *Proceedings of the National Academy of Sciences USA* 103(10), 3546–3551.

Rejeb, I.B., Pastor, V. and Mauch-Mani, B. (2014) Plant responses to simultaneous biotic and abiotic stress: Molecular mechanisms. *Plants* 3(4), 458–475.

Romeis, J., Driesche, R.G.V., Barratt, B.I. and Bigler, F. (2008) Insect-resistant transgenic crops and biological control. In: Romeis, J., Shelton, A. and Kennedy, G. (eds) *Integration of Insect-Resistant Genetically Modified Crops within IPM Programs*. Springer, Dordrecht, pp. 87–117.

Salesse-Smith, C.E., Sharwood, R.E., Busch, F.A., Kromdijk, J., Bardal, V. *et al.* (2018) Overexpression of Rubisco subunits with RAF1 increases Rubisco content in maize. *Nature Plants* 4(10), 802–810.

Sanahuja, G., Banakar, R., Twyman, R.M., Capell, T. and Christou, P. (2011) *Bacillus thuringiensis*: A century of research, development and commercial applications. *Plant Biotechnology Journal* 9(3), 283–300.

Sander, J.D., Goodwin, M.J., Cade, L., Zhang, F., Cifuentes, D. *et al.* (2011) Selection-free zinc-finger-nuclease engineering by context-dependent assembly (CoDA). *Nature Methods* 8(1), 67–69.

Santillán Martínez, M.I., Bracuto, V., Koseoglou, E., Appiano, M., Jacobsen, E. *et al.* (2020) CRISPR/Cas9-targeted mutagenesis of the tomato susceptibility gene *PMR4* for resistance against powdery mildew. *BMC Plant Biology* 20(1), 1–13.

Savary, S., Willocquet, L., Pethybridge, S.J., Esker, P., McRoberts, N. *et al.* (2019) The global burden of pathogens and pests on major food crops. *Nature Ecology & Evolution* 3(3), 430–439.

Schornack, S., Meyer, A., Römer, P., Jordan, T. and Lahaye, T. (2006) Gene-for-gene-mediated recognition of nuclear-targeted AvrBs3-like bacterial effector proteins. *Journal of Plant Physiology* 163(3), 256–272.

Schütte, G., Eckerstorfer, M., Rastelli, V., Reichenbecher, W., Restrepo-Vassalli, S. *et al.* (2017) Herbicide resistance and biodiversity: Agronomic and environmental aspects of genetically modified herbicide-resistant plants. *Environmental Sciences Europe* 29(1), 5.

Segal, D.J., Beerli, R.R., Blancafort, P., Dreier, B., Effertz, K. *et al.* (2003) Evaluation of a modular strategy for the construction of novel polydactyl zinc finger DNA-binding proteins. *Biochemistry* 42(7), 2137–2148.

Shahzad, Z., Gosti, F., Frérot, H., Lacombe, E., Roosens, N. *et al.* (2010) The five AhMTP1 zinc transporters undergo different evolutionary fates towards adaptive evolution to zinc tolerance in *Arabidopsis halleri*. *PLOS Genetics* 6(4), e1000911.

Sharifzadeh, M.S., Abdollahzadeh, G., Damalas, C.A. and Rezaei, R. (2018) Farmers' criteria for pesticide selection and use in the pest control process. *Agriculture* 8(2), 24.

Sharma, P., Kumar, A. and Bhardwaj, R. (2016) Plant steroidal hormone epibrassinolide regulate – heavy metal stress tolerance in *Oryza sativa* L. by modulating antioxidant defense expression. *Environmental and Experimental Botany* 122, 1–9.

Shepherd, M., Turner, J.A., Small, B. and Wheeler, D. (2020) Priorities for science to overcome hurdles thwarting the full promise of the "digital agriculture" revolution. *Journal of the Science of Food and Agriculture* 100(14), 5083–5092.

Shim, J.S., Oh, N., Chung, P.J., Kim, Y.S., Choi, Y.D. *et al.* (2018) Overexpression of *OsNAC14* improves drought tolerance in rice. *Frontiers in Plant Science* 9, 310.

Shinwari, Z.K., Jan, S.A., Nakashima, K. and Yamaguchi-Shinozaki, K. (2020) Genetic engineering approaches to understanding drought tolerance in plants. *Plant Biotechnology Reports* 14(2), 151–162.

Shukla, V.K., Doyon, Y., Miller, J.C., DeKelver, R.C., Moehle, E.A. *et al.* (2009) Precise genome modification in the crop species *Zea mays* using zinc-finger nucleases. *Nature* 459(7245), 437–441.

Singh, B.P., Jayaswal, P.K., Singh, B., Singh, P.K., Kumar, V. *et al.* (2015) Natural allelic diversity in OsDREB1f gene in the Indian wild rice germplasm led to ascertain its association with drought tolerance. *Plant Cell Reports* 34(6), 993–1004.

Singla, J. and Krattinger, S.G. (2016) Biotic stress resistance genes in wheat. In: Wrigley, C., Corke, H., Seetharaman, K. and Faubion, J. (eds) *Encyclopedia of Food Grains*, 2nd edn. Academic Press, New York, pp. 388–392.

Sinha, S., Sandhu, K., Bisht, N., Naliwal, T., Saini, I. *et al.* (2019) Ascertaining the paradigm of secondary metabolism enhancement through gene level modification in therapeutic plants. *Journal of Young Pharmacists* 11(4), 337–343.

Somssich, M. (2019) A short history of plant transformation. *PeerJ Preprints*. Available at: https://peerj.com/preprints/27556v2.pdf

Song, A., Zhu, X., Chen, F., Gao, H., Jiang, J. *et al.* (2014) A chrysanthemum heat shock protein confers tolerance to abiotic stress. *International Journal of Molecular Sciences* 15(3), 5063–5078.

Steiner, B., Michel, S., Maccaferri, M., Lemmens, M., Tuberosa, R. *et al.* (2019) Exploring and exploiting the genetic variation of fusarium head blight resistance for genomic-assisted breeding in the elite durum wheat gene pool. *Theoretical and Applied Genetics* 132(4), 969–988.

Sun, M., Shen, Y., Yang, J., Cai, X., Li, H. *et al.* (2020) MiR535 negatively regulates cold tolerance in rice. *Molecular Breeding* 40(1), 14.

Sun, Y., Jiao, G., Liu, Z., Zhang, X., Li, J. *et al.* (2017) Generation of high-amylose rice through CRISPR/Cas9-mediated targeted mutagenesis of starch branching enzymes. *Frontiers in Plant Science* 8, 298.

Teramoto, S., Takayasu, S., Kitomi, Y., Arai-Sanoh, Y., Tanabata, T. *et al.* (2020) High-throughput three-dimensional visualization of root system architecture of rice using X-ray computed tomography. *Plant Methods* 16, 66.

Thitisaksakul, M., Tananuwong, K., Shoemaker, C.F., Chun, A., Tanadul, O. *et al.* (2015) Effects of timing and severity of salinity stress on rice (*Oryza sativa* L.) yield, grain composition, and starch functionality. *Journal of Agricultural and Food Chemistry* 63(8), 2296–2304.

Tian, S., Jiang, L., Cui, X., Zhang, J., Guo, S. *et al.* (2018) Engineering herbicide-resistant watermelon variety through CRISPR/Cas9-mediated base-editing. *Plant Cell Reports* 37(9), 1353–1356.

Tian, Z., Wang, J.-W., Li, J. and Han, B. (2021) Designing future crops: Challenges and strategies for sustainable agriculture. *Plant Journal* 105(5), 1165–1178.

Townsend, J.A., Wright, D.A., Winfrey, R.J., Fu, F., Maeder, M.L. *et al.* (2009) High-frequency modification of plant genes using engineered zinc-finger nucleases. *Nature* 459(7245), 442–445.

Ueta, R., Abe, C., Watanabe, T., Sugano, S.S., Ishihara, R. *et al.* (2017) Rapid breeding of parthenocarpic tomato plants using CRISPR/Cas9. *Scientific Reports* 7(1), 507.

Van Oort, P.A. and Zwart, S.J. (2018) Impacts of climate change on rice production in Africa and causes of simulated yield changes. *Global Change Biology* 24(3), 1029–1045.

Varkonyi-Gasic, E., Wang, T., Voogd, C., Jeon, S., Drummond, R.S.M. *et al.* (2019) Mutagenesis of kiwifruit CENTRORADIALIS-like genes transforms a climbing woody perennial with long juvenility and axillary flowering into a compact plant with rapid terminal flowering. *Plant Biotechnology Journal* 17(5), 869–880.

Varshney, R.K., Bohra, A., Yu, J., Graner, A., Zhang, Q. *et al.* (2021) Designing future crops: Genomics-assisted breeding comes of age. *Trends in Plant Science* 26(6), 631–649.

Veillet, F., Perrot, L., Chauvin, L., Kermarrec, M.-P., Guyon-Debast, A. *et al.* (2019) Transgene-free genome editing in tomato and potato plants using *Agrobacterium*-Mediated delivery of a CRISPR/Cas9 cytidine base editor. *International Journal of Molecular Sciences* 20(2), 402.

Wang, F., Wang, C., Liu, P., Lei, C., Hao, W. *et al.* (2016) Enhanced rice blast resistance by CRISPR/Cas9-targeted mutagenesis of the ERF transcription factor gene OsERF922. *PLOS One* 11(4), e0154027.

Wang, L., Samac, D.A., Shapir, N., Wackett, L.P., Vance, C.P. *et al.* (2005) Biodegradation of atrazine in transgenic plants expressing a modified bacterial *Atrazine chlorohydrolase* (*atzA*) gene. *Plant Biotechnology Journal* 3(5), 475–486.

Wei, J., Guo, Y., Liang, G., Wu, K., Zhang, J. *et al.* (2015) Cross-resistance and interactions between bt toxins Cry1Ac and Cry2Ab against the cotton bollworm. *Scientific Reports* 5, 7714.

Wu, J., Yan, G., Duan, Z., Wang, Z., Kang, C. *et al.* (2020) Roles of the *Brassica napus* DELLA protein BnaA6.RGA, in modulating drought tolerance by interacting with the ABA signaling component BnaA10.ABF2. *Frontiers in Plant Science* 11, 577.

Xie, X., Qin, G., Si, P., Luo, Z., Gao, J. *et al.* (2017) Analysis of *Nicotiana tabacum* PIN genes identifies *NtPIN4* as a key regulator of axillary bud growth. *Physiologia Plantarum* 160(2), 222–239.

Xu, J., Kang, B.C., Naing, A.H., Bae, S.J., Kim, C.K. *et al.* (2020) CRISPR/Cas9-mediated editing of 1-a minocyclopropane-1-carboxylate oxidase1 enhances petunia flower longevity. *Plant Biotechnology Journal* 18, 287–297.

Xu, Z.S., Feng, K. and Xiong, A.S. (2019) CRISPR/Cas9-mediated multiply targeted mutagenesis in orange and purple carrot plants. *Molecular Biotechnology* 61(3), 191–199.

Yadava, P., Abhishek, A., Singh, R., Singh, I., Kaul, T. *et al.* (2016) Advances in maize transformation technologies and development of transgenic maize. *Frontiers in Plant Science* 7, 1949.

Yang, Q.-Q., Zhang, C.-Q., Chan, M.-L., Zhao, D.-S., Chen, J.-Z. *et al.* (2016) Biofortification of rice with the essential amino acid lysine: Molecular characterization, nutritional evaluation, and field performance. *Journal of Experimental Botany* 67(14), 4285–4296.

Zhai, Y., Yu, K., Cai, S., Hu, L., Amoo, O. *et al.* (2020) Targeted mutagenesis of *BnTT8* homologs controls yellow seed coat development for effective oil production in *Brassica napus* L. *Plant Biotechnology Journal* 18(5), 1153–1168.

Zhang, X., Wang, W., Wang, M., Zhang, H.Y. and Liu, J.H. (2016) The miR396b of *Poncirus trifoliata* functions in cold tolerance by regulating ACC oxidase gene expression and modulating ethylene-polyamine homeostasis. *Plant & Cell Physiology* 57(9), 1865–1878.

Zhang, Y., Bai, Y., Wu, G., Zou, S., Chen, Y. *et al.* (2017) Simultaneous modification of three homoeologs of *TaEDR1* by genome editing enhances powdery mildew resistance in wheat. *Plant Journal* 91(4), 714–724.

Zhang, Y., Pribil, M., Palmgren, M. and Gao, C. (2020) A CRISPR way for accelerating improvement of food crops. *Nature Food* 1(4), 200–205.

Zhao, C., Liu, B., Piao, S., Wang, X., Lobell, D.B. *et al.* (2017) Temperature increase reduces global yields of major crops in four independent estimates. *Proceedings of the National Academy of Sciences* 114(35), 9326–9331.

Zhu, X.G., Long, S.P. and Ort, D.R. (2010) Improving photosynthetic efficiency for greater yield. *Annual Review of Plant Biology* 61, 235–261.

3 Biotechnological Tools to Improve Soil Health and Fertility

Silvina M.Y. Lopez[1,2], Virginia Martínez Alcántara[2,3], Juan M. Reparaz[3] and Pedro A. Balatti[2,3,4*]

[1]Consejo Nacional de Investigaciones Científicas y Técnicas (CONICET), Godoy Cruz, Argentina; [2]Centro de Investigaciones de Fitopatología (CIDEFI CIC-UNLP), La Plata Pcia de Buenos Aires, Argentina; [3]Cátedra de Microbiología Agrícola – Facultad de Ciencias Agrarias y Forestales – Universidad Nacional de La Plata, La Plata, Argentina; [4]Comisión de Investigaciones Científicas de la Provincia de Buenos Aires (CICPBA), La Plata, Argentina

Abstract

In this chapter, soils are presented as living environments, with microbes and eukaryotic cells believed to be intermingled in such a way that they can be considered a continuum of cells connected by many types of interactions, much like a complex tissue. This implies that any environmental change in the soil might affect plant growth and yield, but also that anything cultivated will profoundly impact the soil. Therefore, biotechnology can be used to generate environmental conditions for the soil biota to improve its activity, thereby altering the capacity of soils to act as a source of nutrients and as an environment with low levels of pathogens that can promote growth. To achieve this, biotechnological tools such as metagenomics, nanotechnology, and the selection of plants might be utilized to create healthy soils for the sustainable management and production of crops.

Keywords: environmental conditions, biotechnological tools, nanotechnology, sustainable management

3.1 Introduction

Soils are complex matrixes which are rather difficult to study due to their composition, conformation, and diversity. They host different communities of micro- and macroorganisms that change rapidly in response to environmental changes (Aqeel *et al.*, 2023). Therefore, soil microbial diversity is threatened and might be lost or reduced by land use and management techniques that together have a direct impact on soil health and, because of this, on soil productivity. Therefore, sustainable agricultural practices that respect ecosystem-based approaches should be implemented to promote healthy soils, which might lead to sustainable production systems.

One of the key features of soils is their quality and/or health. For many years soils were mostly defined based on their structure, organic and mineral contents as well as some other physicochemical features (Miransari, 2013), which is the matrix for sustaining the life of micro- and macroorganisms, plants, and animals.

*Corresponding author: pbalatti@gmail.com

DOI: 10.1079/9781800627659.0003

Plant growth and yield demand 16 macro- and micronutrients that are mostly provided by soils. If these are available in adequate amounts and in an economically sustainable environment, plant growth as well as yield is maximized (Nielsen *et al.*, 2015). However, among the chemical properties of soils, nutrient availability is a function of the oxidation state; which is modified by, among other things, the redox potential and pH of soils, which also is under the influence of microbial activity (Husson, 2013). Soils are a crucial component of ecosystems (Nielsen *et al.*, 2015), because the soil microbiota is responsible for essential soil processes such as decomposition, mineralization of organic matter, nutrient cycling, plant growth promotion, pathogen control, and degradation of recalcitrant compounds, all crucial activities for ecosystem regulation (Swift *et al.*, 1979; Bardgett *et al.*, 2005; Dominati *et al.*, 2010; Bardgett and Wardle, 2010; Nielsen *et al.*, 2011; Bissett *et al.*, 2013) and plant production.

Based on this and other findings, nowadays soils are considered a living environment where microorganism activity, which is governed by energy flow, plays a key role in soil and plant health (Fig. 3.1) (Miransari, 2013).

3.2 Health of Living Soil

Since soils are considered as a living system, a new concept has emerged regarding soil health, defined by Doran and Zeiss (2000) as "the capacity of the soil to function as a vital living system, within ecosystem and land-use boundaries, to sustain plant and animal productivity, maintain or enhance water and air quality, and promote plant and animal health."

Considering that soils are alive, one of the key features for soil functioning is energy, which drives soil microorganism activity in many ways, including modifying the pH, redox potential, physical and chemical factors of soil environment (Szent-Gyorgyi, 2012). Thus, soil health considers not only physical and chemical features but also biological ones. Although soils appear to be highly resilient, mostly because they have a microbiome with many communities that are metabolically redundant, they are often quite fragile (Clunes *et al.*, 2022). In fact, soils and their microbial communities are affected by degradation processes triggered by land use, chemical pollution, and climate change that occur all over the world. This is vitally important considering that there is not much land left for agricultural use.

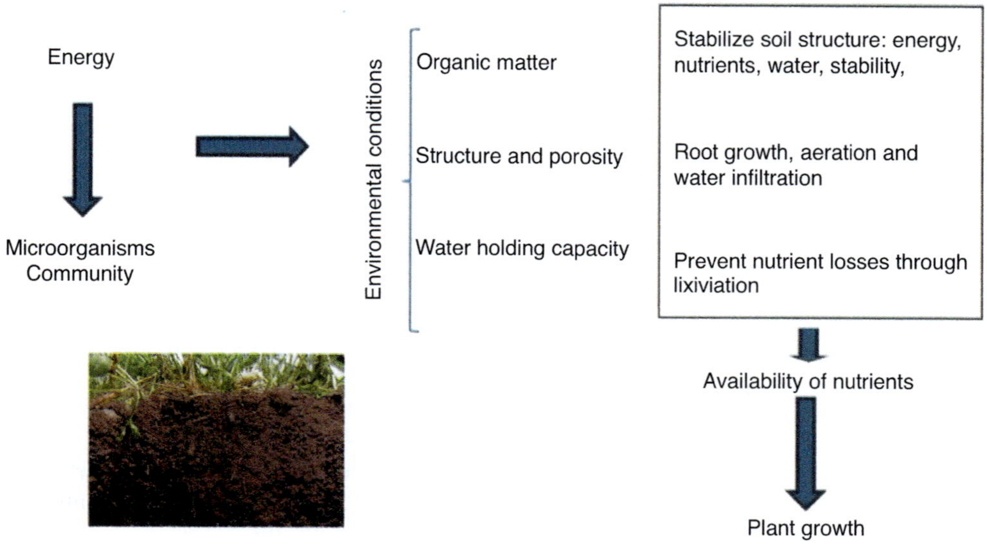

Fig. 3.1. Soil microbiome. The microbial community of soil is dependent on energy, in the form of organic matter. The microbiome activity finally influences nutrient availability and therefore plant growth.

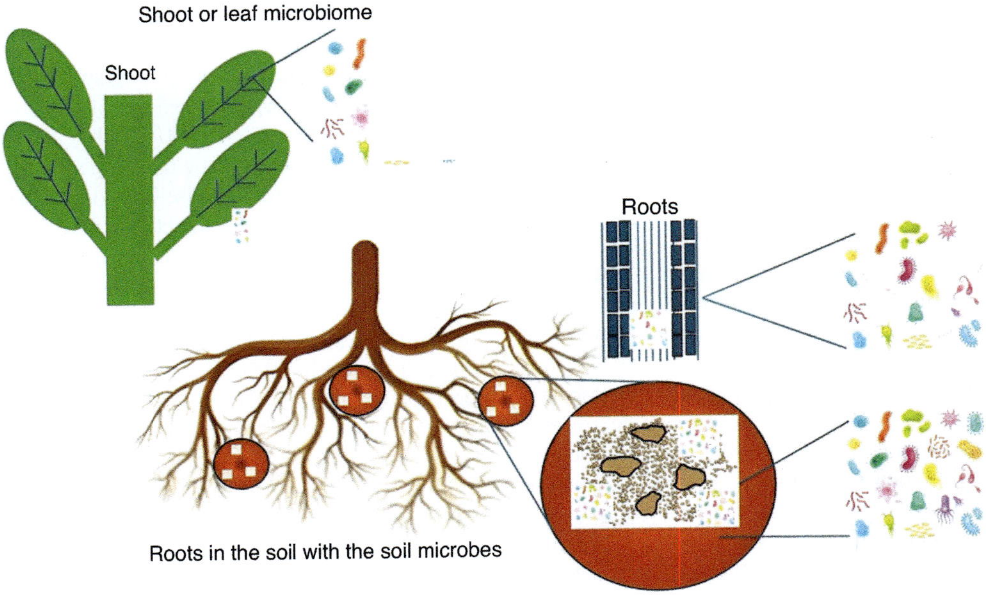

Fig. 3.2. Network (tissue) of prokaryotic and eukaryotic cells in soils. The cells are connected and interact by sending signals and modfying the environment, creating a continuum of cells from the soil to the upper plant parts. In each of these environments, microbes play crucial roles.

The term 'microbiome' was defined as the 'set of resident microorganisms that inhabit a given host/environment' (Lederberg and McCray, 2001) or as the community of microorganisms that should also include the 'theater of activity,' which together makes the environment. The latter would include the entire spectrum of molecules produced by pathogenic and nonpathogenic microorganisms, such as their structural elements (nucleic acids, proteins, lipids, polysaccharides) and metabolites (signaling molecules, toxins, organic and inorganic molecules) (Berg *et al.*, 2020; Medina *et al.*, 2021). We know that each environment hosts a group of diverse organisms, creating its own microbiome that plays a crucial biological role (Fig. 3.2). Microorganisms such as bacteria, actinomycetes, fungi, algae, and protozoa, as well as macroorganisms and plant roots are actively interacting, which is very important for soil health. One of the main attributes of microbiomes is their diversity, their microbial composition indicated by the richness, relative abundance, and phylogenetic structure of taxa assembled (Allison and Martiny, 2008). These provide an array of different metabolic functions that are crucial for the environment (Trivedi *et al.*, 2020).

In general, soil management is aimed at maintaining or increasing soil health in terms of nutrients and pathogens, meaning that more nutrients will be available and the pathogen population of soil-borne ones is reduced; such effects are promoted by a wide array of mechanisms (Abawi and Widmer, 2000; Bailey and Lazarovits, 2003; Ghorbani *et al.*, 2008). Management practices mostly associated with soil health include crop rotations, organic amendments with cover crops, zero or reduced tillage, all of which lead to increased microbial biomass and activity, that affect both the availability of nutrients and the capacity of the soil to suppress diseases (Chander *et al.*, 1998; O'Donnell *et al.*, 2001; Welbaum *et al.*, 2004; Bonilla *et al.*, 2012; Page *et al.*, 2013).

Knowledge regarding functional aspects of the soil microbiome is evolving rapidly, in part due to considering the soil biota as a crucial component of soils but also because many new analytical tools are available, facilitating studies regarding diversity as well as our understanding of the functional role of prokaryotic and eukaryotic microbes in the soil.

In terms of interaction and connection in the soil, there might be a large network or tissue formed by many different types of organisms. The physical matrix of clay, sand, and organic matter hosts a microbiome formed by bacteria, fungi, algae, protozoa, etc. Prokaryotic and eukaryotic organisms live together in close contact, meaning that they affect each other in many different ways. In a similar way, rhizospheric soil has a microbiome which is much richer in quantity and diversity, because roots excrete a richer milieu of organic compounds, among them flavonoids, secondary metabolites, and compounds that either attract or repulse mostly flagellated mobile bacteria. Because of this, bacterial populations in the rhizosphere frequently are in higher numbers. This is important not only due to the role rhizospheric organisms play, but also because these are the bacteria in intimate contact with plant roots that can alter plant growth the most. However, they also move into the plant, becoming the most important source of bacteria for the rhizoendobiome of plants that is formed by eukaryotic and prokaryotic organisms living endophytically in plant roots (López *et al.*, 2018b). Some of the latter population of bacteria might move upward within the plant to the shoots, leaves, flowers or fruits. Each of these organs contains endophytic microorganisms including bacteria and fungi that, as described, have their origins in the soil (López *et al.*, 2020).

Taking all this together, we envisage soils as a network or tissue made up of a complex array of eukaryotic and prokaryotic cells that work together, whose genomes encode a wide array of metabolic pathways that not only allow the selective growth of organisms but also play many different biological roles. Such a tissue is connected in many different ways by different types of signals that move in and out of it. So we can consider that live soils have a huge tissue-like structure connected all the way from the soil to the top of the plant. All these cells constitute the genome of the soil matrix that has sequences of microbial organisms (bacteria, fungi, etc.) as well as plant cells. In a way, this is not that different from the concept stated by Turner *et al.* (2013) which considered the community of beneficial microorganisms of plants (microbiome of plants) as part of the plant genotype. Such a concept implies that any change, within such an environment, might modify the soil by inducing or provoking changes in the plants and/or the organisms. This means that changes in the plant communities will undoubtedly have a profound impact on the soil microbiome and therefore on soil health.

Considering this relationship, it is evident that soils can be improved in at least two ways: by actions that modify the soil itself or by cultivating plants that release different exudates and contain different endosymbiotic organisms.

Any time we sow a different plant species or inoculate organisms or apply an agrochemical, we are altering the environment and, by extension, all the activities within the soil's genome network including the bacteria interacting with the plant organs, roots, leaves, fruits, etc. This means we can develop innovative biotechnological tools to improve soil health and productivity.

As stated above, the soil biota is formed by many types of organisms that perform different metabolic activities. Their number is the result of the response of each microbial species to the environment. Regarding this, two additional findings have substantially changed our way of looking at bacteria. One is that bacteria form biofilms, meaning on one hand that they act together in a milieu created by the deposition and accumulation of several layers of bacteria that are firmly adhered to different surfaces, which might lead to metabolic activities particularly on soil particles (Cai *et al.*, 2019). On the other hand, the quorum sensing system has been described (Chong *et al.*, 2012). This is a signaling pathway bacteria and fungi use to communicate between cells in such a way that, under certain circumstances, they behave like saprophytic bacteria. If they proliferate and increase their numbers, the quorum sensing response might lead to the expression of a different set of genes. As a result of this, saprophytic bacteria might behave as pathogens.

3.3 Suppressive Soils Are Healthy Soils?

Disease-suppressive soils were defined by Baker and Cook (1974) as those soils where pathogens cannot establish or persist or do not cause disease. While general suppressiveness has been attributed to the microbiome and most frequently has been associated with competition for resources, specific suppressiveness of soils has been attributed to defined groups or communities of organisms that somehow interfere with the pathogen cycle (Kinkel *et al.*, 2011). Whether a general or suppressive soil is identified, the question arises whether this is due to a different community of microorganisms hosted by the soil or to the environmental conditions of each soil that determine microbial growth and therefore a change in the biological balance.

3.4 Microbes Play a Crucial Role in Nutrient Cycling

The soil microbiome plays a critical part in soil nutrient availability due to its key role in cycling nutrients, which additionally are dependent on soil characteristics such as pH and redox potential. Nutrients including N, P, K, S, Fe, Zn, Co, and Mn are affected by soil bacteria and environmental conditions.

Nitrogen-fixing bacteria fix otherwise unavailable N_2 either alone or in symbiosis, increasing in this way the input of N (Sepp *et al.*, 2023). N is probably the best known system that has been studied in detail and has been in use for many years (López *et al.*, 2018a). In addition to this, several species of bacteria are involved in nitrogen cycling within soils, due to their ability to modify the oxidation state of N. In the case of P, several bacteria and fungi have been described that solubilize P in the soil, which is important because of stochiometric reasons as excess applied P precipitates and becomes unavailable for plants. Many microbial organisms including bacteria, fungi, and algae exhibit P solubilization and mineralization ability. Soil bacteria that solubilize or mineralize phosphorus include *Pseudomonas* spp., *Agrobacterium* spp., and *Bacillus circulans* (Babalola and Glick, 2012), *Azotobacter* (Kumar *et al.*, 2015), *Bacillus* (Jahan *et al.*, 2013; Raj *et al.*, 2014), *Burkholderia* (Rahi *et al.*, 2010; Zhao *et al.*, 2014; Istina *et al.*, 2015), *Enterobacter*, *Erwinia* (Chakraborty *et al.*, 2009), *Kushneria* (Zhu *et al.*, 2011), *Paenibacillus* (Bidondo *et al.*, 2011), *Ralstonia*, *Rhizobium* (Tajini *et al.*, 2012), *Rhodococcus*, *Serratia*, *Bradyrhizobium*, *Salmonella*, *Sinomonas*, and *Thiobacillus* (Postma *et al.*, 2010; Raj *et al.*, 2014). Also, P is solubilized by fungi, among them *Alternaria*, *Aspergillus*, *Cephalosporium*, *Cladosporium*, *Curvularia*, *Cunninghamella*, *Chaetomium*, *Fusarium*, *Glomus*, *Helminthosporium*, *Micromonospora*, *Mortierella*, *Myrothecium*, *Paecilomyces*, *Penicillium*, *Phoma*, *Pichia fermentans*, *Populospora*, *Pythium*, *Rhizoctonia*, *Rhizopus*, *Saccharomyces*, *Schizosaccharomyces*, *Sclerotium*, *Torula*, *Trichoderma*, and *Yarrowia* (Srinivasan *et al.*, 2012; Sharma *et al.*, 2013). In general, P solubilization is achieved by releasing organic acids and phosphatases. Interestingly, low levels of P increase the development of superficial roots of plants, thus increasing the organic content of soils and having an impact on plant root growth.

Also, K availability can be enhanced by plant growth-promoting rhizobacteria (PGPR) like *Bacillus mucilaginous* and *Penibacilllus mucilaginosus*, which solubilize K from minerals by mechanisms associated with acidolysis chelation, redox reactions and synthesis of oxalates, lactates, and gluconates (Marschner and Rengel, 2007; Uroz *et al.*, 2009). Another element necessary for plant growth is S, whose biogeochemical cycle is governed by microbial activity that is drastically changed by the presence of organisms, oxygenic conditions, and pH. This means that S is used as a source or acceptor of electrons in soils, depending on environmental conditions. Meyer and Kuever (2007) described a set of enzymes present in a collection of sulfur bacteria like *Thiobacillus*, *Thiotrix*, Chorobiaceae, *Beggiatoa*, and Chromatiaceae in soils.

Fe is another nutrient whose solubility in soils is dependent on the environment; in humid soils Fe is highly available. One strategy used to increase Fe availability is the synthesis of siderophores, a quite frequent feature of PGPR. *Bacillus subtilis* increase mobility of Fe in the rhizosphere by acidification of the environment, resulting in an increase in Fe uptake (Zhang *et al.*, 2009). Zn is another relevant element whose availability for plants is regulated by fungi (Miransari, 2013). He *et al.* (2010) showed that Zn-tolerant bacteria such as *Bacillus cereus*, *B. subtilis*, *Flavobacterium*, and *Pseudomonas aeruginosa* might increase available Zn in soils and therefore uptake by plants. Copper and

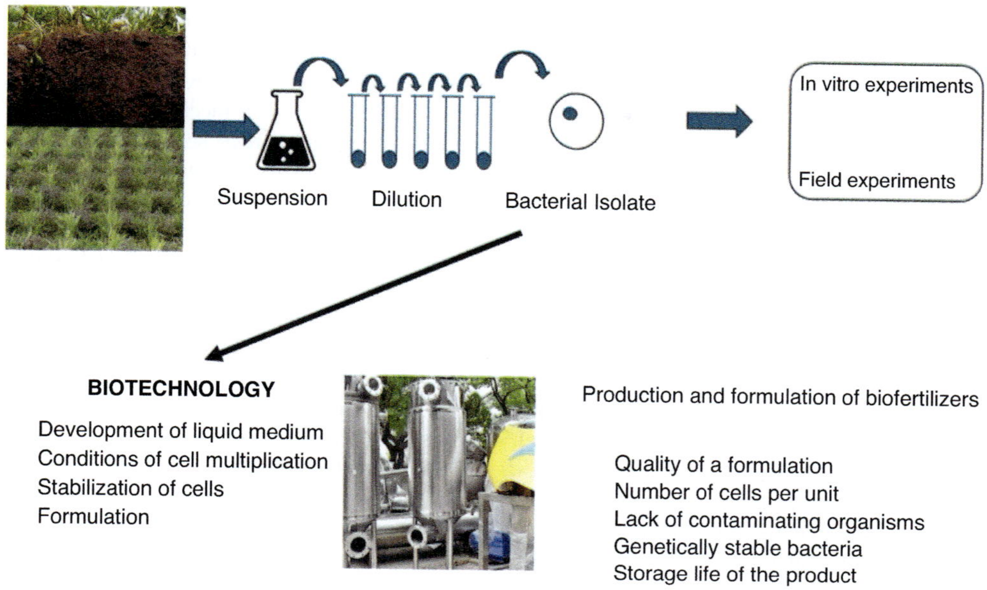

Fig. 3.3. A typical procedure for the isolation of bacteria and/or fungi from soils or plants for the development of a biofertilizer. The organisms are assayed *in vitro* and in the field. A critical biotechnological step is developing the media and other conditions for the multiplication of the organisms. Then, cells are stabilized and formulated.

manganese are two other nutrients affected by microbial activities that somehow affect the synthesis and release of root exudates by plants.

Therefore, we might increase nutrient availability in soils by inoculating organisms in high numbers, hoping that if they survive, they will behave as observed in the preliminary studies. However, a question remains regarding the impact adding an incredibly large population of an organism to the soil has on plants and the soil microbial community. The problem is that, currently, we lack the ability to measure the impact of such a large input of organisms on the soil microbial community.

3.5 Nanotechnology Is a Biotechnological Tool to Improve Soil Health

Regarding biotechnology, the most widely used strategy is formulation of inoculants with organisms based on bacteria with outstanding capacity or ability to promote growth (Fig. 3.3). In most cases, single-species inoculants that fix nitrogen, promote plant growth or biocontrol plant pathogens have been used to inoculate legumes and quite often the response is rather poor. Probably this is because in nature, the communities of microbes act together, which may result in synergistic effects. Will the application of specific populations of organisms solve soil problems in terms of nutrient cycling? It might but the environmental conditions of soil are probably the key factors.

Another biotechnological impact on soil microbial activity can be provoked by means of nanomaterials. Nanotechnology is an area whose impact on soil management has increased substantially, since it improves the physicochemical properties of materials applied to soils. This might affect the activity of soil microbes that promote plant growth (Rajput *et al.*, 2021; Khan *et al.*, 2021). Mahmoud *et al.* (2016) applied a *Staphylococcus aureus* and n-Fe_3O_4-Phth-*S. aureus* complex which promoted efficient removal and extraction of Pb, Ni, and Cu. Thus, nanotechnology can be used to remove heavy metals in heavily polluted soils, a common problem in many areas of the world that is related to the origin of soils.

Another use of nanomaterials might be to improve nutrient delivery in soils, which has an impact on soil enzyme activity, as described by Zhao *et al.* (2021). These authors used carbon nanoparticles to improve soil fertility, nutrient uptake, and maize growth. As expected, the introduction of nanomaterials in soils might affect soil structure and activity of microbes in the rhizosphere. The latter activity is a function of the characteristics and concentrations of nanomaterials, as well as the environment these materials generate and the communities that form the microbiome.

Nanomaterials have unique properties that modify soil biological activities, so nanomaterial application can impact plant growth by enhancing synthesis of plant hormones, altering gene expression, influencing antioxidant activities, and altering carbohydrate, fatty acid, and amino acid synthesis in the soil, as stated by Al-Khayri *et al.* (2023).

In summary, nanotechnology appears to be a potent tool to impact on soil health, by improving nutrient availability and thus the size and activity of soil microbial communities. In addition, nanomaterials can be used to sequester heavy metal ions. Therefore, nanotechnology is already playing and will continue to play a key role in the future in developing tools for improving soil quality.

3.6 Disease-Suppressive Soils

As mentioned earlier, disease-suppressive soils are those where disease development is at a minimal level even in the presence of pathogens and susceptible host plants (Siegel-Hertz *et al.*, 2018). These types of soils help to develop a strategy to use such mechanisms in agriculture. To do this, we should identify and isolate organisms with the capacity to control soil-borne pathogens. Then, we can introduce them to soils by means of commercial formulations of disease-suppressing microbes or by engineering the indigenous microbiome, through practices that sustain the populations and activities of these microbes. General suppressive soils typically have a balance in soil conditions and are more dependent on long-term microbiota diversity (Berendsen *et al.*, 2012).

There is room to recover and/or increase the suppressive capacity of soils. To do this, it is necessary to identify as many strategies as possible and these are basically the mechanisms microbes use for the suppression of soil-borne diseases. They include parasitism against pathogens by beneficial microbial communities, production of metabolites, toxins, and antibiotics, competition for nutrients/resources/substrates, activation of disease resistance genes in the host plant by beneficial microorganisms, and improvement of plant nutrition and soil health (Jayaraman *et al.*, 2021). A soil with functional microbial biodiversity increases ecosystem resilience, making the soil less vulnerable to short-term changes in the environment (Jayaraman *et al.*, 2021). A proper understanding of biotic lines of defense and how the environment impacts on them is crucial to developing management systems, aimed at shaping this type of plant defense against phytopathogens (Nwokolo *et al.*, 2021).

3.7 Biotechnology Is Aimed at Improving the Suppressive Capacity of Soils

There is a growing body of evidence and knowledge on the various approaches to understanding and promoting soil disease suppression by soil management, as well as plant and soil type (Yin *et al.*, 2013; Löbmann *et al.*, 2016), use of compost (Hadar and Papadopoulou, 2012; Lutz *et al.*, 2020), use of marine microalgae derivatives (Hamed *et al.*, 2018; Gonçalves, 2021), and evaluation of microbial communities through omics studies (Kopecky *et al.*, 2019; Deng *et al.*, 2020).

3.7.1 Organic matter

Based on the described findings, any practice that triggers an increase in diversity and/or quantity of microbes of the soil microbiome will lead to suppressive soils. There are many different alternative practices, like management of soils in a sustainable way so that the level of organic matter remains high. As mentioned earlier, the soil organic matter provides energy for cycling of nutrients, retaining water, and stabilizing soils. The way to maintain high levels of organic matter in the soil is to reduce disturbances. Any management practice that disturbs soils provokes a change in the structure of the microbial community and available energy (Nielsen *et al.*, 2015). Basically, disease management including crop rotation and/or covering crops can help to maintain resources and energy in the soil, which enhances microbial activity. Different cover plants might enhance the growth of different microbial communities. Organic applications and conservation tillage impact on soil microbial activity and therefore nitrogen cycling and plant growth.

A high content of organic matter added to the soil is usually associated with a lower incidence of soil-borne pathogens (Bonanomi *et al.*, 2007; Chen and Nelson, 2008; Alfano *et al.*, 2011), which is related to a higher level of available energy, as well as physical and chemical features of soils. When disease-suppressing microorganisms, such as *Trichoderma hamatum* or *Bacillus subtilis*, are added, the plant's defense system is improved through induced systemic resistance (Barnett *et al.*, 2006; Yogev *et al.*, 2010; Hadar and Papadopoulou, 2012). These microbes provide compounds such as toxic or stimulant volatile agents that are released and cause changes in the physical and chemical properties of the soil and further affect the development of pathogens (Tilston *et al.*, 2002; López *et al.*, 2021).

3.7.2 Biopolymers

Chitin and a derivative, chitosan, are obtained by enzymatic or chemical deacetylation of raw chitin, extracted from food waste of crustacean shells, fungi, fish, and mollusks (Maddaloni *et al.*, 2020). Chitin/chitosan can be used in different forms such as solutions, hydrogels, nanomicroparticles and their derivatives, which have different biochemical properties and various levels of deacetylation of raw chitin. Chitin and/or chitosan temporarily increase root growth and reduce the incidence of nematode infection in plant roots (Gooday, 1994; Green *et al.*, 2006; Manucharova *et al.*, 2007; Hjort *et al.*, 2010; Mendes *et al.*, 2011; Radwan *et al.*, 2012), while chitin stimulates hydrolysis of the hyphae of pathogenic chitinous fungi (Cretoiu *et al.*, 2013).

In summary, the addition of organic matter could encourage highly suppressive soils (disease reduction of more than 80%) but this is limited to only 12% of all cases (Bonanomi *et al.*, 2007). Furthermore, the suppressive capacity of applications varies significantly depending on the type of soil-borne pathogen (Bonanomi *et al.*, 2007; Jambhulkar *et al.*, 2015).

Therefore, metagenomic studies regarding the impact environmental changes have on microbial community composition and diversity are crucial, as we might then be able to determine the environmental conditions that promote or inhibit growth of microbial communities.

The microbial composition of communities can change under different environmental disturbances, depending on their resistance, resilience, functional redundancy, and functional similarity, which are associated with the alteration of soil microbial processes (Allison and Martiny, 2008) such as nutrient cycling. Considering that most microbial groups are sensitive and not immediately resilient to disturbances (Allison and Martiny, 2008), the assessment of microbial composition is important to determine or predict the change of soil microbial processes, in response to different disturbances (Allison and Martiny, 2008). Metagenomic studies can also indicate the identity of the taxa or clades linked to microbial responses to disturbances.

Trichoderma is a soil fungus that promotes plant growth. Even though the fungus lives in the soil, plant growth promotion or disease control only occurs when it is supplemented at high numbers. This effect lasts for a certain period, whose length is directly related to the rate at which the *Trichoderma* population decreases in soils. So the question is, should we add *Trichoderma* or should we generate the conditions to promote the proliferation of natural soil populations? This might be particularly complex within the soil microbiome, where eukaryotic and prokaryotic cells are interacting. In any case, as mentioned earlier, metagenomics might show not only the evolution of organism populations but also their response to soil management.

Another biotechnological impact on soils can be achieved by crop rotation. Nowadays, geneticists develop new plant materials basically looking mainly at plant characteristics. Plant microbes are important since we know they might be providing a larger metabolic capacity than the plant itself. Furthermore, part of those organisms might be released to the soil. Biotechnology should allow us to study, on one side, the plant microbes with the desired capacity that we can somehow introduce into soils. However, identification of the signals that trigger growth and development of specific bacterial soil populations that play a role in pathogen suppression would also be an important research aim.

3.7.3 Omics

A biotechnological tool that has been and will continue to affect soil biology is metagenomics. This is basically a study aimed at identifying all the organisms contained by soils. Briefly, it consists in extracting soil genomic DNA to make a library and sequence the *16S* and the *18S rDNA*. This technique allows us to determine the genotypic potential of soils, though it is not saying much regarding the functionality of soils. However, metagenomic studies allow us to determine the composition of soil microbiomes, which will reflect the environmental impact on microbial communities of soils.

Biotechnology has provided us with many tools to study microbiomes, including high-throughput sequencing, terminal restriction fragment length polymorphisms (T-RFLP), automated ribosomal intergenic spacer analysis (ARISA), denaturing gradient gel electrophoresis (DGGE), DNA probes, etc., all of which provide much more detailed and precise information regarding microbial changes in soils (Tiquia *et al.*, 2002; Edel-Hermann *et al.*, 2004; Mazzola, 2004; Van Elsas and Costa, 2007; Sugiyama *et al.*, 2010). One of the most important characteristics of soil microbiomes is their response to environmental changes that might be followed by means of metagenomics, transcriptomics, proteomics, and metabolomics (Fig. 3.4), and each of these techniques will be telling us completely different things.

Next-generation sequencing provides a better overview of soil microbial communities during microbiome manipulation, but sometimes does not provide information related to the highest taxonomic resolution of soil microbial communities. It is estimated that

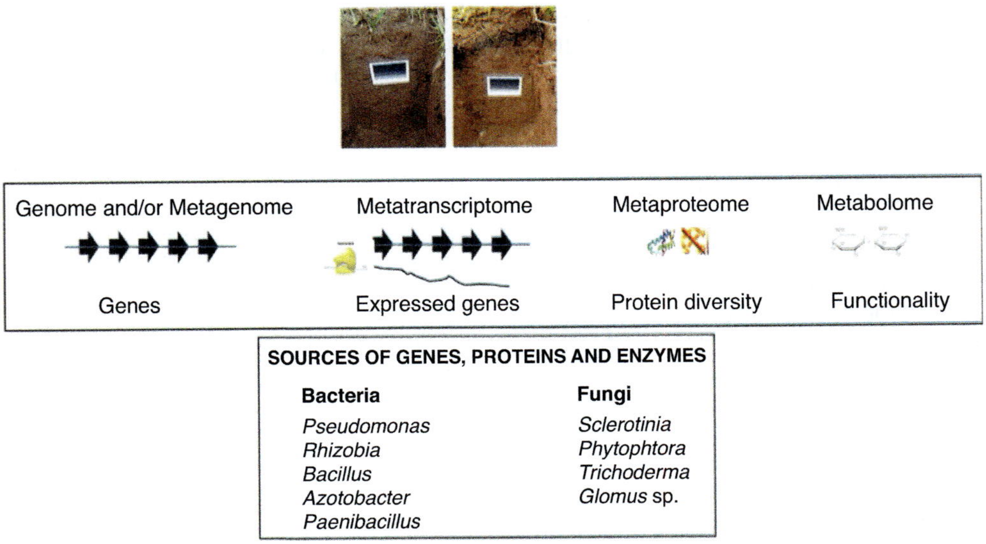

Fig. 3.4. The omics available for soil analysis. The bacteria and fungi included are those that have been used as sources of genes, proteins or enzymes involved in disease suppression. There are very many other organisms available.

only approximately 1% of all microorganisms are culturable and identifiable using traditional sequencing technologies (Solden *et al.*, 2016).

The field of metagenomics emerged due to the need to understand the diversity and functional role of nonculturable microbes (Handelsman *et al.*, 1998; Chen and Pachter, 2005), to answer questions related to the ecological dynamics of microbial communities and to understand the mechanisms of interactions and community coevolution (Segata *et al.*, 2013). Metagenomic studies have been aimed at promoting plant growth by studying endophytic microorganisms (López *et al.*, 2018a, 2020) and studying the role of functional genes for plant and soil health (Bramhachari *et al.*, 2017; Fadiji and Babalola, 2020; Adeleke *et al.*, 2021). The second type of plant health research involves understanding how soil microbes suppress plant pathogens (López *et al.*, 2020). Understanding which taxa are involved in soil suppression using a metagenomic approach could reveal their diverse functions and shed light on the biological mechanism of suppression (Crandall *et al.*, 2020).

With the development of high-throughput sequencing techniques, entire microbiomes will be identified from many different soils. The next step will be to leverage this information so that it does not merely remain as a description of microbiome composition but rather brings us to a closer understanding of their functional capacities, not just their antagonistic interactions that might lead to suppression of plant diseases.

It is necessary to understand the role of microbes in soil disease suppression and the potential that biological formulations might have for suppressing plant diseases. Furthermore, integrating multi-omics approaches will help us to determine which are the microbial actors in soils, the way they are distributed through time and space, and their roles or functions whether acting alone or together (Crandall *et al.*, 2020).

3.8 Conclusion

Soil microbiomes play a crucial role in soil quality and health. However, their activity is dependent on the energy present, so any input of energy, whether naturally or by amending soils, will have an impact on soil microbe communities.

Furthermore, the biological properties of soil may depend on whether the microbial community is sensitive, resilient, or functionally redundant in the face of a given disturbance or environmental change.

Soil microbes form a close connected network of prokaryotic and eukaryotic cells, which includes the rhizosphere, the roots of plants and the plant endobiome, so any change in any one of these organisms will provoke an environmental change and a change in soil biological activity.

Biotechnology is constantly developing new tools to study soil biology. However, although the genome capacity of the soil's organisms is important, but their function in the soils is more, so many metabolomic studies of soils under different conditions are necessary.

Nanotechnology is a powerful technique particularly to improve fertilization and interactions of applied nutrients with microbes so it could have a high impact on soil health and nutrient content.

The genetics of cover plants could have a great effect on soil nutrient capacity and health; breeding should consider the analysis of the microbial genome carried by plants either to modify it or develop commercial formulations to inoculate plants. This will enhance the development of sustainable agriculture.

References

Abawi, G.S. and Widmer, T.L. (2000) Impact of soil health management practices on soilborne pathogens, nematodes and root diseases of vegetable crops. *Applied Soil Ecology* 15(1), 37–47.

Adeleke, B.S., Babalola, O.O. and Glick, B.R. (2021) Plant growth-promoting root-colonizing bacterial endophytes. *Rhizosphere* 20, 100433.

Alfano, G., Lustrato, G., Lima, G., Vitullo, D. and Ranalli, G. (2011) Characterization of composted olive mill wastes to predict potential plant disease suppressiveness. *Biological Control* 58(3), 199–207.

Al-Khayri, J.M., Rashmi, R., Surya Ulhas, R., Sudheer, W.N., Banadka, A. *et al.* (2023) The role of nanoparticles in response of plants to abiotic stress at physiological, biochemical, and molecular levels. *Plants* 12(2), 292.

Allison, S.D. and Martiny, J.B.H. (2008) Resistance, resilience, and redundancy in microbial communities. *Proceedings of the National Academy of Sciences* 105(suppl 1), 11512–11519.

Aqeel, M., Ran, J., Hu, W., Irshad, M.K., Dong, L. *et al.* (2023) Plant-soil-microbe interactions in maintaining ecosystem stability and coordinated turnover under changing environmental conditions. *Chemosphere* 318, 137924.

Babalola, O.O. and Glick, B.R. (2012) The use of microbial inoculants in African agriculture: Current practice and future prospects. *Journal of Food, Agriculture and Environment* 10, 540–549.

Bailey, K.L. and Lazarovits, G. (2003) Suppressing soil-borne diseases with residue management and organic amendments. *Soil and Tillage Research* 72(2), 169–180.

Baker, K.F. and Cook, R.J. (1974) *Biological Control of Plant Pathogens*. W.H. Freeman, New York.

Bardgett, R.D. and Wardle, D.A. (2010) *Aboveground-Belowground Linkages: Biotic Interactions, Ecosystem Processes, and Global Change*. Oxford University Press, Oxford.

Bardgett, R.D., Usher, M.B. and Hopkins, D.W. (eds) (2005) *Biological Diversity and Function in Soils*. Cambridge University Press, Cambridge.

Barnett, S.J., Roget, D.K. and Ryder, M.H. (2006) Suppression of *Rhizoctonia solani* AG-8 induced disease on wheat by the interaction between Pantoea, Exiguobacterium, and Microbacteria. *Soil Research* 44(4), 331–342.

Berendsen, R.L., Pieterse, C.M. and Bakker, P.A. (2012) The rhizosphere microbiome and plant health. *Trends in Plant Science* 17(8), 478–486.

Berg, G., Rybakova, D., Fischer, D., Cernava, T., Vergès, M.-C.C. *et al.* (2020) Microbiome definition revisited: Old concepts and new challenges. *Microbiome* 8(1), 103.

Bidondo, L.F., Silvani, V., Colombo, R., Pérgola, M., Bompadre, J. *et al.* (2011) Pre-symbiotic and symbiotic interactions between Glomus intraradices and two *Paenibacillus* species isolated from AM propagules *In Vitro* and *in Vivo* assays with soybean (AG043RG) as plant host. *Soil Biology and Biochemistry* 43(9), 1866–1872.

Bissett, A., Brown, M.V., Siciliano, S.D. and Thrall, P.H. (2013) Microbial community responses to anthropogenically induced environmental change: Towards a systems approach. *Ecology Letters* 16, 128–139.

Bonanomi, G., Antignani, V., Pane, C. and Scala, F. (2007) Suppression of soilborne fungal diseases with organic amendments. *Journal of Plant Pathology* 89, 311–324.

Bonilla, N., Gutiérrez-Barranquero, J.A., Vicente, A. and Cazorla, F.M. (2012) Enhancing soil quality and plant health through suppressive organic amendments. *Diversity* 4(4), 475–491.

Bramhachari, P.V., Nagaraju, G.P. and Kariali, E. (2017) Metagenomic approaches in understanding the mechanism and function of PGPRs: Perspectives for sustainable agriculture. *Agriculturally Important Microbes for Sustainable Agriculture* I, 163–182.

Cai, P., Sun, X., Wu, Y., Gao, C., Mortimer, M. *et al.* (2019) Soil biofilms: Microbial interactions, challenges, and advanced techniques for ex-situ characterization. *Soil Ecology Letters* 1, 85–93.

Chakraborty, U., Chakraborty, B.N., Basnet, M. and Chakraborty, A.P. (2009) Evaluation of *Ochrobactrum anthropi* TRS-2 and its talc based formulation for enhancement of growth of tea plants and management of brown root rot disease. *Journal of Applied Microbiology* 107(2), 625–634.

Chander, K., Goyal, S., Nandal, D.P. and Kapoor, K.K. (1998) Soil organic matter, microbial biomass and enzyme activities in a tropical agroforestry system. *Biology and Fertility of Soils* 27, 168–172.

Chen, K. and Pachter, L. (2005) Bioinformatics for whole-genome shotgun sequencing of microbial communities. *PLOS Computational Biology* 1(2), e24.

Chen, M.H. and Nelson, E.B. (2008) Seed-colonizing microbes from municipal biosolids compost suppress *Pythium ultimum* damping-off on different plant species. *Phytopathology* 98(9), 1012–1018.

Chong, T.M., Koh, C.L., Sam, C.K., Choo, Y.M., Yin, W.F. *et al.* (2012) Characterization of quorum sensing and quorum quenching soil bacteria isolated from Malaysian tropical montane forest. *Sensors* 12(4), 4846–4859.

Clunes, J., Valle, S., Dörner, J., Martínez, O., Pinochet, D. *et al.* (2022) Soil fragility: A concept to ensure a sustainable use of soils. *Ecological Indicators* 139, 108969.

Crandall, S.G., Gold, K.M., Jiménez-Gasco, M.D.M., Filgueiras, C.C. and Willett, D.S. (2020) A multi-omics approach to solving problems in plant disease ecology. *PLOS One* 15(9), e0237975.
Cretoiu, M.S., Korthals, G.W., Visser, J.H. and Elsas, J.D. (2013) Chitin amendment increases soil suppressiveness toward plant pathogens and modulates the actinobacterial and oxalobacteraceal communities in an experimental agricultural field. *Applied and Environmental Microbiology* 79(17), 5291–5301.
Deng, X., Zhang, N., Shen, Z., Zhu, C., Li, R. *et al.* (2020) Rhizosphere bacteria assembly derived from fumigation and organic amendment triggers the direct and indirect suppression of tomato bacterial wilt disease. *Applied Soil Ecology* 147, 103364.
Dominati, E., Patterson, M. and Mackay, A. (2010) A framework for classifying and quantifying the natural capital and ecosystem services of soils. *Ecological Economics* 69(9), 1858–1868.
Doran, J.W. and Zeiss, M.R. (2000) Soil health and sustainability: Managing the biotic component of soil quality. *Applied Soil Ecology* 15(1), 3–11.
Edel-Hermann, V., Dreumont, C., Pérez-Piqueres, A. and Steinberg, C. (2004) Terminal restriction fragment length polymorphism analysis of ribosomal RNA genes to assess changes in fungal community structure in soils. *FEMS Microbiology Ecology* 47(3), 397–404.
Elsas, J.V. and Costa, R. (2007) Molecular assessment of soil microbial communities with potential for plant disease suppression. In: Punja, Z., Boer, S. and Sanfaçon, H. (eds) *Biotechnology and Plant Disease Management*. CAB International, Wallingford, pp. 498–517.
Fadiji, A.E. and Babalola, O.O. (2020) Metagenomics methods for the study of plant-associated microbial communities: A review. *Journal of Microbiological Methods* 170, 105860.
Ghorbani, R., Wilcockson, S., Koocheki, A. and Leifert, C. (2008) Soil management for sustainable crop disease control: A review. *Environmental Chemistry Letters* 6, 149–162.
Gonçalves, A.L. (2021) The use of microalgae and cyanobacteria in the improvement of agricultural practices: A review on their biofertilising, biostimulating and biopesticide roles. *Applied Sciences* 11(2), 871.
Gooday, G.W. (1994) Physiology of microbial degradation of chitin and chitosan. *Biochemistry of Microbial Degradation* 279–312.
Green, S.J., Inbar, E., Michel, Jr. F.C.J., Hadar, Y. and Minz, D. (2006) Succession of bacterial communities during early plant development: Transition from seed to root and effect of compost amendment. *Applied and Environmental Microbiology* 72(6), 3975–3983.
Hadar, Y. and Papadopoulou, K.K. (2012) Suppressive composts: Microbial ecology links between abiotic environments and healthy plants. *Annual Review of Phytopathology* 50(1), 133–153.
Hamed, S.M., Abd El-Rhman, A.A., Abdel-Raouf, N. and Ibraheem, I.B. (2018) Role of marine macroalgae in plant protection and improvement for sustainable agriculture technology. *Beni-Suef University Journal of Basic and Applied Sciences* 7(1), 104–110.
Handelsman, J., Rondon, M.R., Brady, S.F., Clardy, J. and Goodman, R.M. (1998) Molecular biological access to the chemistry of unknown soil microbes: A new frontier for natural products. *Chemistry & Biology* 5(10), R245–R249.
He, C.Q., Tan, G.E., Liang, X.D.U.W., Du, W., Chen, Y.L. *et al.* (2010) Effect of Zn-tolerant bacterial strains on growth and Zn accumulation in *Orychophragmus violaceus*. *Applied Soil Ecology* 44(1), 1–5.
Hjort, K., Bergström, M., Adesina, M.F., Jansson, J.K., Smalla, K. *et al.* (2010) Chitinase genes revealed and compared in bacterial isolates, DNA extracts and a metagenomic library from a phytopathogen-suppressive soil. *FEMS Microbiology Ecology* 71(2), 197–207.
Husson, O. (2013) Redox potential (Eh) and pH as drivers of soil/plant/microorganism systems: A transdisciplinary overview pointing to integrative opportunities for agronomy. *Plant and Soil* 362(1–2), 389–417.
Istina, I.N., Widiastuti, H., Joy, B. and Antralina, M. (2015) Phosphate-solubilizing microbe from saprists peat soil and their potency to enhance oil palm growth and P uptake. *Procedia Food Science* 3, 426–435.
Jahan, M., Nassiri Mahallati, M., Amiri, M.B. and Ehyayi, H.R. (2013) Radiation absorption and use efficiency of sesame as affected by biofertilizers inoculation in a low input cropping system. *Industrial Crops and Products* 43, 606–611.
Jambhulkar, P.P., Sharma, M., Lakshman, D. and Sharma, P. (2015) Natural mechanisms of soil suppressiveness against diseases caused by Fusarium, Rhizoctonia, Pythium, and Phytophthora. In: Meghvansi, M. and Varma, A. (eds) *Organic Amendments and Soil Suppressiveness in Plant Disease Management*. Springer, Cham, pp. 95–123.

Jayaraman, S., Naorem, A.K., Lal, R., Dalal, R.C., Sinha, N.K. *et al.* (2021) Disease-suppressive soils-beyond food production: A critical review. *Journal of Soil Science and Plant Nutrition* 21(2), 1437–1465.

Khan, S.T., Adil, S.F., Shaik, M.R., Alkhathlan, H.Z., Khan, M. *et al.* (2021) Engineered nanomaterials in soil: Their impact on soil microbiome and plant health. *Plants* 11(1), 109.

Kinkel, L.L., Bakker, M.G. and Schlatter, D.C. (2011) A coevolutionary framework for managing disease-suppressive soils. *Annual Review of Phytopathology* 49(1), 47–67.

Kopecky, J., Samkova, Z., Sarikhani, E., Kyselková, M., Omelka, M. *et al.* (2019) Bacterial, archaeal and micro-eukaryotic communities characterize a disease-suppressive or conducive soil and a cultivar resistant or susceptible to common scab. *Scientific Reports* 9(1), 14883.

Kumar, A., Sharma, S. and Mishra, S. (2015) Evaluating effect of arbuscular mycorrhizal fungal consortia and *Azotobacter chroococcum* in improving biomass yield of *Jatropha curcas*. *Plant Biosystems* 150(5), 1056–1064.

Lederberg, J. and McCray, A.T. (2001) 'Ome sweet 'Omics – A genealogical treasury of words. *Scientist* 15(7), 8–8.

Löbmann, M.T., Vetukuri, R.R., Zinger, L., Alsanius, B.W., Grenville-Briggs, L.J. *et al.* (2016) The occurrence of pathogen suppressive soils in Sweden in relation to soil biota, soil properties, and farming practices. *Applied Soil Ecology* 107, 57–65.

López, S.M., Pastorino, G.N. and Balatti, P.A. (2021) Volatile organic compounds profile synthesized and released by endophytes of tomato (*Solanum lycopersici* L.) and their antagonistic role. *Archives of Microbiology* 203(4), 1383–1397.

López, S.M., Sánchez, M.D.M., Pastorino, G.N., Franco, M.E., García, N.T. *et al.* (2018a) Nodulation and delayed nodule senescence: Strategies of two *Bradyrhizobium japonicum* isolates with high capacity to fix nitrogen. *Current Microbiology* 75(8), 997–1005.

López, S.M.Y., Pastorino, G.N., Franco, M.E.E., Medina, R., Lucentini, C.G. *et al.* (2018b) Microbial endophytes that live within the seeds of two tomato hybrids cultivated in Argentina. *Agronomy* 8(8), 136.

López, S.M.Y., Pastorino, G.N., Fernández-González, A.J., Franco, M.E., Fernández-López, M. *et al.* (2020) The endosphere bacteriome of diseased and healthy tomato plants. *Archives of Microbiology* 202(10), 2629–2642.

Lutz, S., Thuerig, B., Oberhaensli, T., Mayerhofer, J., Fuchs, J.G. *et al.* (2020) Harnessing the microbiomes of suppressive composts for plant protection: From metagenomes to beneficial microorganisms and reliable diagnostics. *Frontiers in Microbiology* 11, 1810.

Maddaloni, M., Vassalini, I. and Alessandri, I. (2020) Green routes for the development of chitin/chitosan sustainable hydrogels. *Sustainable Chemistry* 1(3), 325–344.

Mahmoud, M.E., Abdou, A.E., Mohamed, S.M. and Osman, M.M. (2016) Engineered *Staphylococcus aureus* via immobilization on magnetic Fe3O4-phthalate nanoparticles for biosorption of divalent ions from aqueous solutions. *Journal of Environmental Chemical Engineering* 4(4), 3810–3824.

Manucharova, N.A., Vlasenko, A.N. and Stepanov, A.L. (2007) Temperature as an autoecological factor of chitinolytic microbial complex formation in soils . *Biology Bulletin* 34, 163–169.

Marschner, P. and Rengel, Z. (2007) Contributions of rhizosphere interactions to soil biological fertility. In: Abbott, L. and Murphy, D. (eds) *Soil Biological Fertility: A Key to Sustainable Land Use in Agriculture*. Springer, Dordrecht, pp. 81–98.

Mazzola, M. (2004) Assessment and management of soil microbial community structure for disease suppression. *Annual Review of Phytopathology* 42(1), 35–59.

Medina, R., Franco, M.E., Cruz Cabral, L., Bahima, J.V., Patriarca, A. *et al.* (2021) The secondary metabolites profile of *Stemphylium lycopersici*, the causal agent of tomato grey leaf spot, is complex and includes host and non-host specific toxins. *Australasian Plant Pathology* 50(1), 105–115.

Mendes, R., Kruijt, M., Bruijn, I., Dekkers, E., Voort, M. *et al.* (2011) Deciphering the rhizosphere microbiome for disease-suppressive bacteria. *Science* 332(6033), 1097–1100.

Meyer, B. and Kuever, J. (2007) Molecular analysis of the distribution and phylogeny of dissimilatory adenosine-5'-phosphosulfate reductase-encoding genes (aprBA) among sulfur-oxidizing prokaryotes. *Microbiology* 153(10), 3478–3498.

Miransari, M. (2013) Soil microbes and the availability of soil nutrients. *Acta Physiologiae Plantarum* 35(11), 3075–3084.

Nielsen, U.N., Ayres, E., Wall, D.H. and Bardgett, R.D. (2011) Soil biodiversity and carbon cycling: A review and synthesis of studies examining diversity–function relationships. *European Journal of Soil Science* 62(1), 105–116.

Nielsen, U.N., Wall, D.H. and Six, J. (2015) Soil biodiversity and the environment. *Annual Review of Environment and Resources* 40(1), 63–90.

Nwokolo, N.L., Enebe, M.C., Chigor, C.B., Chigor, V.N. and Dada, O.A. (2021) The contributions of biotic lines of defence to improving plant disease suppression in soils: A review. *Rhizosphere* 19, 100372.

O'Donnell, A.G., Seasman, M., Macrae, A., Waite, I. and Davies, J.T. (2001) Plants and fertilisers as drivers of change in microbial community structure and function in soils. *Plant and Soil* 232, 135–145.

Page, K., Dang, Y. and Dalal, R. (2013) Impacts of conservation tillage on soil quality, including soil-borne crop diseases, with a focus on semi-arid grain cropping systems. *Australasian Plant Pathology* 42(3), 363–377.

Postma, J., Nijhuis, E.H. and Someus, E. (2010) Selection of phosphorus solubilizing bacteria with biocontrol potential for growth in phosphorus rich animal bone charcoal. *Applied Soil Ecology* 46(3), 464–469.

Radwan, M.A., Farrag, S.A., Abu-Elamayem, M.M. and Ahmed, N.S. (2012) Extraction, characterization, and nematicidal activity of chitin and chitosan derived from shrimp shell wastes. *Biology and Fertility of Soils* 48(4), 463–468.

Rahi, P., Pathania, V., Gulati, A., Singh, B., Bhanwra, R.K. *et al.* (2010) Stimulatory effect of phosphate-solubilizing bacteria on plant growth, stevioside and rebaudioside-A contents of *Stevia rebaudiana* Bertoni. *Applied Soil Ecology* 46(2), 222–229.

Raj, D.P.R.S., Linda, R. and Babyson, R.S. (2014) Molecular characterization of phosphate solubilizing bacteria (PSB) and plant growth promoting rhizobacteria (PGPR) from pristine soils. *International Journal of Innovative Science, Engineering and Technology* 1, 317–324.

Rajput, V.D., Singh, A., Singh, V.K., Minkina, T.M. and Sushkova, S. (2021) Impact of nanoparticles on soil resource. In: Amrane, A., Mohran, D., Nguyen, T., Assadi, A. and Yasin, G. (eds) *Nanomaterials for Soil Remediation*. Elsevier, New York, pp. 65–85.

Segata, N., Boernigen, D., Tickle, T.L., Morgan, X.C., Garrett, W.S. *et al.* (2013) Computational meta'omics for microbial community studies. *Molecular Systems Biology* 9(1), 666.

Sepp, S.-K., Vasar, M., Davison, J., Oja, J., Anslan, S. *et al.* (2023) Global diversity and distribution of nitrogen-fixing bacteria in the soil. *Frontiers in Plant Science* 14, 1100235.

Sharma, S.B., Sayyed, R.Z., Trivedi, M.H. and Gobi, T.A. (2013) Phosphate solubilizing microbes: Sustainable approach for managing phosphorus deficiency in agricultural soils. *SpringerPlus* 2, 587.

Siegel-Hertz, K., Edel-Hermann, V., Chapelle, E., Terrat, S., Raaijmakers, J.M. *et al.* (2018) Comparative microbiome analysis of a Fusarium wilt suppressive soil and a fusarium wilt conducive soil from the Châteaurenard region. *Frontiers in Microbiology* 9, 568.

Solden, L., Lloyd, K. and Wrighton, K. (2016) The bright side of microbial dark matter: Lessons learned from the uncultivated majority. *Current Opinion in Microbiology* 31, 217–226.

Srinivasan, R., Yandigeri, M.S., Kashyap, S. and Alagawadi, A.R. (2012) Effect of salt on survival and P-solubilization potential of phosphate solubilizing microorganisms from salt affected soils. *Saudi Journal of Biological Sciences* 19(4), 427–434.

Sugiyama, A., Vivanco, J.M., Jayanty, S.S. and Manter, D.K. (2010) Pyrosequencing assessment of soil microbial communities in organic and conventional potato farms. *Plant Disease* 94(11), 1329–1335.

Swift, M.J., Heal, O.W., Anderson, J.M. and Anderson, J.M. (1979) *Decomposition in Terrestrial Ecosystems*, Vol. 5. University of California Press, Berkeley.

Szent-Gyorgyi, A. (2012) *Introduction to a Submolecular Biology*. Elsevier, New York.

Tajini, F., Trabelsi, M. and Drevon, J.J. (2012) Combined inoculation with *Glomus intraradices* and *Rhizobium tropici* CIAT899 increases phosphorus use efficiency for symbiotic nitrogen fixation in common bean (*Phaseolus vulgaris* L.). *Saudi Journal of Biological Sciences* 19(2), 157–163.

Tilston, E.L., Pitt, D. and Groenhof, A.C. (2002) Composted recycled organic matter suppresses soil-borne diseases of field crops. *New Phytologist* 154(3), 731–740.

Tiquia, S.M., Lloyd, J., Herms, D.A., Hoitink, H.A. and Michel Jr, F.C. Jr. (2002) Effects of mulching and fertilization on soil nutrients, microbial activity and rhizosphere bacterial community structure determined by analysis of TRFLPs of PCR-amplified 16S rRNA genes. *Applied Soil Ecology* 21(1), 31–48.

Trivedi, P., Leach, J.E., Tringe, S.G., Sa, T. and Singh, B.K. (2020) Plant-microbiome interactions: From community assembly to plant health. *Nature Reviews. Microbiology* 18(11), 607–621.

Turner, T.R., James, E.K. and Poole, P.S. (2013) The plant microbiome. *Genome Biology* 14, 1–10.

Uroz, S., Calvaruso, C., Turpault, M.P. and Frey-Klett, P. (2009) Mineral weathering by bacteria: Ecology, actors and mechanisms. *Trends in Microbiology* 17(8), 378–387.

Welbaum, G.E., Sturz, A.V., Dong, Z. and Nowak, J. (2004) Managing soil microorganisms to improve productivity of agro-ecosystems. *Critical Reviews in Plant Sciences* 23(2), 175–193.

Yin, C., Hulbert, S.H., Schroeder, K.L., Mavrodi, O., Mavrodi, D. *et al.* (2013) Role of bacterial communities in the natural suppression of *Rhizoctonia solani* bare patch disease of wheat (*Triticum aestivum* L.). *Applied and Environmental Microbiology* 79(23), 7428–7438.

Yogev, A., Raviv, M., Hadar, Y., Cohen, R., Wolf, S. *et al.* (2010) Induced resistance as a putative component of compost suppressiveness. *Biological Control* 54(1), 46–51.

Zhang, H., Sun, Y., Xie, X., Kim, M.S., Dowd, S.E. *et al.* (2009) A soil bacterium regulates plant acquisition of iron via deficiency-inducible mechanisms. *Plant Journal* 58(4), 568–577.

Zhao, F., Xin, X., Cao, Y., Su, D., Ji, P. *et al.* (2021) Use of carbon nanoparticles to improve soil fertility, crop growth and nutrient uptake by corn (*Zea mays* L.). *Nanomaterials* 11(10), 2717.

Zhao, K., Penttinen, P., Zhang, X., Ao, X., Liu, M. *et al.* (2014) Maize rhizosphere in Sichuan, China, hosts plant growth promoting *Burkholderia cepacia* with phosphate solubilizing and antifungal abilities. *Microbiological Research* 169(1), 76–82.

Zhu, F., Qu, L., Hong, X. and Sun, X. (2011) Isolation and characterization of a phosphate-solubilizing halophilic bacterium *Kushneria* sp. YCWA18 from Daqiao Saltern on the coast of Yellow Sea of China. *Evidence-Based Complementary and Alternative Medicine* 2011(1), 615032.

4 Harnessing Plant–Microbial Interaction for Sustainable Crop Production Using Biotechnology

Iqra Owais[1], Tahleel and Muhammad Khashi u Rahman[2]*

[1]Department of Botany, Emerson University, Multan, Pakistan; [2]Institute for Agribiotechnology Research (CIALE) and Microbiology and Genetics Department, University of Salamanca, Salamanca, Spain

Abstract

Plants interact and establish symbiotic associations with diverse microorganisms. The mutual interactions between plants and these microbes play a significant role in stimulating plant growth and making them resilient during stressful conditions. Using advanced biotechnological tools, plant-associated microbes and their microbial products can be converted into essential plant probiotics, thereby reducing the input of chemical formulations. This makes them organic and eco-friendly candidates for maintaining plant and soil health. With advances in the biotechnological fields, we are now able to enhance plant–microbial interactions using different genetic engineering techniques. Genetically engineering plants and beneficial microbes using clustered regularly interspaced short palindromic repeats with associated protein 9 (CRISPR-Cas9) technology can introduce advantageous mutations in the genes encoding receptors, help plants and microbes recognize broader ranges of signaling molecules more effectively, and strengthen plant–microbial interactions. In this chapter, we discuss the complexity of plant–microbial interactions and how we can enhance these interactions using advanced biotechnology to sustain crop and food security.

Keywords: symbiotic associations, CRISPR-Cas9, food security, genetic engineering

4.1 Introduction

Food security and sustainable agriculture are closely related concepts that ensure consistent accessibility and affordability of nutritious food. Sustainable agricultural practices mainly focus on building more resilient food systems that effectively address the present and future challenges of global food demands. These farming practices are necessary for producing high yields of nutrient-rich and climate-resilient crops. A sustainable agriculture system also promises the conservation of resources like water and microbial diversity in the soil and contributes to food security (Khashi u Rahman *et al.*, 2021).

However, the production of enough food to meet the needs of a constantly growing global population is one of the most concerning issues. Over the next 40 years, the world population is expected to increase by 35% (Takahashi *et al.*, 2020). As conventional farming practices slowly degrade soil structure and fertility, we

*Corresponding author: khashiurahman@usal.es

 Plant Biotechnology and Sustainable Agriculture (ed. Tariq Aftab)
DOI: 10.1079/9781800627659.0004

need to develop an eco-friendly and sustainable agriculture system by improving the plant-associated beneficial microbiome. Plants have diverse microbial assemblages including plant growth-promoting rhizobacteria (PGPR), arbuscular mycorrhizal fungi (AMF), archaea, and others that develop symbiotic associations with the plant. Both microbes and plants receive advantages from this symbiotic or mutualistic relationship (Lima *et al.*, 2009). These microbial communities directly or indirectly stimulate plant growth and development by improving nitrogen fixation, mineral solubilization, nutrient acquisition, disease resistance, and stress tolerance (Compant *et al.*, 2019).

With biotechnological advances in genomic and metabolomic studies, we can stimulate plant–microbe interactions using various genome editing techniques such as zinc finger nucleases (ZFNs) and transcription activator-like effector nucleases (TALENs). Also, CRISPR-Cas9 (clustered regularly interspaced short palindromic repeats with associated protein 9) is an advanced and effective genome editing technology that allows for precise gene insertion, deletion, or alteration at any genome locus (Pratheesh *et al.*, 2020). These genetic engineering techniques have potential applications in precisely modifying the genomes of plant-beneficial microbes that bring multiple benefits to food and agricultural industries. We can also develop organic plant probiotics from beneficial microbes using biotechnological approaches. These eco-friendly microbe-based plant probiotics function as biofertilizers, biopesticides, and biostimulants for plants, reduce the reliance on chemical fertilizers, and play a significant role in maintaining the structure and biodiversity of agricultural land. Biotechnologically prepared plant probiotics minimize environmental degradation and contribute to long-term food security by ensuring the availability of resources for sustainable agriculture (Fig. 4.1).

In this chapter, our primary aim is to comprehend the mechanisms of assemblage of the plant-associated microbiome, and how we can use CRISPR-Cas9 to enhance plant–microbial interactions and improve agronomic traits. We will also discuss how we can produce plant probiotics from symbiotic microbes using advanced

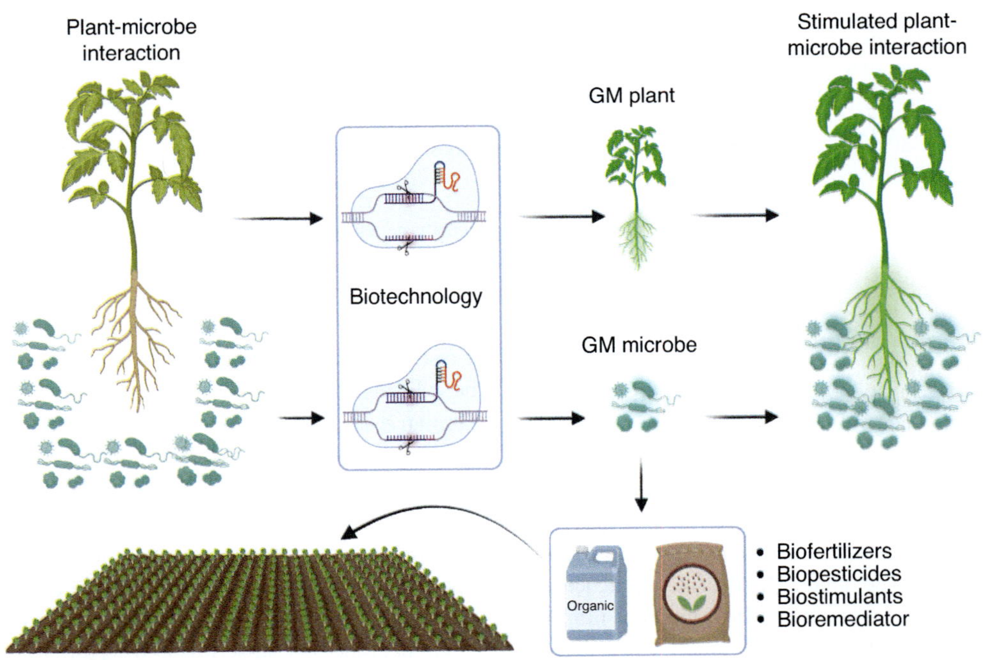

Fig. 4.1. Diagram of biotechnologically improved plant–microbe interactions for sustainable crop production. Figure created via BioRender.com.

biotechnology for sustainable crop production, effectively contributing to global food security.

4.2 Plant-Associated Microbiome

Plants harbor diverse communities of microorganisms in the rhizosphere (the region around the plant roots), phyllosphere (aerial parts of the plant), and endosphere (internal tissues of the plant). These plant-associated beneficial microbes include AMF, PGPR, and nitrogen-fixing and other PGP endosymbionts (Mishra *et al.*, 2021). These abundant microbial communities are collectively known as microbiota while the collective genomes of microbiota in a particular habitat are called the microbiome (Schlaeppi and Bulgarelli, 2015).

The plant microbiome encompasses all the interactions between itself and the host plant. These interactions can be parasitic, commensalistic, or mutualistic and play an essential role in nutrient cycling, decomposition, and plant health. In parasitic interactions, microbes develop pathogenic relationships with the host and adversely impact plants' health and development (Morgan *et al.*, 2005). In commensalistic interaction, only one of the two partners benefits from symbiosis, while the other remains unaffected (Newton *et al.*, 2010). A mutualistic relationship involves mutual benefit, with plants offering photosynthetic carbon to microbes in exchange for mineral nutrients and protection (Powell and Klironomos, 2007). Mutualistic plant-associated microbes include nitrogen-fixing bacteria that are free-living symbionts mainly belonging to *Rhizobium*, *Azorhizobium*, *Mesorhizobium*, and *Frankia* species, located in the root nodules. These bacteria symbiotically fix nitrogen in higher plants and make it easily accessible to them (Bhattacharyya and Jha, 2012). Other microbes including AMF, belonging to the phylum *Glomeromycota*, colonize the root cortex and enhance plant growth by stimulating nutrient acquisition from the soil (Soumare, 2015). Mutualistic plant–microbial interactions take part in the primary and secondary metabolism of the plant and significantly promote plant growth and overall fitness.

One of the most important components of the plant-associated microbiome is PGPR which promotes plant growth and development by different mechanisms. In the absence of pathogenic microbes, some members of free-living PGPR belonging to *Azospirillum*, *Burkholderia*, *Bacillus*, and *Herbaspirillum* can directly fix nitrogen (Vessey, 2003). These rhizospheric microbes benefit plants with an additional supply of nitrogen along with the production of plant growth regulators and significantly contribute to plant growth and development (Ramos Solano *et al.*, 2008). Other PGPR produce siderophores that chelate iron, solubilize phosphorus, and make these nutrients easily accessible to plants. These microbes promote plant growth by suppressing the activity of phytopathogens by producing antimicrobial compounds or stimulating the plant's defense mechanism (Reddy, 2012). Some PGPR belonging to the genera *Azotobacter*, *Azospirillum*, *Arthrobacter*, *Bacillus*, *Burkholderia*, *Flavobacterium*, *Erwinia*, *Micrococcus*, *Agrobacterium*, *Serratia*, and *Pseudomonas* effectively augment plant and soil sustainability (Gray and Smith, 2005).

Apart from their growth-promoting activity, PGPR play a significant role in stimulating plants' resilience to various biotic stresses, such as insect pests and microbial diseases, and abiotic stresses, including drought, salinity, and heavy metal contamination.

4.2.1 Role of PGPR in biotic stress management

Phytopathogens, including pathogenic bacteria, fungi, viruses, nematodes, and insect pests, affect global crop yield and pose serious threats to food sustainability. The use of chemical pesticides and insecticides can worsen plant and soil health. Plants have developed various innate defense strategies including accumulation of reactive oxygen species (ROS), callose deposition, defense hormone signaling, cell wall rigidification, and regulation of genes encoding pathogenesis-related proteins that protect them from pathogenic attacks (Nürnberger *et al.*, 2004).

Apart from plants' innate immune responses, the application of PGPR is an alternative and promising approach to mitigate the harmful impact of biotic stresses and make plants more resilient to pathogenic attacks. These microbes, including some strains of *Pseudomonas fluorescens, Bacillus subtilis, B. amyloliquefaciens, B. mycoides, B. pumilus*, and *B. cereus*, are reported to significantly reduce the impact of phytopathogens in two ways: direct and indirect (Kumar *et al.*, 2018). Direct mechanisms include production of antimicrobial compounds, lytic enzyme secretion, siderophore production, phosphate solubilization, allelochemical production, and competition with other microbes, while indirect mechanisms include stimulation of plant secondary metabolite production and induction of the plant's immune response against these pathogens (Fadiji and Babalola, 2020).

Some PGPR synthesize antimicrobial compounds such as lipopeptides, polyketides, polymyxins, flavonoids, quinones, alkaloids, and phenols that restrict pathogenic activity (Hamid *et al.*, 2021). Several antibiotic compounds including 2,4-diacetylphloroglucinol, tropolone, tensin, pyrrolnitrin, phenazine, oomycin, hydrogen cyanide, oligomycin A, and cyclic lipopeptides are produced by some members of *Streptomyces, Stenotrophomonas*, and *Bacillus* (Compant *et al.*, 2005). These compounds influence plant metabolic activities and membrane structures, and restrict synthesis of the pathogenic cell walls, thereby reducing their pathogenic activity (Maksimov *et al.*, 2011). Lytic enzymes such as β−1,3-glucanase, chitinase, protease, and cellulase produced by rhizobacteria degrade the cell walls and suppress the growth and development of invading filamentous hyphae and other microbes (Gao *et al.*, 2010). These microbes also compete with other pathogenic microbes for resources in the soil and serve as potential biocontrol agents (Oukala *et al.*, 2021). Some PGPR produce siderophores that protect plants from phytopathogens by competing for iron, reducing iron availability for pathogen growth (Cesa-Luna *et al.*, 2020).

Plant growth-promoting rhizobacteria and other nonpathogenic microbes can stimulate the defense responses of plants through induced systemic resistance (ISR) mediated by the jasmonic acid or ethylene signaling pathways (Ward *et al.*, 1991). Various strains of *Pseudomonas* and *Bacillus* elicit systemic disease resistance in plants through a variety of signal translocation pathways and ISR-related gene expression, enhancing the defense responses against broader ranges of pathogenic microbes (Choudhary *et al.*, 2007). *Pseudomonas fluorescens*, a PGP bacterium, maintains plant health by producing antifungal compounds (Haas and Défago, 2005). These beneficial microbes stimulate the production of plant secondary metabolites, including low molecular antimicrobial compounds such as phytoalexins that boost the plant's resistance to biotic stress (Gao *et al.*, 2010). The incorporation of PGPR into integrated disease management programs can help farmers develop more resilient cropping systems that support environmental sustainability and agricultural productivity (Han *et al.*, 2021).

4.2.2 Role of PGPR in abiotic stress management

Agricultural productivity is suppressed by several abiotic stress factors including drought, salinity, alkalinity, nutrient deprivation, and soil contamination that negatively impact plant growth through ion toxicity, hormonal and nutritional imbalance, and physiological and metabolic changes (Gao *et al.*, 2020). Plants harboring PGPR show enhanced tolerance to these stress conditions. These microbes regulate the expression of stress-responsive genes and make plants more resilient in stressful environments (Subiramani *et al.*, 2020).

Drought and salinity are the two major factors responsible for the decline in agricultural yields. Prolonged unavailability of water and elevated levels of salts in the soil cause ion toxicity, and disrupt the water balance and normal physiological processes in plants. Synthesis of 1-aminocyclopropane-1-carboxylate (ACC) deaminase, indole-3-acetic acid, cytokinins, gibberellins, and abscisic acid by several members of the genera *Azospirillum, Bacillus, Burkholderia, Xanthobacter, Pseudomonas*, and *Enterobacter* has been proved to stimulate plant tolerance to drought and salinity stress (Kumar *et al.*, 2019; Khashi u Rahman *et al.*, 2020). Consistent drought periods trigger PGPR to produce extracellular polymeric substances, osmolytes, phytohormones, and antioxidants that induce physiological changes in the roots and stimulate stress tolerance (Yang *et al.*,

2009). Production of the phytohormone abscisic acid (ABA) by PGPR during drought stress induces the closure of stomata to reduce water loss, aids in osmotic regulation, and optimizes water transport efficiency (Percival Zhang *et al.*, 2006). Some strains of *Streptomyces* were reported to confer enhanced salt tolerance in rice through the activity of ACC deaminase to transform ethylene precursor into ammonia and α-ketobutyrate, thus lowering ethylene levels in the plant (Jaemsaeng *et al.*, 2018).

Plant growth-promoting rhizobacteria play a pivotal role in alleviating heavy metal toxicity and promoting efficient mineral absorption. The activity of *Sphingomonas macrogoltabidus*, *Alyssum murale*, *Microbacterium liquefaciens*, and *Thlaspi caerulescens* is of immense importance in reclaiming soil contaminated with heavy metals (Abou-Shanab *et al.*, 2003). Abiotic stresses induce the production and accumulation of ROS in plants which causes oxidative damage to the cellular components, leading to cell death. Some PGPR have been found to enhance the activity of antioxidant enzymes, such as glutathione reductase, ascorbate peroxidase, superoxide dismutase, and catalase, and nonenzymatic antioxidants such as α-tocopherol, ascorbate, and flavonoids, and to protect plants from oxidative damage (Nivetha *et al.*, 2021). By employing these mechanisms, PGPR serve as an excellent strategy to alleviate the impact of abiotic stresses.

4.3 How Plants Assemble Their Microbiome

As an exceptionally rich microbial reservoir, soil habitat plays a key role in facilitating the selection of host microbiota. Microbial activities and the physiochemical properties of the soil are greatly affected by the root activities of the host plant. In the rhizosphere, plant roots secrete a variety of organic compounds including amino acids, sugars, organic acids, phenols, secondary metabolites, and flavonoids that function as chemoattractants for microbes in the soil surrounding the roots (Rahman *et al.*, 2020; Vives-Peris *et al.*, 2020). These exuded compounds directly influence the microbial communities in the rhizosphere and selectively attract or repel the microbes (Leach *et al.*, 2017). These compounds are the source of early communication with the soil microbes and initiate the recruitment process. They also synthesize signaling molecules that trigger plant responses necessary for colonization (van West *et al.*, 2002).

These observations suggest that the selection of beneficial microbes in the rhizosphere is not only a plant-driven mechanism but a complex evolutionary process influenced by the interplay between the host plant and its associated microbiome (Andreote *et al.*, 2014). Effective colonization of microbes in a particular niche involves the ability of microbes to surpass host immune defenses and endure different abiotic stresses.

Microbes have adapted several traits such as chemotaxis, exopolysaccharide production, and production of cell wall-degrading enzymes for successful invasion, translocation, and colonization in the host plant (Piromyou *et al.*, 2015). These microbes produce microbe-associated molecular patterns (MAMPs) or pathogen-associated molecular patterns (PAMPs) that are recognized by the receptor-like kinase membrane proteins known as pattern recognition receptors (PRRs) of the host plant (Vandenkoornhuyse *et al.*, 2015). Inside the host plant, symbiotic microbes use xylem vessels as their main transport route to spread and colonize nutrient-rich intercellular spaces of the vegetative parts (Compant *et al.*, 2010). However, the mechanisms by which host plants and environment shape microbiome composition and interactions in the rhizosphere, phyllosphere, and endosphere are still largely unexplored. Considering the biological potential of this mutualistic relationship between the host plant and its associated microbiome, we can improve the integrated metabolism of these interactions to achieve the targets of agricultural sustainability and food security.

4.4 Using CRISPR-Cas9 to Improve Plant–Microbe Interaction

Recent advances in genomics have introduced various tools for genome editing that hold immense potential for enhancing plant–microbe associations. These interactions are crucial in making crops more climate resilient and sustainable. Genome editing has revolutionized the agriculture and breeding system by producing genetically modified plants with desirable agronomic traits.

CRISPR-Cas9 offers enhanced plant traits that govern microbial colonization and symbiosis through targeted genome editing. CRISPR-Cas9 is the most efficient and reliable genome editing technique that can precisely introduce advantageous mutations or delete the genetic components responsible for undesirable traits. One of the main components of the CRISPR-Cas9 system is Cas-protein enzymes that act as molecular scissors to cleave the targeted DNA, and another is guide RNA (gRNA) that ensures the correct binding of Cas9 machinery to the targeted DNA (Shalem *et al.*, 2014). Using CRISPR-Cas9 to target particular genes from the microbe or the host plant can exponentially improve plant–microbial interactions, thereby contributing to sustainable crop production and food security.

Plants release certain types of root exudates that favor the proliferation of a specific beneficial microbiome in the rhizosphere. Some plants were found to elevate the secretion of malate, citrate, or oxalate during stress conditions (Yu *et al.*, 2023), which act as important carbon sources for plant-beneficial microbes. For example, some beneficial strains of *Burkholderia* spp. were reported to utilize oxalate as a carbon source (Kost *et al.*, 2014). This selective recruitment of beneficial microbes by liberating special chemical compounds is a key strategy adopted by plants to boost symbiotic associations.

The efficiency of Cas9-mediated gene editing can be significantly enhanced by some chemical alterations in the gRNA (Lee *et al.*, 2017). Engineering microbes to produce elevated levels of phytohormones, enhancing mineral nutrient solubilization and nitrogen-fixing activity contributes to overall plant growth and vigor. CRISPR-mediated genetic mutations in fungal microbes such as *Beauveria bassiana*, *Purpureocillium lilacinum*, and *Trichoderma* spp. make them effective biocontrol agents which inhibit the growth of nematodes and insect larvae (Jiao *et al.*, 2019). CRISPR-Cas9 can precisely delete or disable the genes responsible for the pathogenicity of microbes or the susceptibility factor of the host plant to induce disease resistance (Das *et al.*, 2019).

Node factors released by nitrogen-fixing bacteria are recognized by lysin motif receptor-like kinases (LysM-RLKs) on the surface of legume root hairs. These receptors then initiate signal transduction pathways for nodule formation in the roots (Geurts and Bisseling, 2002). Using CRISPR-Cas9 to edit the plant genes responsible for recognizing LysM-RLKs can enhance the efficiency of node factor recognition and the establishment of mutualistic relationships, potentially extending the benefits of rhizobial nitrogen fixation in legume and nonlegume crops. Similarly, PGPR and other microbes produce MAMPs such as flagellin-sensing 2 (FLS2) receptor and chitin elicitor receptor kinase 1 (CERK1) that are recognized by the PRRs in the plants, leading to symbiotic plant–microbe association (Oleńska *et al.*, 2020).

Enhancing the sensitivity of PRRs using CRISPR-Cas9 can exponentially improve the ability of plants to recognize and respond to beneficial PGP bacteria, leading to stronger plant–microbe interaction.

4.5 Development of Plant-Beneficial Microbes Into Probiotics

One of the major causes of soil degradation is excessive use of chemical fertilizers, pesticides, herbicides, and other products that contain hazardous chemical compounds. Continued reliance on agrochemicals can deteriorate soil health and severely affect the soil microflora, leading to reduced crop productivity. The use of organic microbial products as an alternative to chemical fertilizers has gained the interest of many scientists. The ubiquitous nature of microbes and their effective applications in agriculture have made them a potential candidate for resolving issues related to crop production and food security. Plants' complex symbiotic interactions with diverse varieties of microbes in the soil play a promising role in stimulating and sustaining nutrient cycling, soil structure, stress tolerance, and ecosystem functioning.

The products of symbiotic microbes can be made into organic plant probiotics using different biotechnological techniques. These microbe-based probiotics act as biofertilizers, biostimulators, biopesticides, bioherbicides, bioremediators, and biocontrol agents according to plant and soil needs (Sharma *et al.*, 2012). Free-living PGPR, including members of several genera, e.g., *Azotobacter*, *Bacillus*, *Pseudomonas*, *Rhizobium*, *Xanthomonas*, *Arthrobacter*, and *Enterobacter*, act as biofertilizers and biostimulants by enhancing nutrient acquisition, mineral solubilization, plant growth, and tolerance to various biotic and abiotic stresses (Nath Yadav, 2017). These PGPR along with *Aspergillus* and *Penicillium* function as phosphate solubilizers and convert insoluble forms of phosphate, i.e., HPO_4 and H_2PO_4, into more soluble forms which are easily accessible to plants (Anand *et al.*, 2016). Various members of mycorrhizal fungi from *Scutellospora*, *Entrophospora*, *Sclerocystis*, *Gigaspora*, and *Glomus* colonize plant roots and form symbiotic associations, contributing to the enhanced acquisition of P, Ca, Zn, Mg, and S from the soil to stimulate plant growth (Sullia, 1991; Rahman and Siddique, 2021).

An environment-friendly approach to plant disease management is the use of symbiotic microbes as biopesticides. These microbial pesticides can be used as effective biocontrol agents against insects and pests that are vectors of deadly plant pathogens. Microbial products obtained from some members of *Burkholderia*, *Pseudomonas*, *Saccharopolyspora*, *Streptomyces*, *Beauveria bassiana*, *Isaria*, *Hirsutella*, and *Verticillium* spp. can be manufactured into bioinsecticides and biopesticides to prevent plant diseases and yield loss (Thakur *et al.*, 2020). Microbe-based plant probiotics can degrade organic pollutants in contaminated soil. Microbes such as *Aspergillus niger* CSR3, *Burkholderia cepacia* G, and *Microbacterium lactium* YJ7 play an important role in the reclamation of soil contaminated with heavy metals (Tiwari *et al.*, 2023). Harnessing the potential of biotechnologically engineered microbes can develop effective plant probiotics with novel metabolites.

The production of microbe-based formulations has a promising future in creating a climate-resilient and sustainable agriculture system. With advances in the field of biotechnology, several plant probiotics are currently available on the market; some of them are listed in Table 4.1. However, the efficiency of these microbes remains an unresolved challenge to eliminate chemical input in the field.

4.6 Conclusion

The intricate dynamics of mutualistic relationships between plants and soil microbes hold tremendous potential for plant growth and environmental sustainability. The plant-associated symbiotic microbiome stimulates plant health, nutrient acquisition, and resistance to different biotic and abiotic stresses. Several essential bioactive compounds synthesized by plant–microbe interactions reduce the need for chemical fertilizers and pesticides, thereby mitigating environmental impact and promoting soil health. Beneficial microbes and their products can be converted into organic plant probiotics that replenish the nutrient requirements of plants and contribute to soil health.

Recent advances in omics technologies have helped us better understand and improve microbe-mediated strategies for plant growth promotion and biocontrol activities. Innovations in biotechnology such as genetically engineered microbes and plants, precision microbiome management, and bioinformatics-driven insights offer new and more sustainable farming practices that optimize crop yields while reducing reliance on chemical inputs.

Considering the significance of microbes and their influence on plant health, future studies need to delve deeper into comprehending plant–microbial associations to ensure agricultural sustainability and address the challenges of global food security.

Table 4.1. Commercially available microbe-based plant probiotics.

Product name	Microbes	Manufacturer	Properties	References
Avanex®	*Epichloë coenophiala*	PGG Wrightson Seeds Ltd	Insecticide	Pennell *et al.* (2016)
BioEnsure®	*Muscodor albus*	Adaptive Symbiotic Technologies	Biofertilizer	Sharma *et al.* (2021)
BioBoost	*Delftia acidovorans*	Brett-Young Seeds Ltd	Biofertilizer	Song *et al.* (2012)
AQ-10 Biofungicide	*Ampelomyces quisqualis* M-10	Ecogen	Biofungicide	Song *et al.* (2012)
Biopromotor	*Azospirillum* spp.	Manidharma Biotech	Biofertilizer	Song *et al.* (2012)
MaxQ®	*Epichloë*	Grasslanz Technology Ltd	Bioinsecticide	Pandey *et al.* (2022)
Yates Nature's Way Caterpillar Killer	*Bacillus thuringiensis*	Dulux Pty Ltd	Bioinsecticide	Glare and O'Callaghan (2019)
Invade Bioshield Grass-Grub	*Serratia entomophila*	Biostart	Bioinsecticide	Glare and O'Callaghan (2019)
YieldShield	*Bacillus pumilus* GB34	Bayer CropScience	Biofungicide	Song *et al.* (2012)
MADEX 3	*Cydia pomonella* granulosis virus	Key Industries Ltd	Bioinsecticide	Glare and O'Callaghan (2019)
Poncho Votivo	*Bacillus firmus*	Bayer New Zealand Ltd	Bioinsecticide and nematicide	Glare and O'Callaghan (2019)
Rootonic	*Piriformospora indica*	European Patent Office	Biofertilizer	Smriti and Ajit (2014)
DIPEL ES	*Bacillus thuringiensis*	Valent BioSciences	Bioinsecticide and pesticide	Glare and O'Callaghan (2019)

References

Abou-Shanab, R., Angle, J., Delorme, T., Chaney, R., Berkum, P. *et al.* (2003) Rhizobacterial effects on nickel extraction from soil and uptake by *Alyssum murale*. *New Phytologist* 158(1), 219–224.

Anand, K., Kumari, B. and Mallick, M. (2016) Phosphate solubilizing microbes: An effective and alternative approach as biofertilizers. *International Journal of Pharmacy and Pharmaceutical Sciences* 8, 37–40.

Andreote, F.D., Gumiere, T. and Durrer, A. (2014) Exploring interactions of plant microbiomes. *Scientia Agricola* 71(6), 528–539.

Bhattacharyya, P.N. and Jha, D.K. (2012) Plant growth-promoting rhizobacteria (PGPR): Emergence in agriculture. *World Journal of Microbiology & Biotechnology* 28(4), 1327–1350.

Cesa-Luna, C., Baez, A., Quintero-Hernández, V., Cruz-Enríquez, J., Castañeda-Antonio, M.D. *et al.* (2020) The importance of antimicrobial compounds produced by beneficial bacteria on the biocontrol of phytopathogens. *Acta Biológica Colombiana* 25(1), 140–154.

Choudhary, D.K., Prakash, A. and Johri, B.N. (2007) Induced systemic resistance (ISR) in plants: Mechanism of action. *Indian Journal of Microbiology* 47(4), 289–297.

Compant, S., Duffy, B., Nowak, J., Clément, C. and Barka, E.A. (2005) Use of plant growth-promoting bacteria for biocontrol of plant diseases: Principles, mechanisms of action, and future prospects. *Applied and Environmental Microbiology* 71(9), 4951–4959.

Compant, S., Clément, C. and Sessitsch, A. (2010) Plant growth-promoting bacteria in the rhizo- and endosphere of plants: Their role, colonization, mechanisms involved and prospects for utilization. *Soil Biology and Biochemistry* 42(5), 669–678.

Compant, S., Samad, A., Faist, H. and Sessitsch, A. (2019) A review on the plant microbiome: Ecology, functions, and emerging trends in microbial application. *Journal of Advanced Research* 19, 29–37.

Das, A., Sharma, N. and Prasad, M. (2019) CRISPR-Cas9: A novel weapon in the arsenal to combat plant diseases. *Frontiers in Plant Science* 9.

Fadiji, A.E. and Babalola, O.O. (2020) Elucidating mechanisms of endophytes used in plant protection and other bioactivities with multifunctional prospects. *Frontiers in Bioengineering and Biotechnology* 8.

Gao, F.-K., Dai, C.-C. and Liu, X.-Z. (2010) Mechanisms of fungal endophytes in plant protection against pathogens. *African Journal of Microbiology Research* 4, 1346–1351.

Gao, H., Yang, W., Li, C., Zhou, X., Gao, D. *et al.* (2020) Gene expression and K^+ uptake of two tomato cultivars in response to sub-optimal temperature. *Plants* 9(1), 65.

Geurts, R. and Bisseling, T. (2002) Rhizobium nod factor perception and signalling. *Plant Cell* 14, S239–S249.

Glare, T.R. and O'Callaghan, M. (2019) Microbial biopesticides for control of invertebrates: Progress from New Zealand. *Journal of Invertebrate Pathology* 165, 82–88.

Gray, E.J. and Smith, D.L. (2005) Intracellular and extracellular PGPR: Commonalities and distinctions in the plant–bacterium signaling processes. *Soil Biology and Biochemistry* 37(3), 395–412.

Haas, D. and Défago, G. (2005) Biological control of soil-borne pathogens by fluorescent pseudomonads. Nature Reviews. *Microbiology* 3(4), 307–319.

Hamid, S., Lone, R. and Mohamed, H.I. (2021) Production of antibiotics from PGPR and their role in biocontrol of plant diseases. In: Mohamed, H.I., El-Beltagi, H.E.-D.S. and Abd-Elsalam, K.A. (eds) *Plant Growth-Promoting Microbes for Sustainable Biotic and Abiotic Stress Management*. Springer, Cham, pp. 441–461.

Han, Z., Di, C., Rahman, M.K. u, Gao, D., Wu, F. *et al.* (2021) Repeated application of rice straw stabilizes soil bacterial community composition and inhibits clubroot disease. *Agriculture* 11(2), 108.

Jaemsaeng, R., Jantasuriyarat, C. and Thamchaipenet, A. (2018) Molecular interaction of 1-aminocycl opropane-1-carboxylate deaminase (ACCD)-producing endophytic *Streptomyces* sp. GMKU 336 towards salt-stress resistance of *Oryza sativa* L. cv KDML105. *Scientific Reports* 8(1), 1950.

Jiao, Y., Li, Y., Li, Y., Cao, H., Mao, Z. *et al.* (2019) Functional genetic analysis of the leucinostatin biosynthesis transcription regulator lcsL in *Purpureocillium lilacinum* using CRISPR-Cas9 technology. *Applied Microbiology and Biotechnology* 103(15), 6187–6194.

Khashi u Rahman, M., Ma, C., Wu, F. and Zhou, X. (2020) Changes in cucumber *Rhizosphere bacillus* spp. abundance and community structure during long-term continuous monocropping. *Polish Journal of Environmental Studies* 30(2), 1795–1800.

Khashi u Rahman, M., Hussain, Z., Zhou, X., Ali, I., Wu, F. *et al.* (2021) Intercropping: A substitute but identical of biofertilizers. In: Dar, G.H. (ed.) *Microbiota and Biofertilizers, Vol 2: Ecofriendly Tools for Reclamation of Degraded Soil Environs*. Springer, Cham, pp. 293–309.

Kost, T., Stopnisek, N., Agnoli, K., Eberl, L. and Weisskopf, L. (2014) Oxalotrophy, a widespread trait of plant-associated *Burkholderia* species, is involved in successful root colonization of lupin and maize by *Burkholderia phytofirmans*. *Frontiers in Microbiology* 4.

Kumar, A., Patel, J.S., Meena, V.S. and Ramteke, P.W. (2019) Plant growth-promoting rhizobacteria: Strategies to improve abiotic stresses under sustainable agriculture. *Journal of Plant Nutrition* 42(11–12), 1402–1415.

Kumar, A., Singh, V.K., Tripathi, V., Singh, P.P. and Singh, A.K. (2018) Plant growth-promoting rhizobacteria (PGPR): Perspective in agriculture under biotic and abiotic stress. In: Prasad, R., Gill, S.S. and Tuteja, N. (eds) *Crop Improvement Through Microbial Biotechnology*. Elsevier, New York, pp. 333–342.

Leach, J.E., Triplett, L.R., Argueso, C.T. and Trivedi, P. (2017) Communication in the phytobiome. *Cell* 169(4), 587–596.

Lee, K., Mackley, V.A., Rao, A., Chong, A.T., Dewitt, M.A. *et al.* (2017) Synthetically modified guide RNA and donor DNA are a versatile platform for CRISPR-Cas9 engineering. *eLife* 6, e25312.

Lima, P.T., Faria, V.G., Patraquim, P., Ramos, A.C., Feijó, J.A. *et al.* (2009) Plant-microbe symbioses: New insights into common roots. *BioEssays* 31(11), 1233–1244.

Maksimov, I.V., Abizgil'dina, R.R. and Pusenkova, L.I. (2011) Plant growth promoting rhizobacteria as alternative to chemical crop protectors from pathogens (review). *Applied Biochemistry and Microbiology* 47(4), 333–345.

Mishra, S., Bhattacharjee, A. and Sharma, S. (2021) An ecological insight into the multifaceted world of plant-endophyte association. *Critical Reviews in Plant Sciences* 40(2), 127–146.

Morgan, J.A.W., Bending, G.D. and White, P.J. (2005) Biological costs and benefits to plant-microbe interactions in the rhizosphere. *Journal of Experimental Botany* 56(417), 1729–1739.

Nath Yadav, A. (2017) Plant growth promoting bacteria: Biodiversity and multifunctional attributes for sustainable agriculture. *Advances in Biotechnology & Microbiology* 5(5).

Newton, A.C., Fitt, B.D.L., Atkins, S.D., Walters, D.R. and Daniell, T.J. (2010) Pathogenesis, parasitism and mutualism in the trophic space of microbe-plant interactions. *Trends in Microbiology* 18(8), 365–373.

Nivetha, N., Lavanya, A.K., Vikram, K.V., Asha, A.D., Sruthi, K.S. *et al.* (2021) PGPR-mediated regulation of antioxidants: Prospects for abiotic stress management in plants. In: Singh, H.B., Vaishnav, A. and Sayyed, R.Z. (eds) *Antioxidants in Plant-Microbe Interaction*. Springer, Singapore, pp. 471–497.

Nürnberger, T., Brunner, F., Kemmerling, B. and Piater, L. (2004) Innate immunity in plants and animals: Striking similarities and obvious differences. *Immunological Reviews* 198(1), 249–266.

Oleńska, E., Małek, W., Wójcik, M., Swiecicka, I., Thijs, S. *et al.* (2020) Beneficial features of plant growth-promoting rhizobacteria for improving plant growth and health in challenging conditions: A methodical review. *Science of the Total Environment* 743, 140682.

Oukala, N., Aissat, K. and Pastor, V. (2021) Bacterial endophytes: The hidden actor in plant immune responses against biotic stress. *Plants* 10(5), 1012.

Pandey, S.S., Jain, R., Bhardwaj, P., Thakur, A., Kumari, M. *et al.* (2022) Plant probiotics – endophytes pivotal to plant health. *Microbiological Research* 263, 127148.

Pennell, C.G.L., Popay, A.J., Rolston, M.P., Townsend, R.J., Lloyd-West, C.M. *et al.* (2016) Avanex unique endophyte technology: Reduced insect food source at airports. *Environmental Entomology* 45(1), 101–108.

Percival Zhang, Y.-H., Himmel, M.E. and Mielenz, J.R. (2006) Outlook for cellulase improvement: Screening and selection strategies. *Biotechnology Advances* 24(5), 452–481.

Piromyou, P., Songwattana, P., Greetatorn, T., Okubo, T., Kakizaki, K.C. *et al.* (2015) The type iii secretion system (T3SS) is a determinant for rice-endophyte colonization by non-photosynthetic *Bradyrhizobium*. *Microbes and Environments* 30(4), 291–300.

Powell, J. and Klironomos, J. (2007) The ecology of plant–microbial mutualisms. In: Paul, E.A. (ed.) *Soil Microbiology, Ecology and Biochemistry*, 3rd edn. Academic Press, San Diego, pp. 257–281.

Pratheesh, P.T., Lal, S., Tuvikene, R., Manickam, S. and Sudheer, S. (2020) The contribution of microbial biotechnology to sustainable development in agriculture and allied sectors. In: Rastegari, A.A., Yadav, A.N. and Yadav, N. (eds) *New and Future Developments in Microbial Biotechnology and Bioengineering*. Elsevier, New York, pp. 17–28.

Rahman, M.K. u, Tan, S., Ma, C., Wu, F. and Zhou, X. (2020) Exogenously applied ferulic acid and p-coumaric acid differentially affect cucumber rhizosphere *Trichoderma* spp. community structure and abundance. *Plant, Soil and Environment* 66(9), 461–467.

Rahman, M.K. u and Siddique, F. (2021) Physiological and molecular strategies of plant adaptationin phosphorus-deficient soils. In: Aftab, T. and Hakeem, K.R. (eds) *Frontiers in Plant-Soil Interaction*. Academic Press, New York, pp. 367–380.

Ramos Solano, B., Barriuso, J. and Gutiérrez Mañero, F.J. (2008) Physiological and molecular mechanisms of plant growth promoting rhizobacteria (PGPR). In: Ahmad, I., Pichtel, J. and Hayat, S. (eds) *Plant-Bacteria Interactions*. John Wiley & Sons, New York, pp. 41–54.

Reddy, P.P. (ed.) (2012) Plant growth-promoting rhizobacteria (PGPR). In: *Recent Advances in Crop Protection*. Springer India, New Delhi, pp. 131–158.

Schlaeppi, K. and Bulgarelli, D. (2015) The plant microbiome at work. *Molecular Plant-Microbe Interactions* 28(3), 212–217.

Shalem, O., Sanjana, N.E., Hartenian, E., Shi, X., Scott, D.A. *et al.* (2014) Genome-scale CRISPR-Cas9 knockout screening in human cells. *Science* 343(6166), 84–87.

Sharma, H., Rai, A.K., Dahiya, D., Chettri, R. and Nigam, P.S. (2021) Exploring endophytes for in vitro synthesis of bioactive compounds similar to metabolites produced in vivo by host plants. *AIMS Microbiology* 7(2), 175–199.

Sharma, S., Gupta, R., Dugar, G. and Srivastava, A.K. (2012) Impact of application of biofertilizers on soil structure and resident microbial community structure and function. In: Maheshwari, D.K. (ed.) *Bacteria in Agrobiology: Plant Probiotics*. Springer, Berlin, pp. 65–77.

Smriti, S. and Ajit, V. (2014) From *Piriformospora indica* to rootonic: A review. *African Journal of Microbiology Research* 8(32), 2984–2992.

Song, D., Ibrahim, S. and Hayek, S. (2012) Recent application of probiotics in food and agricultural science. In: Rigobelo, E. (ed.) *Probiotics*. Intech, Manhattan, pp. 1–34.

Soumare, A. (2015) Role of arbuscular mycorrhizal fungi and nitrogen fixing bacteria on legume growth under various environmental stresses. *International Journal of Biosciences* 7(4), 31–46.

Subiramani, S., Ramalingam, S., Muthu, T., Nile, S.H. and Venkidasamy, B. (2020) Development of abiotic stress tolerance in crops by plant growth-promoting rhizobacteria (PGPR). In: Kumar, M., Kumar, V. and Prasad, R. (eds) *Phyto-Microbiome in Stress Regulation*. Springer, Singapore, pp. 125–145.

Sullia, S.B. (1991) Use of vesicular - arbuscular mycorrhiza (VAM) as biofertilizer for horticultural plants in developing countries. In: Prakash, J. and Pierik, R.L.M. (eds) *Horticulture – New Technologies and Applications: Proceedings of the International Seminar on New Frontiers in Horticulture, organized by Indo-American Hybrid Seeds*, Springer, Bangalore, India, November 25–28, 1990, pp. 49–53.

Takahashi, K., Muraoka, R. and Otsuka, K. (2020) Technology adoption, impact, and extension in developing countries' agriculture: A review of the recent literature. *Agricultural Economics* 51(1), 31–45.

Thakur, N., Kaur, S., Tomar, P., Thakur, S. and Yadav, A.N. (2020) Microbial biopesticides: Current status and advancement for sustainable agriculture and environment. In: Rastegari, A.A., Yadav, A.N. and Yadav, N. (eds) *New and Future Developments in Microbial Biotechnology and Bioengineering*. Elsevier, New York, pp. 243–282.

Tiwari, P., Kang, S. and Bae, H. (2023) Plant-endophyte associations: Rich yet under-explored sources of novel bioactive molecules and applications. *Microbiological Research* 266, 127241.

Vandenkoornhuyse, P., Quaiser, A., Duhamel, M., Le Van, A. and Dufresne, A. (2015) The importance of the microbiome of the plant holobiont. *New Phytologist* 206(4), 1196–1206.

van West, P., Morris, B.M., Reid, B., Appiah, A.A., Osborne, M.C. *et al.* (2002) Oomycete plant pathogens use electric fields to target roots. *Molecular Plant-Microbe Interactions* 15(8), 790–798.

Vessey, J.K. (2003) Plant growth promoting rhizobacteria as biofertilizers. *Plant and Soil* 255(2), 571–586.

Vives-Peris, V., Ollas, C., Gómez-Cadenas, A. and Pérez-Clemente, R.M. (2020) Root exudates: From plant to rhizosphere and beyond. *Plant Cell Reports* 39(1), 3–17.

Ward, E.R., Uknes, S.J., Williams, S.C., Dincher, S.S., Wiederhold, D.L. *et al.* (1991) Coordinate gene activity in response to agents that induce systemic acquired resistance. *Plant Cell* 3(10), 1085–1094.

Yang, J., Kloepper, J.W. and Ryu, C.M. (2009) Rhizosphere bacteria help plants tolerate abiotic stress. *Trends in Plant Science* 14(1), 1–4.

Yu, H., Gao, D., Rahman, M., Chen, S. and Wu, F. (2023) L-phenylalanine in potato onion (*Allium cepa* var. *aggregatum* G. Don) root exudates mediates neighbor detection and trigger physio-morphological root responses of tomato. *Frontiers in Plant Science* 14, 1056629.

5 Metabolic Engineering of Tea for Enhanced Health Benefits

Misbah Naz[1,2], Muhammad Rahil Afzal[1,2], Adalberto Benavides-Mendoza[3], Kashif Akhtar[4], Nazeer Ahmed[1,2] and Shi Yuanzhi[5*]

[1]State Key Laboratory of Green Pesticide, Guiyang, China; [2]Key Laboratory of Green Pesticide and Agricultural Bioengineering, Ministry of Education, Guizhou University, Guiyang, China; [3]Department of Horticulture, Antonio Narro Autonomous Agrarian University, Saltillo, Mexico; [4]State Key Laboratory for Conservation and Utilization of Subtropical Agro-bioresources, Guangxi Key Laboratory of Sugarcane Biology, College of Life Science and Technology, Guangxi University, Nanning, China; [5]Key Laboratory of Tea Biology and Resource Utilization of Tea, Tea Research Institute, Chinese Academy of Agricultural Sciences, Hangzhou, China

Abstract

Tea, derived from the leaves of *Camellia sinensis*, is one of the most widely consumed beverages worldwide, renowned for its diverse health benefits attributed to its rich phytochemical composition. However, the natural variability in tea plant genetics and environmental factors limits the consistent production of desired bioactive compounds. Metabolic engineering is emerging as a promising approach to enhance the health-promoting properties of tea by modulating metabolic pathways responsible for the synthesis of key phytochemicals. This chapter outlines recent advances in metabolic engineering strategies applied to tea plants, focusing on the manipulation of pathways leading to the biosynthesis of catechins, theanine, caffeine, and other antioxidants. Techniques such as gene editing, transgenic approaches, and traditional breeding methods are discussed in the context of their application for improving tea quality and health benefits. Furthermore, the influence of external factors such as environmental conditions and cultivation practices on metabolic engineering outcomes is examined. Insights into the potential health benefits conferred by engineered tea varieties, including enhanced antioxidant activity, flavor profiles, and disease resistance, are elucidated. This comprehensive overview underscores the importance of metabolic engineering in tailoring tea plants for optimized phytochemical composition and heightened health-promoting properties, thereby paving the way for the development of novel tea cultivars with enhanced nutritional and therapeutic value.

Keywords: *Camellia sinensis*, metabolic engineering, health benefits, metabolic pathways, novel tea cultivars

5.1 Introduction

Tea, derived from the leaves of the *Camellia sinensis* plant, has been drunk for centuries and is one of the most widely consumed beverages globally (Issaoui *et al.*, 2020; Chan *et al.*, 2024). Beyond its cultural significance and social rituals, tea is renowned for its numerous health benefits

*Corresponding author: shiyuanzhi@caas.cn

DOI: 10.1079/9781800627659.0005

attributed to its rich composition of bioactive compounds, including polyphenols, catechins, amino acids, vitamins, and minerals (Kumar *et al.*, 2024). The consumption of tea has been associated with various health-promoting effects, such as antioxidant, antiinflammatory, antimicrobial, and anticancer properties, as well as potential benefits for cardiovascular health, weight management, and cognitive function (Joshi *et al.*, 2024). These health benefits have sparked growing interest in optimizing the nutritional composition of tea through metabolic engineering approaches to enhance its therapeutic potential and meet the evolving demands of consumers for functional foods and beverages (Cui *et al.*, 2024).

Metabolic engineering, a multidisciplinary field at the intersection of biology, chemistry, and engineering, offers powerful tools and strategies to manipulate the metabolic pathways of plants for the production of desired compounds (Şimşek *et al.*, 2024). In recent years, metabolic engineering has emerged as a promising approach for enhancing the health benefits of tea by modulating the biosynthesis of bioactive compounds and improving their abundance, bioavailability, and efficacy (Mohanta *et al.*, 2024). This chapter provides an overview of the current state of research in the metabolic engineering of tea for enhanced health benefits, highlighting key achievements, challenges, and future directions (Sanjarnia *et al.*, 2024).

Research in the metabolic engineering of tea has primarily focused on three main areas: enhancement of specific bioactive compounds, modulation of flavor profiles, and improvement of agronomic traits and stress tolerance. One of the most extensively studied bioactive compounds in tea is catechins, a class of polyphenolic antioxidants with potent health-promoting properties (Priya and Manjegowda, 2024). Catechins, particularly epigallocatechin gallate (EGCG), have been associated with numerous health benefits, including anticancer, antiinflammatory, and cardioprotective effects. Several studies have employed metabolic engineering techniques to increase the levels of catechins in tea leaves through the overexpression of key biosynthetic genes or transcription factors involved in catechin biosynthesis (Li *et al.*, 2024), for example successfully enhanced catechin production in tea plants by simultaneously expressing chalcone synthase and anthocyanidin reductase genes, leading to the development of tea cultivars with fortified antioxidant properties (Wakamatsu *et al.*, 2024; Yahaya *et al.*, 2024).

In addition to catechins, researchers have also explored the modulation of flavor compounds in tea to improve its sensory characteristics and consumer acceptance. Flavor is a critical determinant of tea quality, influencing consumer preferences and market demand (McBey and Nadathur, 2024). Metabolic engineering strategies have been employed to manipulate the levels of volatile and nonvolatile compounds responsible for tea aroma and taste, such as terpenoids, methylxanthines, and amino acids (Sukumaran and Gopi, 2023). Zeng *et al.* (2023) demonstrated the modulation of terpene biosynthesis in tea plants using metabolic engineering techniques to create cultivars with distinct flavor profiles. By precisely controlling the expression of key genes involved in terpene biosynthesis, the researchers were able to enhance the aroma and taste of tea, offering consumers a more diverse and appealing range of options (Wei *et al.*, 2023). Furthermore, metabolic engineering holds promise for improving the agronomic traits and stress tolerance of tea plants to enhance their resilience to biotic and abiotic stresses (Olanrewaju *et al.*, 2024). Climate change, pests, and diseases pose significant challenges to tea cultivation, affecting yield, quality, and sustainability. By manipulating the expression of stress-responsive genes and metabolic pathways, researchers aim to develop tea cultivars with enhanced resistance to environmental stresses, such as drought, heat, cold, and pathogens (Ahmed *et al.*, 2018). One study employed CRISPR-Cas9-mediated genome editing to introduce targeted mutations in stress-related genes in tea plants, resulting in improved stress tolerance and yield stability (Li *et al.*, 2023).

Despite significant advances, several challenges remain in the metabolic engineering of tea for enhanced health benefits. Regulatory hurdles, public acceptance, and consumer perception of genetically modified organisms (GMOs) present barriers to the commercialization of genetically engineered tea cultivars. Regulatory frameworks vary across countries, and stringent regulations may hinder the adoption of genetically modified tea plants (Su *et al.*, 2023). Moreover, consumer awareness and attitudes towards GMOs can influence market acceptance and demand for genetically engineered tea products (Bawa and Anilakumar, 2013). Kamau and Karitu (2023) investigated consumer awareness, perceptions, and acceptance of genetically modified tea in China, highlighting the importance of addressing public concerns and building trust to facilitate the adoption of genetically engineered tea cultivars.

Another challenge is maintaining agronomic traits and yield while manipulating metabolic pathways to enhance specific bioactive compounds or stress tolerance. Genetic modifications may inadvertently affect plant morphology, growth, and productivity, leading to unintended consequences and trade-offs (Varzakas *et al.*, 2018). Comprehensive phenotypic analysis and field trials are essential to assess the impact of metabolic engineering interventions on agronomic performance and yield stability. Additionally, emerging technologies such as genome editing and marker-assisted selection offer opportunities to precisely target desired traits while minimizing off-target effects and accelerating breeding programs (Vallarino *et al.*, 2020; Kamau and Karitu, 2023).

In the future, research in the metabolic engineering of tea is poised to address these challenges and unlock new opportunities for the development of genetically improved tea cultivars with enhanced health benefits and resilience to environmental stresses (Hazra *et al.*, 2022). Emerging technologies, such as CRISPR-Cas9 genome editing, metabolic modeling, and high-throughput phenotyping, will continue to revolutionize the field, enabling precise manipulation of metabolic pathways and accelerated breeding programs (Yang *et al.*, 2021). Integration of omics approaches, such as genomics, transcriptomics, proteomics, and metabolomics, will enhance our understanding of tea metabolism and facilitate the development of tailored breeding strategies for improved cultivars (Faryad *et al.*, 2021). Furthermore, collaborative efforts among researchers, industry stakeholders, policymakers, and consumers are essential to overcome regulatory hurdles, address public concerns, and promote the adoption of genetically engineered tea cultivars (Sultan and El-Qassem, 2021).

In conclusion, metabolic engineering holds immense potential for enhancing the health benefits, flavor, and resilience of tea cultivars to meet the evolving demands of consumers and the tea industry. By manipulating the metabolic pathways of tea plants, researchers aim to optimize the nutritional composition, sensory characteristics, and stress tolerance of tea, offering consumers a diverse range of functional foods and beverages with improved quality and therapeutic potential (Choi and Lee, 2023). Despite challenges, ongoing research and technological advances offer promising avenues for the development of genetically improved tea cultivars, contributing to the sustainability and resilience of tea production in the face of global challenges such as climate change and food security (Söderholm, 2020).

5.2 Bioactive Compounds in Tea and Their Health Benefits

5.2.1 Catechins and their antioxidant properties

Catechins are a group of flavonoids that are abundant in tea and are known for their potent antioxidant properties (Musial *et al.*, 2020). The major catechins found in tea are epicatechin (EC), epigallocatechin (EGC), epicatechin gallate (ECG), and epigallocatechin gallate (EGCG). Among these, EGCG is the most biologically active and extensively studied catechin (Almatroodi *et al.*, 2020).

Catechins exert their antioxidant effects through several mechanisms, including scavenging reactive oxygen species (ROS), chelating metal ions, and inhibiting prooxidant enzymes (Bernatoniene and Kopustinskiene, 2018). These antioxidant properties contribute to the potential of catechins in preventing oxidative stress-related diseases, such as cardiovascular diseases, cancer, and neurodegenerative disorders. Additionally, catechins have been shown to exhibit antiinflammatory, antimicrobial, and antiobesity effects, further contributing to their health-promoting properties (Sheng *et al.*, 2023).

5.2.2 Theaflavins and their antiinflammatory effects

Theaflavins are a class of polyphenolic compounds formed during the enzymatic oxidation of catechins during the processing of black tea. Theaflavins, including theaflavin (TF), theaflavin-3-gallate (TF3G), theaflavin-3'-gallate (TF3'G), and theaflavin-3,3'-digallate (TFDG), are potent antioxidants and exhibit antiinflammatory properties (Ansari *et al.*, 2023). They have been shown to inhibit the production of proinflammatory cytokines, such as interleukin-6 (IL-6) and tumor necrosis factor-alpha (TNF-α), by modulating various signaling pathways, including the nuclear factor-kappa B (NF-XB) and mitogen-activated protein kinase (MAPK) pathways (Luo *et al.*, 2010). Additionally, theaflavins have been reported to have neuroprotective effects, potentially reducing the risk of neurodegenerative diseases like Alzheimer's and Parkinson's (Afzal *et al.*, 2022).

Table 5.1 provides a snapshot of various strategies for metabolic engineering of tea plants to enhance health benefits, with a focus on increasing catechin content. Each strategy has its advantages and considerations, and a combination of approaches may be necessary to achieve optimal results.

Table 5.1. Metabolic engineering of tea plants to enhance health benefits, with a focus on increasing catechin content.

Strategy	Description	References
Genetic modification	Introduce genes encoding enzymes involved in catechin biosynthesis, such as flavonoid biosynthetic pathway genes, to enhance catechin production. Examples include overexpression of genes encoding flavanone 3-hydroxylase (F3H) and flavan-3-ol synthase (FLS).	Pang *et al.* (2013)
Genome editing	Utilize CRISPR-Cas9 or other genome editing tools to precisely edit endogenous genes involved in catechin biosynthesis, optimizing their expression levels or modifying enzyme activity to increase catechin accumulation without introducing foreign DNA.	Wang *et al.* (2024)
Elicitation	Apply chemical elicitors, such as methyl jasmonate or salicylic acid, to induce the plant's defense mechanisms and stimulate catechin production as a response to stress. This approach enhances catechin levels without genetic modification.	Mubeen *et al.* (2023)
Environmental optimization	Optimize growing conditions, such as light intensity, temperature, and soil nutrients, to promote catechin accumulation. Shading, controlled temperature regimes, and balanced fertilization can influence secondary metabolite biosynthesis in tea plants.	Ahmed *et al.* (2018)
Phytochemical profiling	Utilize omics technologies, such as metabolomics, to identify key regulatory genes and pathways associated with catechin biosynthesis. By profiling metabolite composition, researchers can pinpoint targets for metabolic engineering to enhance catechin content.	Vom Endt *et al.* (2002)
Trait stacking	Combine genetic modifications or environmental optimizations aimed at enhancing catechin production with other desirable traits, such as disease resistance or yield potential, to develop tea cultivars with improved health benefits and agronomic performance.	Singha and Singha (2024)

5.2.3 Other beneficial metabolites in tea

Apart from catechins and theaflavins, tea contains various other bioactive compounds that contribute to its health-promoting properties. Theanine, a unique nonproteinogenic amino acid found in tea, has been shown to induce a state of relaxed alertness and improve cognitive function (Nunes *et al.*, 2015). Caffeine, a methylxanthine present in tea, acts as a stimulant and has been associated with enhancing cognitive performance, increasing energy levels, and promoting thermogenesis. Tea also contains vitamins, such as vitamin C and folate, which act as antioxidants and play crucial roles in various metabolic processes (Acaroz *et al.*, 2019). Additionally, tea is a rich source of essential minerals, including fluoride which is beneficial for dental health. The synergistic effects of these diverse bioactive compounds contribute to the overall health benefits associated with tea consumption (Samanta, 2022).

5.3 Strategies for Metabolic Engineering of Tea

5.3.1 Overexpression of biosynthetic genes

One of the primary strategies employed in metabolic engineering of tea involves the overexpression of key biosynthetic genes involved in the production of target compounds (Liu *et al.*, 2021). This approach aims to increase the flux through specific metabolic pathways, leading to enhanced accumulation of desired metabolites. For instance, overexpression of the CsANR gene, encoding anthocyanidin reductase, in tea plants resulted in an increase in the accumulation of epicatechin and total catechins by up to two-fold compared to control plants (Zhao *et al.*, 2023b). Similarly, Yang *et al.* (2022) reported enhanced levels of catechins, including EGCG, EGC, and ECG, in transgenic tea plants overexpressing the CsDFR gene, encoding dihydroflavonol 4-reductase. In another study, Li *et al.* (2022b) reported an increase in theanine content in transgenic tea plants overexpressing the CsaLDC gene, encoding the enzyme L-arginine decarboxylase, which catalyzes the formation of ethylamine, a precursor for theanine biosynthesis.

5.3.2 Regulation of transcription factors

Transcription factors play crucial roles in regulating the expression of biosynthetic genes involved in the production of various metabolites (Vom Endt *et al.*, 2002). Therefore, manipulating the activity or expression levels of these transcription factors can provide an effective strategy for metabolic engineering in tea plants (Li *et al.*, 2017). Overexpression of the CsMYB5 transcription factor in tea plants leads to increased expression of flavonoid biosynthetic genes and elevated levels of catechins, particularly EGCG and EGC (Luo *et al.*, 2018). Similarly, Wei *et al.* (2023) identified and characterized a MYB transcription factor, CsMYB5, that regulates the expression of caffeine biosynthetic genes, offering a potential target for metabolic engineering strategies (Yang *et al.*, 2012; Mohajer *et al.*, 2023). Fang *et al.* (2021) identified and characterized a MYB transcription factor, CsTF13, that regulates the expression of genes involved in theanine biosynthesis, providing a potential target for enhancing theanine production in tea plants.

5.3.3 Metabolic flux modelling and *in silico* approaches

Metabolic flux modeling and *in silico* approaches have gained increasing attention in metabolic engineering as they enable the prediction and optimization of metabolic pathways for enhanced production of target compounds (Das *et al.*, 2020). Tariq *et al.* (2023) employed a metabolic flux modeling approach to identify potential bottlenecks and optimize the flux distribution in the flavonoid biosynthetic pathway of tea plants. This approach guided the selection of genetic modifications that led to the production of novel flavonoid derivatives with enhanced antioxidant and antiinflammatory activities. Similarly, a study employed a genome-scale metabolic model of tea plants to predict the impact of gene knockouts and overexpression on the production of key metabolites, including catechins and theanine. This *in silico* approach provided valuable insights for designing metabolic engineering strategies and prioritizing target genes for manipulation (Lee *et al.*, 2012).

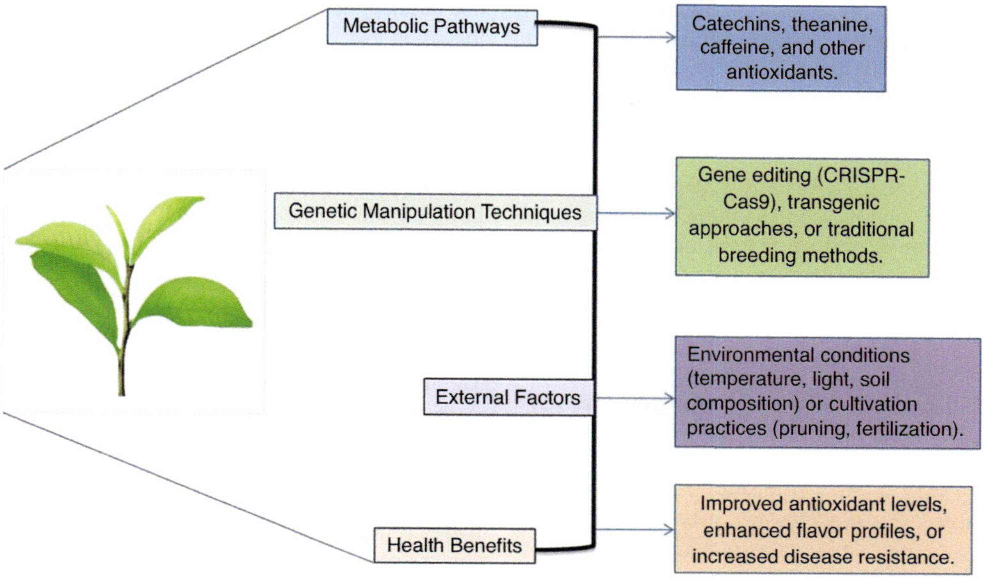

Fig. 5.1. The process of metabolic engineering in tea plants and how it leads to enhanced health benefits.

Metabolic engineering in tea plants involves the targeted modification of biosynthetic pathways to enhance the production of beneficial secondary metabolites, such as catechins, theanine, and flavonoids (Fig. 5.1). Through the use of tools like genome-scale metabolic models and CRISPR-Cas9 gene editing, researchers can identify key regulatory genes and enzymes involved in these pathways. By overexpressing or silencing specific genes, the metabolic flux can be redirected to increase the accumulation of compounds with antioxidant, antiinflammatory, and neuroprotective properties. These enhanced phytochemicals not only improve the nutritional and medicinal value of tea but also contribute to its potential as a functional beverage for promoting human health.

5.4 Case Studies in Metabolic Engineering of Tea

Tea, renowned for its health benefits and diverse flavors, has garnered considerable attention in metabolic engineering (Pang *et al.*, 2013). Through innovative approaches, researchers have aimed to enhance specific compounds in tea leaves, leading to the development of novel cultivars with improved nutritional profiles and flavor profiles (Yadav and Ahuja, 2007). Here, we delve into three significant case studies showcasing the advances in metabolic engineering of tea (Villanueva-Bermejo *et al.*, 2017).

5.4.1 Development of catechin-fortified tea cultivars

Catechins are a type of natural phenolic compound found in tea, particularly abundant in green tea, that have been linked to various health benefits, including antioxidant and antiinflammatory properties (Tang *et al.*, 2019). Researchers have employed metabolic engineering techniques to enhance catechin levels in tea cultivars (Li *et al.*, 2024). Several studies have reported the successful manipulation of biosynthetic pathways to increase catechin content in tea plant leaves. By overexpressing key enzymes involved in catechin biosynthesis, researchers have developed cultivars with elevated levels of these beneficial compounds, offering enhanced health benefits to consumers. Overall, the development of catechin-fortified tea cultivars involves a combination of traditional breeding methods, biotechnology, agronomic practices, and molecular techniques (Singh and Hazarika, 2020). By selecting or engineering tea plants with higher catechin content, researchers aim to produce teas with enhanced health benefits and market value. However, it is essential to consider factors such as regulatory approval, consumer acceptance, and environmental sustainability throughout the development process (Guo *et al.*, 2017).

5.4.2 Modulation of flavor compounds

Flavor is a crucial determinant of tea quality and influences consumer preferences (Cho *et al.*, 2005). Metabolic engineering strategies have been employed to modulate the levels of flavor compounds in tea leaves, enhancing aroma and taste profiles (Zhao *et al.*, 2023a). Research by Liu *et al.* (2023) exemplifies this approach, where the manipulation of genes involved in terpene biosynthesis led to the development of tea cultivars with distinct flavor characteristics. By precisely controlling the expression of key genes, researchers achieved targeted modulation of flavor compounds, catering to diverse consumer preferences and expanding the market for specialty teas (Sharma *et al.*, 2019).

5.4.3 Enhancement of other beneficial metabolites

Beyond catechins and flavor compounds, tea contains many other beneficial metabolites with potential health-promoting effects (Tounekti *et al.*, 2013). Metabolic engineering endeavors have focused on enhancing the levels of these compounds to further augment the nutritional value of tea (Issaoui *et al.*, 2020). Studies such as those by Zhou *et al.* (2024) have explored the manipulation of pathways involved in the biosynthesis of amino acids and vitamins in tea plants. By enhancing the expression of relevant genes, researchers succeeded in elevating the levels of essential nutrients in tea leaves, offering enhanced health benefits to consumers. Understanding the biosynthesis and regulation of these secondary metabolites is crucial for enhancing tea quality and developing applications in various industries, including food, beverages, pharmaceuticals, and cosmetics (Ding *et al.*, 2024). These case studies underscore the potential of metabolic engineering in tailoring the biochemical composition of tea, paving the way for the development of novel cultivars with improved nutritional profiles and flavor attributes (Jiang *et al.*, 2024).

5.5 Challenges and Future Prospects

The field of metabolic engineering in tea faces several challenges and holds promising future prospects for the development of improved cultivars (Hassan *et al.*, 2024). In summary, while metabolic engineering in tea faces challenges related to genetic complexity, regulatory constraints, and environmental factors, ongoing advances in breeding techniques, genome editing, and omics technologies offer promising avenues for developing improved tea cultivars with enhanced secondary metabolite profiles and resilience to environmental stressors (Malyukova and Velikiy, 2024; Torrente-Murciano *et al.*, 2024).

5.5.1 Regulatory hurdles and public acceptance

Two major challenges in the metabolic engineering of tea are navigating regulatory frameworks and gaining public acceptance (Asiminicesei *et al.*, 2024). GMOs can face stringent regulations, and consumer perception of genetically engineered products can influence market acceptance (Naveen and Sontakke, 2024). Overcoming these hurdles requires robust regulatory strategies and effective communication to build public trust (Ryan *et al.*, 2024). Research by Arhin *et al.* (2024) highlights the importance of engaging stakeholders and addressing concerns regarding safety and environmental impact to facilitate the acceptance of genetically modified tea cultivars.

5.5.2 Maintaining agronomic traits and yield

Another challenge is ensuring that metabolic engineering interventions do not compromise essential agronomic traits or yield (Khan *et al.*, 2024). Manipulating metabolic pathways can sometimes lead to unintended consequences, such as altered plant morphology or reduced productivity (Barupal and Kumar, 2024). Research by Wang *et al.* (2021) emphasizes the importance of comprehensive phenotypic analysis and field trials to assess the impact of metabolic engineering on agronomic performance and yield stability (Singha and Singha, 2024). Strategies such as genome editing and marker-assisted selection can help mitigate these challenges by precisely targeting desired traits while minimizing off-target effects (Wang *et al.*, 2024).

Maintaining agronomic traits and yield while enhancing secondary metabolite accumulation in tea plants requires a holistic approach that integrates breeding strategies, molecular tools, agronomic practices, and environmental management techniques (Haq *et al.*, 2024). By prioritizing trait stacking, utilizing advanced breeding technologies, optimizing cultural practices, and ensuring stress tolerance, it is possible to develop tea cultivars that meet the dual goals of productivity and quality in tea production (Hina *et al.*, 2024).

5.6 Emerging Technologies and Future Directions

Emerging technologies and future directions in the field of metabolic engineering and breeding of tea plants hold immense potential for revolutionizing tea cultivation, enhancing product quality, and meeting evolving consumer demands (Raza *et al.*, 2024). These advances encompass a wide range of approaches, from genomics and biotechnology to precision agriculture and sustainable practices (Mokrani and Nabti, 2024). In the following discussion, we explore some of these emerging technologies and their potential implications for the future of tea production (Jauhar, 2006).

One of the most promising areas of advancement is the integration of genomics and omics technologies into tea breeding programs. High-throughput sequencing techniques, such as next-generation sequencing (NGS), enable researchers to decipher the tea plant's entire genome rapidly (Sekhon, 2014). This genomic information provides insights into the genetic basis of important traits, including yield, quality, and stress tolerance. By identifying genes associated with secondary metabolite biosynthesis and agronomic traits, breeders can develop molecular markers for marker-assisted selection (MAS) and genomic selection, facilitating the breeding of tea cultivars with desired characteristics (Faller, 2022). Furthermore, omics technologies, including transcriptomics, proteomics, and metabolomics, offer comprehensive insights into the molecular mechanisms underlying tea plant metabolism (Li *et al.*, 2022a). By profiling gene expression patterns, protein abundance, and metabolite composition, researchers can unravel the complex regulatory networks governing secondary metabolite biosynthesis and stress responses in tea plants (Zhang *et al.*, 2020). This knowledge can inform targeted metabolic engineering strategies aimed at enhancing specific metabolites, such as catechins, while maintaining agronomic traits and yield.

In addition to genomics and omics approaches, biotechnological tools, such as genetic engineering and genome editing, hold promise for accelerating the development of improved tea cultivars. CRISPR-Cas9 technology, in particular, allows precise modifications to the tea plant's genome, enabling targeted gene editing without the introduction of foreign DNA (Kumar *et al.*, 2018; Zhang *et al.*, 2020). This opens up opportunities to engineer tea plants with enhanced traits, including increased catechin content, disease resistance, and environmental resilience. However, it is essential to address regulatory considerations and ensure transparent communication with stakeholders regarding the safety and acceptance of genetically modified tea cultivars (Raza *et al.*, 2024).

Precision agriculture techniques, leveraging remote sensing, geographic information systems (GIS), and sensor technologies, offer new opportunities for optimizing tea cultivation practices. By monitoring environmental parameters, soil conditions, and plant health in real time, growers can implement site-specific management strategies to maximize yield and quality while minimizing resource inputs and environmental impacts (Söderholm, 2020). This data-driven approach enables precision irrigation, fertilization, and pest management tailored to the unique needs of each tea plantation, contributing to sustainable tea production.

Furthermore, the adoption of climate-smart agriculture practices is crucial for mitigating the effects of climate change on tea cultivation. By developing climate-resilient tea cultivars adapted to changing environmental conditions, breeders can ensure the long-term sustainability of tea production (Faller, 2022). This involves incorporating traits such as heat tolerance, drought resistance, and disease tolerance into new cultivars through breeding and biotechnological approaches.

Overall, the integration of emerging technologies, including genomics, biotechnology, precision agriculture, and climate-smart practices, holds great promise for the future of tea cultivation (Liu *et al.*, 2021). By harnessing these advances, researchers and growers can develop tea cultivars that are not only high-yielding and resilient but also rich in health-promoting compounds such as catechins, meeting the demands of both producers and consumers in a rapidly evolving global market (Nunes *et al.*, 2015). However, it is essential to address regulatory, ethical, and socioeconomic considerations to ensure the responsible and equitable adoption of these technologies across the tea industry (Villanueva-Bermejo *et al.*, 2017).

5.7 Conclusion

In conclusion, metabolic engineering holds immense potential for enhancing the nutritional quality, flavor, and stress tolerance of tea cultivars. Despite regulatory and agronomic challenges, ongoing research and technological advancements offer promising avenues for the development of genetically improved tea varieties. By addressing regulatory concerns, optimizing agronomic performance, and harnessing emerging technologies, the future of metabolic engineering in tea looks bright, paving the way for sustainable and resilient tea production to meet the evolving demands of consumers and the tea industry.

References

Acaroz, U., Arslan-Acaroz, D. and Ince, S. (2019) A wide perspective on nutrients in beverages. In: Grumezescu, A. and Holban, A. (eds) *Nutrients in Beverages*. Elsevier, New York, pp. 1–39.

Afzal, O., Dalhat, M.H., Altamimi, A.S.A., Rasool, R., Alzarea, S.I. *et al.* (2022) Green tea catechins attenuate neurodegenerative diseases and cognitive deficits. *Molecules* 27(21), 7604.

Ahmed, S., Griffin, T., Cash, S.B., Han, W.-Y., Matyas, C. *et al.* (2018) Global climate change, ecological stress, and tea production. In: Han, W., Li, X. and Ahammed, G. (eds) *Stress Physiology of Tea in the Face of Climate Change*. Springer, Singapore, pp. 1–23.

Almatroodi, S.A., Almatroudi, A., Khan, A.A., Alhumaydhi, F.A., Alsahli, M.A. *et al.* (2020) Potential therapeutic targets of epigallocatechin gallate (EGCG), the most abundant catechin in green tea, and its role in the therapy of various types of cancer. *Molecules* 25(14), 3146.

Ansari, M.H.R., Saher, S., Parveen, R., Khan, W., Khan, I.A. *et al.* (2023) Role of gut microbiota metabolism and biotransformation on dietary natural products to human health implications with special reference to biochemoinformatics approach. *Journal of Traditional and Complementary Medicine* 13(2), 150–160.

Arhin, I., Yeboah, E., Liu, X., Liu, A., Chen, X. *et al.* (2024) Integrating farmers' perception of sustainable agricultural technologies towards the development of sustainable tea production in China. *International Journal of Agricultural Sustainability* 22(1), 2303886.

Asiminicesei, D.-M., Fertu, D.I. and Gavrilescu, M. (2024) Impact of heavy metal pollution in the environment on the metabolic profile of medicinal plants and their therapeutic potential. *Plants* 13(6), 913.

Barupal, M. and Kumar, A. (2024) CRISPR/Cas systems for enhancing photosynthesis: Climate resilience and food production. In: Kumar, A., Arora, S., Ogita, S., Yau, Y. and Mukherjee, K. (eds) *Gene Editing in Plants: CRISPR-Cas and Its Applications*. Springer, Singapore, pp. 477–519.

Bawa, A.S. and Anilakumar, K.R. (2013) Genetically modified foods: Safety, risks and public concerns-a review. *Journal of Food Science and Technology* 50(6), 1035–1046.

Bernatoniene, J. and Kopustinskiene, D.M. (2018) The role of catechins in cellular responses to oxidative stress. *Molecules* 23(4), 965.

Chan, K.-T., Wu, H.-Y., Tin, W.-Y., But, P.P.-H., Cheung, S.C.-H. *et al.* (2024) Ethnopharmacology of five flowers herbal tea, a popular traditional beverage in Hong Kong and South China. *Journal of Ethnobiology and Ethnomedicine* 20(1), 36.

Cho, H.Y., Chung, S.J., Kim, H.S. and Kim, K.O. (2005) Effect of sensory characteristics and non-sensory factors on consumer liking of various canned tea products. *Journal of Food Science* 70(8), s532–s538.

Choi, K.R. and Lee, S.Y. (2023) Systems metabolic engineering of microorganisms for food and cosmetics production. *Nature Reviews Bioengineering* 1(11), 832–857.

Cui, R., Zhang, C., Pan, Z.H., Hu, T.G. and Wu, H. (2024) Probiotic-fermented edible herbs as functional foods: A review of current status, challenges, and strategies. *Comprehensive Reviews in Food Science and Food Safety* 23(2), e13305.

Das, M., Patra, P. and Ghosh, A. (2020) Metabolic engineering for enhancing microbial biosynthesis of advanced biofuels. *Renewable and Sustainable Energy Reviews* 119, 109562.

Ding, K., Lv, W., Ren, H., Xiong, F., Zhang, Y. *et al.* (2024) Small world but large differences: Cultivar-specific secondary metabolite-mediated phyllosphere fungal homeostasis in tea plant (*Camellia sinensis*). *Plant and Soil* 502, 725–743.

Faller, A.C. (2022) *Genetic Characterization of Tea and Its Quality Attributes, Tea as a Food Ingredient*. CRC Press, Boca Raton, pp. 347–374.

Fang, K., Xia, Z., Li, H., Jiang, X., Qin, D. *et al.* (2021) Genome-wide association analysis identified molecular markers associated with important tea flavor-related metabolites. *Horticulture Research* 8(1), 42.

Faryad, A., Aziz, F., Tahir, J., Kousar, M., Qasim, M. *et al.* (2021) Integration of OMICS technologies for crop improvement. *Protein and Peptide Letters* 28(8), 896–908.

Guo, F., Guo, Y., Wang, P., Wang, Y. and Ni, D. (2017) Transcriptional profiling of catechins biosynthesis genes during tea plant leaf development. *Planta* 246(6), 1139–1152.

Haq, S.A.U., Bashir, T., Roberts, T.H. and Husaini, A.M. (2024) Ameliorating the effects of multiple stresses on agronomic traits in crops: Modern biotechnological and omics approaches. *Molecular Biology Reports* 51(1), 41.

Hassan, S., Karaila, G.K., Singh, P., Meenatchi, R., Venkateswaran, A.S. *et al.* (2024) Implications of fungal nanotechnology for sustainable agriculture – applications and future perspectives. *Biocatalysis and Agricultural Biotechnology* 57, 103110.

Hazra, A., Dasgupta, N., Bera, B., Mahadani, P., Sengupta, C. *et al.* (2022) Omics advances in tea research. In: Rout, G. and Peter, K. (eds) *Omics in Horticultural Crops*. Elsevier, New York, pp. 367–382.

Hina, A., Abbasi, A., Arshad, M., Imtiaz, S., Shahid, S. *et al.* (2024) Utilization of multi-omics approaches for crop improvement. In: Fiaz, S. and Prakash, C. (eds) *Omics-Based Techniques for Global Food Security*. Wiley, Chichester, pp. 91–121.

Issaoui, M., Delgado, A.M., Caruso, G., Micali, M., Barbera, M. *et al.* (2020) Phenols, flavors, and the mediterranean diet. *Journal of AOAC International* 103(4), 915–924.

Jauhar, P.P. (2006) Modern biotechnology as an integral supplement to conventional plant breeding: The prospects and challenges. *Crop Science* 46(5), 1841–1859.

Jiang, X., Zhao, J., Gao, D., Zhang, X., Qiu, H. *et al.* (2024) Metabolomic and genome-wide association studies drive genetic dissection and gene mining in tea plant. *Horticulture Advances* 2(1), 1–13.

Joshi, D.D., Deb, L., Kaul, K., Somkuwar, B.G., Rana, V.S. *et al.* (2024) Relevance of Indian traditional herbal brews for gut microbiota balance. *Indian Journal of Microbiology* 64(4), 1425–1444.

Kamau, J. and Karitu, P. (2023) Consumer perceptions on the introduction of genetically modified food products in Kenya: A qualitative study in Nyeri and Meru counties. *Journal of Agriculture and Horticulture Research* 6(2), 282–287.

Khan, A.A., Iqbal, B., Jalal, A., Khan, K.A., Al-Andal, A. *et al.* (2024) Advanced molecular approaches for improving crop yield and quality: A review. *Journal of Plant Growth Regulation* 43(7), 2091–2103.

Kumar, A., Kumar, R. and Gangwar, S. (2024) Herbal tea: A sip of wellness and serenity. *Agritech* 8, 67.

Kumar, A., Mosa, K.A., Ji, L., Kage, U., Dhokane, D. *et al.* (2018) Metabolomics-assisted biotechnological interventions for developing plant-based functional foods and nutraceuticals. *Critical Reviews in Food Science and Nutrition* 58(11), 1791–1807.

Lee, J.W., Na, D., Park, J.M., Lee, J., Choi, S. *et al.* (2012) Systems metabolic engineering of microorganisms for natural and non-natural chemicals. *Nature Chemical Biology* 8(6), 536–546.

Li, H., Song, K., Li, B., Zhang, X., Wang, D. *et al.* (2023) CRISPR/Cas9 editing sites identification and multi-elements association analysis in *Camellia sinensis*. *International Journal of Molecular Sciences* 24(20), 15317.

Li, J., Wang, Y. and Suh, J.H. (2022a) Multi-omics approach in tea polyphenol research regarding tea plant growth, development and tea processing: Current technologies and perspectives. *Food Science and Human Wellness* 11(3), 524–536.

Li, J., Ren, J., Lei, X., Fan, W., Tang, L. *et al.* (2024) CsREV-CsTCP4-CsVND7 module shapes xylem patterns differentially between stem and leaf to enhance tea plant tolerance to drought. *Cell Reports* 43(4), 113987.

Li, M., Li, Y., Guo, L., Gong, N., Pang, Y. *et al.* (2017) Functional characterization of tea (*Camellia sinensis*) myb4a transcription factor using an integrative approach. *Frontiers in Plant Science* 8, 245151.

Li, M.-Y., Liu, H.-Y., Wu, D.-T., Kenaan, A., Geng, F. *et al.* (2022b) L-theanine: A unique functional amino acid in tea (*Camellia sinensis* L.) with multiple health benefits and food applications. *Frontiers in Nutrition* 9, 853846.

Liu, S.-H., Li, J., Huang, J.-A., Liu, Z.-H. and Xiong, L.-G. (2021) New advances in genetic engineering for l-theanine biosynthesis. *Trends in Food Science & Technology* 114, 540–551.

Liu, Y., Guo, C., Zang, E., Shi, R., Liu, Q. *et al.* (2023) Review on herbal tea as a functional food: Classification, active compounds, biological activity, and industrial status. *Journal of Future Foods* 3(3), 206–219.

Luo, Y., Liu, M., Dai, Y., Yao, X., Xia, Y. *et al.* (2010) Norisoboldine inhibits the production of pro-inflammatory cytokines in lipopolysaccharide-stimulated RAW 264.7 cells by down-regulating the activation of mapks but not NF-κB. *Inflammation* 33(6), 389–397.

Luo, Y., Yu, S., Li, J., Li, Q., Wang, K. *et al.* (2018) Molecular characterization of WRKY transcription factors that act as negative regulators of o-methylated catechin biosynthesis in tea plants (*Camellia sinensis* L.). *Journal of Agricultural and Food Chemistry* 66(43), 11234–11243.

Malyukova, L.S. and Velikiy, A.V. (2024) The regulatory effect of nitrogen on the formation and accumulation of secondary metabolites in different genotypes of *Camellia sinensis* (L.) Kuntze. *Moscow University Soil Science Bulletin* 79(1), 56–64.

McBey, D. and Nadathur, S. (2024) Flavors, taste preferences, and the consumer: Taste modulation and influencing change in dietary patterns for a sustainable earth. In: Nadathur, S., Wanasundara, J. and Scanlin, L. (eds) *Sustainable Protein Sources*. Elsevier, New York, pp. 629–647.

Mohajer, S., Taha, R.M., Hajivand, S., Pirkhezri, M. and Ahmed, A.B.A. (2023) Transient gene expression in molecular farming and functional genomics of tea (*Camellia sinensis*): A review. *Journal of Plant Growth Regulation* 42(7), 4000–4019.

Mohanta, Y.K., Nongbet, A., Panda, J., Chakrabartty, I., Mohanta, T.K. *et al.* (2024) Nanobiotechnology in genetic engineering for abiotic stress resistance in crops. In: Pudake, R., Tripathi, R. and Gill, S. (eds) *Nanotechnology for Abiotic Stress Tolerance and Management in Crop Plants*. Elsevier, New York, pp. 153–164.

Mokrani, S. and Nabti, E. (2024) Recent status in production, biotechnological applications, commercial aspects, and future prospects of microbial enzymes: A comprehensive review. *International Journal of Agricultural Science and Food Technology* 10(1), 006–020.

Mubeen, B., Hasnain, A., Jie, W., Zheng, H., Peijnenburg, W.J. *et al.* (2023) Enhanced production of active photosynthetic and biochemical molecules in *Silybum marianum* L. using biotic and abiotic elicitors in hydroponic culture. *Molecules* 28(4), 1716.

Musial, C., Kuban-Jankowska, A. and Gorska-Ponikowska, M. (2020) Beneficial properties of green tea catechins. *International Journal of Molecular Sciences* 21(5), 1744.

Naveen, A.K. and Sontakke, M. (2024) A review on regulatory aspects, challenges and public perception in acceptance of genetically modified foods. *Food Science and Biotechnology* 33(4), 791–804.

Nunes, A.R., Alves, M.G., Moreira, P.I., Oliveira, P.F. and Silva, B.M. (2015) Impact of green tea (*Camellia sinensis* l.) consumption in diabetes mellitus-induced neurodegeneration. In: Powell, N. (ed.) *Green Tea and Health: Antioxidant Properties, Consumption and Role in Disease Prevention*. Nova Science, New York, pp. 1–33.

Olanrewaju, O.S., Glick, B.R. and Babalola, O.O. (2024) Metabolomics-guided utilization of beneficial microbes for climate-resilient crops. *Current Opinion in Chemical Biology* 79, 102427.

Pang, Y., Abeysinghe, I.S.B., He, J., He, X., Huhman, D. *et al.* (2013) Functional characterization of proanthocyanidin pathway enzymes from tea and their application for metabolic engineering. *Plant Physiology* 161(3), 1103–1116.

Priya, K. and Manjegowda, D.S. (2024) Unravelling the cardiovascular symphony: Decoding the intricate influence of green tea polyphenols from leaf to heart. *Biomedicine* 44(1), 1–12.

Raza, A., Bashir, S., Khare, T., Karikari, B., Copeland, R.G.R. *et al.* (2024) Temperature-smart plants: A new horizon with omics-driven plant breeding. *Physiologia Plantarum* 176(1), e14188.

Ryan, C.D., Henggeler, E., Gilbert, S., Schaul, A.J. and Swarthout, J.T. (2024) Exploring the GMO narrative through labeling: Strategies, products, and politics. *GM Crops & Food* 15(1), 51–66.

Samanta, S. (2022) Potential bioactive components and health promotional benefits of tea (*Camellia sinensis*). *Journal of the American Nutrition Association* 41, 65–93.

Sanjarnia, P., Picchio, M.L., Solis, A.N.P., Schuhladen, K., Fliss, P.M. *et al.* (2024) Bringing innovative wound care polymer materials to the market: Challenges, developments, and new trends. *Advanced Drug Delivery Reviews* 207, 115–217.

Sekhon, B.S. (2014) Nanotechnology in agri-food production: An overview. *Nanotechnology, Science and Applications* 7, 31–53.

Sharma, M., Koul, A., Sharma, D., Kaul, S., Swamy, M.K. *et al.* (2019) Metabolic engineering strategies for enhancing the production of bio-active compounds from medicinal plants. In: Akhtar, M. and Swamy, M. (eds) *Natural Bio-Active Compounds*. Springer, Dordrecht, pp. 287–316.

Sheng, Y., Sun, Y., Tang, Y., Yu, Y., Wang, J. *et al.* (2023) Catechins: Protective mechanism of antioxidant stress in atherosclerosis. *Frontiers in Pharmacology* 14, 1144878.

Şimşek, Ö., Isak, M.A., Dönmez, D., Dalda Şekerci, A., İzgü, T. *et al.* (2024) Advanced biotechnological interventions in mitigating drought stress in plants. *Plants* 13, 717.

Singh, H.R. and Hazarika, P. (2020) Biotechnological approaches for tea improvement. In: Grumezescu, A. and Holban, A. (eds) *Biotechnological Progress and Beverage Consumption*. Elsevier, New York, pp. 111–148.

Singha, S. and Singha, R. (2024) Crop improvement strategies and principles of selective breeding, water-soil-plant-animal nexus in the era of climate change. *IGI Global* 93–113.

Söderholm, P. (2020) The green economy transition: The challenges of technological change for sustainability. *Sustainable Earth* 3, 6.

Su, W., Xu, M., Radani, Y. and Yang, L. (2023) Technological development and application of plant genetic transformation. *International Journal of Molecular Sciences* 24(13), 10646.

Sukumaran, N.P. and Gopi, S. (2023) Biogenesis of plant-derived flavor compounds. In: Gopi, S., Sukumaran, N., Jacob, J. and Thomas, S. (eds) *Natural Flavours, Fragrances, and Perfumes: Chemistry, Production, and Sensory Approach*. Wiley-VCH, Weinheim, pp. 23–46.

Sultan, S. and El-Qassem, A. (2021) Future prospects for sustainable agricultural development. *International Journal of Modern Agriculture and Environment* 1(2), 54–82.

Tang, G.-Y., Meng, X., Gan, R.-Y., Zhao, C.-N., Liu, Q. *et al.* (2019) Health functions and related molecular mechanisms of tea components: An update review. *International Journal of Molecular Sciences* 20, 6196.

Tariq, H., Asif, S., Andleeb, A., Hano, C. and Abbasi, B.H. (2023) Flavonoid production: current trends in plant metabolic engineering and de novo microbial production. *Metabolites* 13(1), 124.

Torrente-Murciano, L., Dunn, J.B., Christofides, P.D., Keasling, J.D., Glotzer, S.C. *et al.* (2024) The forefront of chemical engineering research. *Nature Chemical Engineering* 1(1), 18–27.

Tounekti, T., Joubert, E., Hernández, I. and Munné-Bosch, S. (2013) Improving the polyphenol content of tea. *Critical Reviews in Plant Sciences* 32(3), 192–215.

Vallarino, J.G., Kubiszewski-Jakubiak, S., Ruf, S., Rößner, M., Timm, S. *et al.* (2020) Multi-gene metabolic engineering of tomato plants results in increased fruit yield up to 23. *Scientific Reports* 10(1), 17219.

Varzakas, T., Kandylis, P., Dimitrellou, D., Salamoura, C., Zakynthinos, G. *et al.* (2018) Innovative and fortified food: Probiotics, prebiotics, GMOs, and superfood. In: Ali, M. and Nizar, M. (eds) *Preparation and Processing of Religious and Cultural Foods*. Elsevier, New York, pp. 67–129.

Villanueva-Bermejo, D., Reglero, G. and Fornari, T. (2017) Recent advances in the processing of green tea biomolecules using ethyl lactate. A review. *Trends in Food Science & Technology* 62, 1–12.

Vom Endt, D., Kijne, J.W. and Memelink, J. (2002) Transcription factors controlling plant secondary metabolism: What regulates the regulators. *Phytochemistry* 61(2), 107–114.

Wakamatsu, J., Yamamoto, M., Kikuta, M., Tanaka, W. and Tominaga, R. (2024) Effect of shading on trichome formation and CAPRICE-like gene expression in tea (*Camellia sinensis var. sinensis*) leaves. *Scientia Horticulturae* 330, 113049.

Wang, X., Dong, B., Wang, Z. and Wang, M.J. (2021) Wang et al. Respond. *American Journal of Public Health* 111(4), e23. DOI: 10.2105/AJPH.2021.306164.

Wang, Y., Niu, S., Deng, X., Bai, D., Chen, Z. *et al.* (2024) Genome-wide association study, population structure, and genetic diversity of the tea plant in Guizhou Plateau. *BMC Plant Biology* 24(1), 79.

Wei, J., Yang, Y., Peng, Y., Wang, S., Zhang, J. *et al.* (2023) Biosynthesis and the transcriptional regulation of terpenoids in tea plants (*Camellia sinensis*). *International Journal of Molecular Sciences* 24(8), 6937.

Yadav, S.K. and Ahuja, P.S. (2007) Towards generating caffeine-free tea by metabolic engineering. *Plant Foods for Human Nutrition* 62(4), 185–191.

Yahaya, L.E., Igbinadolor, R.O., Okunade, F.A., Ajewole, A.O., Atanda, J.F. *et al.* (2024) Functionality of green tea (*Camellia sinensis*) blends – effects of spice concentration on composition, antioxidant profile, microbiological and cup quality. *Asian Food Science Journal* 23(3), 28–37.

Yang, C., Wu, P., Cao, Y., Yang, B., Liu, L. *et al.* (2022) Overexpression of dihydroflavonol 4-reductase (CoDFR) boosts flavonoid production involved in the anthracnose resistance. *Frontiers in Plant Science* 13, 1038467.

Yang, D., Liu, Y., Sun, M., Zhao, L., Wang, Y. *et al.* (2012) Differential gene expression in tea (*Camellia sinensis* L.) calli with different morphologies and catechin contents. *Journal of Plant Physiology* 169(2), 163–175.

Yang, Y., Saand, M.A., Huang, L., Abdelaal, W.B., Zhang, J. *et al.* (2021) Applications of multi-omics technologies for crop improvement. *Frontiers in Plant Science* 12, 563953.

Zeng, L., Zhou, X., Fu, X., Hu, Y., Gu, D. *et al.* (2023) Effect of the biosynthesis of the volatile compound phenylacetaldehyde on chloroplast modifications in tea (*Camellia sinensis*) plants. *Horticulture Research* 10(3), uhad003.

Zhang, C., Wang, M., Gao, X., Zhou, F., Shen, C. *et al.* (2020) Multi-omics research in albino tea plants: Past, present, and future. *Scientia Horticulturae* 261, 108943.

Zhao, S., Cheng, H., Xu, P. and Wang, Y. (2023a) Regulation of biosynthesis of the main flavor-contributing metabolites in tea plant (*Camellia sinensis*): A review. *Critical Reviews in Food Science and Nutrition* 63(30), 10520–10535.

Zhao, X., Li, P., Zuo, H., Peng, A., Lin, J. *et al.* (2023b) CsMYBL2 homologs modulate the light and temperature stress-regulated anthocyanin and catechins biosynthesis in tea plants (*Camellia sinensis*). *Plant Journal* 115(4), 1051–1070.

Zhou, X., Hu, L., Hoang, N.H., Thanh, T.L., Zhou, C. *et al.* (2024) The changes in metabolites, quality components, and antioxidant activity of tea (*Camellia sinensis*) infected with *Exobasidium vexans* by applying UPLC-MS/MS-based widely targeted metabolome and biochemical analysis. *Phytopathology* 114(1), 164–176.

6 Molecular Hydrogen Treatments in Food Supplies: Enhanced Growth, Better Yields, and Prolonged Postharvest Storage?

John T. Hancock* **and Jennifer May**
School of Applied Sciences, University of the West of England, Bristol, UK

Abstract

Food security is an issue which will only get more prominent as the human population continues to expand across the globe, and therefore new and innovative approaches to improve crops production and storage need to be tried. The application of molecular hydrogen (H_2) has been suggested to be of benefit to the biomedical industry, but also for the agricultural sector. H_2 can be applied to plants in a variety of ways, and there appear to be several benefits: plants grow better and have enhanced stress responses. There is some evidence that the crops obtained following H_2 treatment are larger and healthier. H_2-based treatments can also be used post harvest and for food processing. Here, examples are given of where H_2 can be used and show benefit in each of these stages. H_2-based treatments need to have a cost/benefit, but there are promising reports that H_2 could be used to enhance various stages of food production in the future.

Keywords: antioxidants, molecular hydrogen, pathogens, plant stress, oxidative stress, reactive nitrogen species, reactive oxygen species

6.1 Introduction

Food insecurity is estimated to affect more than 900 million people worldwide, representing over 11% of the total population (FAO, IFAD, UNICEF, WFP, and WHO, 2023), and has increased in recent years due to the COVID-19 pandemic and political instability (Iannotti *et al.*, 2024). With an increasing global population which is estimated to exceed 9.5 billion by 2050 (Nor A'azizam *et al.*, 2024), this is an ever-increasing issue. The current global agrosystem is becoming increasingly unsustainable, with novel approaches needed to sustain food production in a healthy way. This is also compounded by the threat of continued climate change, with potentially increasing future effects on agriculture worldwide (Vaidya *et al.*, 2024).

Many approaches have been suggested in a bid to increase food security, including focusing on food availability, food quality, food affordability, and sustainability (Iannotti *et al.*, 2024). One aspect of this is a drive for further research into novel methods to improve crop yield and quality, postharvest storage and better food production, aiming to improve both food availability and reduce wastage. Molecular hydrogen has been

*Corresponding author: john.hancock@uwe.ac.uk

DOI: 10.1079/9781800627659.0006

suggested as one such novel method, which may play a future role in improving food security.

Molecular hydrogen (H_2) has been suggested to have profound effects across a range of biological systems, including animals and plants. It has been mooted as wide-ranging therapy for treatment of human disease (Dixon *et al.*, 2013; Ohta, 2014), with conditions including cancer (Noor *et al.*, 2023), diabetes (Zheng *et al.*, 2021), and neurodegenerative disease (Rahman *et al.*, 2021) all being suggested as targets. However, beneficial effects of H_2 are not limited to animal species, with several reports suggesting that it would be good to use in agricultural practice (Zeng *et al.*, 2014; Hancock *et al.*, 2021; Li *et al.*, 2021; Zulfiqar *et al.*, 2021; Alwazeer and Çiğdem, 2022; Wang *et al.*, 2024).

H_2 is a colorless gas, which is lighter than air, a characteristic which needs to be considered if it is going to be used, as discussed below. It is relatively unreactive, although it is explosive in air, a property which has been known for a long time (reviewed by Von Elbe and Lewis, 1942), and exemplified by such events as the Hindenburg airship disaster (DiLisi, 2017). Even if relatively unreactive, to have an effect – or effects – in biological systems, there has to be an interaction with H_2 and the appropriate biological molecules. The exact nature of how H_2 interacts with biological components has been debated, with suggestions ranging from acting as an antioxidant (Kawamura *et al.*, 2020), interacting with hydroxyl radicals (Ohsawa *et al.*, 2007), acting through its redox potential, or having a direct interaction with the polypeptide chains of protein complexes, as discussed previously (Hancock *et al.*, 2022a). It is likely that there is no single mode of action for H_2 in biology, and it is probably a range of interactions all going on simultaneously. There are also short-term effects seen, as well as long-term effects reported, as discussed below. Therefore, H_2 most likely is having effects on the future protein complement of cells, their signaling events and undoubtedly the gene expression which ensues following H_2 application. The molecular biological events initiated by H_2 are beyond the scope of this chapter but will need to be unraveled and understood if H_2 treatments are to be adopted and used in the future.

Interestingly, H_2 is well tolerated by animals, and seems to have no toxic effects. It was suggested (Bjurstedt and Severin, 1948), and then subsequently used as a diving gas back in the 1940s, and the concentration of H_2 used is much higher than that used in biomedical research. Thereby, by extension, any H_2 residue in an agricultural or food product would also be safe to consume. Therefore, using H_2 treatments in agriculture would be assumed to be safe but as discussed below, the creation of residue by-products in any technology to deliver the H_2 should also be considered.

What is certain is that it can be seen in the literature that treatments of plant materials with H_2 is of benefit for food security for the future, and it is examples of this evidence that will be reviewed in this chapter.

6.2 Methods of Molecular Hydrogen Application to Plant Material

Hydrogen is a gas and can be used as such for treating plants, as summarized in Fig. 6.1. Examples of using H_2 fumigation are given below. However, there are a few caveats. First, it is very much lighter than air and so in an expansive volume, the H_2 will not linger at ground level for very long. Therefore, plant treatments may need to be carried out in a confined space if significant plant exposure to H_2 gas is to be achieved. Second, mixing H_2 and O_2 (in the atmosphere) can be explosive.

It is more likely that H_2 could be added to plants as a water which has been enhanced for its H_2 concentration. Water can be bubbled with H_2 gas, and some of this will dissolve to create what is referred to as hydrogen-rich water (HRW). This will have a limited concentration of H_2, which ideally should be quantified before use. However, H_2 does not remain dissolved for very long, and will soon revert to the gaseous phase and deplete the HRW. The kinetics of such depletion were recently published in a perspective article (Russell *et al.*, 2024). As will be seen from the discussion below, some researchers also dilute HRW before use, and may find such diluted solutions are better than the original stock HRW created. A variation of HRW is to bubble hydrogen into a saline solution and create hydrogen-rich saline (HRS), but this probably has limited use in plant science or agriculture, as saline can be stressful to plants (Negrão *et al.*, 2017).

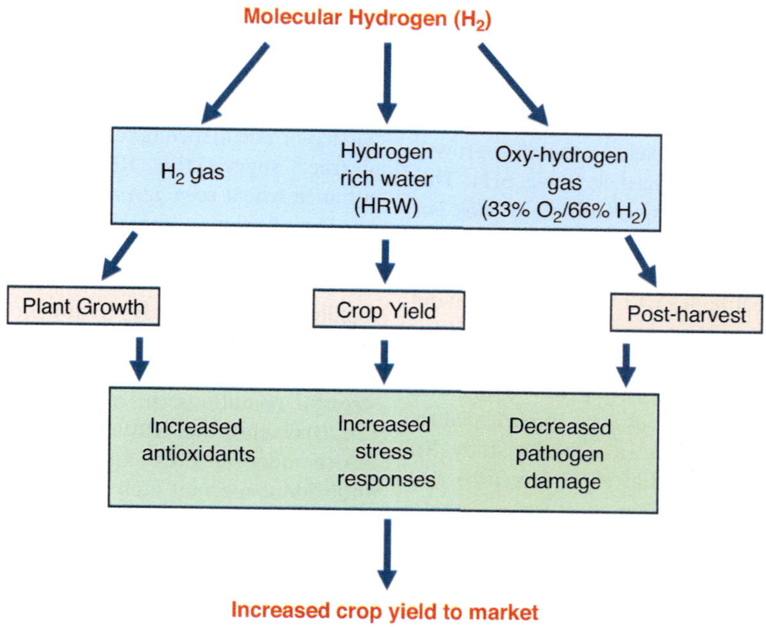

Fig. 6.1. Examples of treatment methods which can be used to apply H_2 to plant material.

One of the ways to ensure that H_2 stays in solution longer is to use nanobubbles, to make what is referred to as hydrogen nanobubble water (HNW). Although there are several techniques for creating nanobubbles (Zhou *et al.*, 2021), a cheap and easy method is to employ a bamboo cane, which was used in a study with chickens (Shin *et al.*, 2016) and by others (Han *et al.*, 2017).

Alternatively, hydrogen gas can be delivered from a donor molecule. The simplest way to create H_2 in solution is to use magnesium metal (Mg), which will react with water and liberate H_2 gas, some of which will dissolve. This is the basis for many 'hydrogen tablets' available commercially. However, this will leave a magnesium ion by-product in the solution, which may not be desirable for agricultural use, although it is well tolerated in the biomedical arena.

Another relatively simple donor molecule for obtaining H_2 is magnesium hydride (MgH_2). This will release H_2 over a longer period than simply using HRW, but it does leave a Mg-based by-product, and plants are not always tolerant of high Mg^{2+} (Guo *et al.*, 2015), although magnesium can be beneficial in alleviating the stress caused by other metals, such as aluminum (Bose *et al.*, 2011). However, MgH_2 has been suggested for use with cut flowers, for example (Li *et al.*, 2020; Nguyen and Lim, 2022).

Other donors are based on nanotechnology and include ammonia borane-loaded hollow mesoporous silica nanoparticle (AB@hMSN). This should give a supply of H_2 over a prolonged period of time, and so avoid a bolus effect of adding HRW. AB@hMSN use has been reported in plants, for example tomato (Wang *et al.*, 2021), and suggested for agriculture more widely (Wang *et al.*, 2024), as well as for use in animals (Liu *et al.*, 2024). However, there will be a by-product here, and it would need to be determined whether any residue of the use of AB@hMSN enters the food chain and what effect it may have on the health of humans and other animals.

Lastly, H_2 can be used as a mixture with oxygen. Electrolysis of water naturally gives a mixture of 66% H_2 and 33% O_2, in what is referred to as oxy-hydrogen gas. Electrolysis is relatively easy to do with an electricity supply, and if required, a membrane can be used to separate the H_2 and O_2, so giving pure H_2 and exhausting the O_2 to the atmosphere. For a relatively recent overview of the methods to administer H_2, see Russell *et al.* (2021).

6.3 Growth Enhancement by Molecular Hydrogen

One of the attributes of the application of H_2 to plants is that it is likely to increase growth of the plant (summarized in Table 6.1). This might include enhancing germination rates. For example, Huang *et al.* (2021) showed increased germination rates in cucumber seeds following HRW treatment, with 75% HRW having maximal effect. They also showed that HRW increased the activity of α-amylase, β-amylase, and total amylase, suggesting that sugar and starch metabolism was at least in part mediating the effects reported. In a similar study, Xu *et al.* (2013) showed that HRW increased rice seed germination during salt stress. Again α/β-amylase activity was increased, as were the antioxidant enzymes superoxide dismutase (SOD), catalase (CAT), and ascorbate peroxidase.

More recently, Huang *et al.* (2023) again showed that H_2 can increase germination in cucumber, but further suggested that the effect was mediated by changes in trehalose metabolism, i.e., trehalose accumulation was increased. Similar data were reported in barley by Song *et al.* (2023), but this time under drought stress. Again, sugar metabolism was increased, and markers of oxidative stress were decreased. HRW, especially at a concentration of 25% HRW, was thought to restore the redox balance of seedlings, particularly by the regulation of the ascorbate-glutathione (ASA-GSH) cycle. In lentil, it was found that HRW increased germination of seed but that this was further enhanced by the addition of plasma-activated water (PAW) (Mansory *et al.*, 2023). In rice it was suggested that HRW alters the water migration of the seed and hence germination was enhanced. On further analysis, several biomolecules had altered accumulation, including soluble dietary fibre, thiamine, and some amino acids. Overall, the authors suggested that HRW increased the nutritional value of the rice and also the eating qualities (Zhu *et al.*, 2022), which is especially pertinent if the rice is being used for food.

Once seeds have germinated, the effects of H_2 can still be seen. Islam *et al.* (2023) treated wheat (*Triticum aestivum* L.) with HRW under drought stress, and found that a 4 hour treatment worked best. HRW enhanced germination but also shoot growth, both as root and shoot weight and length. Although antioxidant levels were also enhanced, it was found that the tissues had higher H_2O_2 levels and NO accumulation. Of particular pertinence here, the authors in their abstract 'suggest that HRW treatment could enhance wheat seed germination, growth, and nutrient absorption, thereby increasing agricultural productivity.'

On scanning the literature it becomes rapidly apparent that H_2 treatment is particularly useful for aiding plants to overcome stress, leading to improved growth of a plant under stressful conditions. In a pre-print, Fu *et al.* (2020) claimed that HRW increased the growth of rice under salt stress, and went on to say that antioxidant enzymes such as SOD and CAT were enhanced following HRW treatment. Later, the same research group published this work separately (Fu *et al.*, 2021). Others have also shown that HRW alleviates salt stress in plants. Xu *et al.* (2013) had also worked on rice, whereas Yang *et al.* (2023) focused on maize.

In rice seedlings, HRW could enhance growth under nitrogen deficiency (Jiang *et al.*, 2023). HRW improved the biomass of the plants but also enhanced the antioxidant capacity, for example by increasing CAT activity. In cucumber, HRW could relieve chilling stress (Wang *et al.*, 2023). Several growth indicators, such as plant height and leaf area, increased following HRW application. Photosynthetic efficiency improved and antioxidants were enhanced, including SOD and CAT. With such wide-ranging effects, it is no surprise that the plants were healthier and grew more vigorously.

At the other end of the temperature scale, Chen *et al.* (2017b) looked at the effect of HRW pretreatment (7 days) on heat stress. Here, the chosen species was also cucumber, and HRW was able to alleviate heat effects on a range of parameters, including photosynthesis, chlorophyll content, electrolyte leakage, and lipid peroxidation. The antioxidant capacity of the plants was also enhanced. Interestingly, there was also an increase in the heat shock protein, HSP70, as well as osmolytes. All in all, HRW had a positive effect on these heat-stressed plants.

Plants are often exposed to heavy metals, and this can cause a stress response (Ghori *et al.*, 2019). Therefore, it is no surprise that there has been interest in whether H_2 treatments can

Table 6.1. Examples of enhanced growth of plants on treatment with molecular hydrogen.

Plant material treated	Method of treatment	Effect seen/reported	Reference(s)
Arabidopsis thaliana	Hydrogen-rich water (HRW)	Drought response	Xie *et al*. (2014)
Cucumber (*Cucumis sativus* L.)	HRW	Increased germination	Huang *et al*. (2021)
Cucumber (*Cucumis sativus* L.)	HRW	Increased germination may be mediated by trehalose	Huang *et al*. (2023)
Cucumber (*Cucumis sativus*)	HRW	Enhanced growth under chilling	Wang *et al*. (2023)
Cucumber (*Cucumis sativus*)	HRW	Enhanced growth under heat stress	Chen *et al*. (2017a)
Cucumber (*Cucumis Sativus* L)	HRW (and CO)	Alleviates drought stress	Chen and Liao (2017)
Cucumber (*Cucumis sativus*)	HRW	Adventitious root induction	Zhu *et al*. (2016)
Rice (*Oryza sativa* L., Wuyunjing 7)	HRW	Increased germination during salt stress	Xu *et al*. (2013)
Rice (brown)	HRW	Increased germination, improved nutritional value	Zhu *et al*. (2022)
Rice	HRW	Increased growth under nitrogen deficiency	Jiang *et al*. (2023)
Rice	HRW	Alleviates heavy metal stress, Cd and Pb	Ma *et al*. (2021)
Rice	HRW	Alleviates salt stress	Fu *et al*. (2020); Fu *et al*. (2021)
Barley (*Hordeum vulgare* L.)	HRW	Increased germination under drought stress	Song *et al*. (2023)
Wheat (*Triticum aestivum* L.)	HRW	Increased growth and antioxidants	Islam *et al*. (2023)
Maize (*Zea mays* L.)	HRW	Alleviates salt stress, better root growth	Yang *et al*. (2023)
Lentil (*L. culinaris*)	HRW, plasma-activated water (PAW)	Increased germination	Mansory *et al*. (2023)
Mung bean	HRW	Promotion of elongation of hypocotyls and roots	Wu *et al*. (2020)
Alfalfa (*Medicago sativa* L.)	HRW	Alleviates Hg stress	Cui *et al*. (2014)
Alfalfa (*M. sativa* L. cv. Victoria)	HRW	Reduced paraquat toxicity	Jin *et al*. (2013)
Alfalfa	HRW	Cd tolerance, gene expression study	Cui *et al*. (2020)
Alfalfa	HRW	Aluminium tolerance	Chen *et al*. (2014)
Chinese cabbage	HRW	Alleviates Cd stress	Wu *et al*. (2015)

Continued

Table 6.1. Continued

Plant material treated	Method of treatment	Effect seen/reported	Reference(s)
Brassica campestris spp. *chinensis* L.	HRW	Cd tolerance, NO mediation	Su *et al.* (2019)
Tetrastigma hemsleyanum	HRW	Cold stress and phenylpropanoid biosynthesis	Liu *et al.* (2022)
Hypsizygus marmoreus (fungus)	HRW	Alleviated stress	Zhang *et al.* (2017)

alleviate such stress. To this end, Ma *et al.* (2021) looked at the use of HRW under cadmium (Cd) and lead (Pb) toxicity. HRW enhanced the growth of rice under these conditions, as assessed by root and shoot biomass. Antioxidants, particularly catalase (CAT) and peroxidase (POD), were also enhanced by HRW, suggesting a mechanism by which the stress is tolerated. Mercury (Hg) stress was alleviated by HRW in alfalfa (*Medicago sativa* L.) (Cui *et al.*, 2014). The authors concluded that HRW reduced the stunting of plant growth caused by the Hg, and decreased the oxidative stress of the tissues. Seedling roots also contained less Hg. In Chinese cabbage HRW was shown to alleviate Cd stress (Wu *et al.*, 2015). Here, there was enhanced growth and less Cd content in the tissues. Upregulation of the genes responsible for Cd uptake were blocked by HRW, whereas a gene needed for enhanced Cd accumulation in root vacuoles was upregulated, for example. Also in this study, oxidative stress was reduced by HRW, with several antioxidant enzymes showing increased activity, including SOD, CAT, and ascorbate peroxidase.

Jin *et al.* (2013) looked at paraquat toxicity in alfalfa and found that HRW alleviated stress in the plant. There was better growth, less lipid peroxidation, and decreased reactive oxygen species (ROS) accumulation. Interestingly, the effects of H_2 appeared to involve the enzyme heme oxygenase-1 (HO-1), with H_2 increasing transcripts for this enzyme, as well as activity. This involvement of HO-1 was further supported by the fact that carbon monoxide (CO) could reverse the effects of H_2.

Carbon monoxide involvement has also been noted by others. For example, Chen and Liao (2017) noted that CO mediated HRW effects under drought conditions when they were investigating the development of adventitious roots. As well as CO, other cell signaling molecules have been shown to be involved in mediating HRW, or H_2, effects in stress responses. Cui *et al.* (2020) looked at the gene expression of alfalfa seedling roots using RNA-Seq, and found that genes encoding proteins involved in sulfur and glutathione metabolism were prominent, suggesting that alterations in this metabolism may be involved in mediating H_2 responses. In a cold stress study of the plant *Tetrastigma hemsleyanum*, it was shown that the phenylpropanoid biosynthesis pathway is important for HRW responses (Liu *et al.*, 2022). Other downstream candidates for mediating responses in plants to H_2 include ROS and NO. In *Arabidopsis*, H_2 led to stomatal closure but this appeared to be mediated by a rise in NO, itself mediated by ROS accumulation (Xie *et al.*, 2014). In cucumber, NO was also involved in mediating H_2 effects, this time inducing adventitious root formation (Zhu *et al.*, 2016). Su *et al.* (2019) reported upregulated NO production caused by increased activity of the NO-generating enzyme nitrate reductase (NR). On the other hand, aluminum-induced reduction in root growth in alfalfa appeared to be mediated by a decrease in NO accumulation (Chen *et al.*, 2014). Plant hormones are likely to be involved too. Wu *et al.* (2020) suggested that in mung bean seedlings, hypocotyl and root elongation induced by HRW was mediated by endogenous levels of auxin and gibberellin.

Although not a plant, HRW was shown to enhance growth of the fungus *Hypsizygus marmoreus* while grown under a variety of stresses (Zhang *et al.*, 2017), such as $CdCl_2$, NaCl, and H_2O_2. HRW improved mycelial growth and biomass under these growth conditions. In a similar manner to that seen in plants, malondialdehyde (MDA) was reduced by HRW, and antioxidants were increased. Therefore, the molecular biological responses here were not different from plants, and indeed similar to that seen in animals (Köktürk *et al.*, 2022)

It is clear that H_2 treatment, mainly reported using HRW, will have effects on the growth of many plant species, many of which are part of, and therefore enhance, the human food supply. One concern within global food insecurity is becoming increasingly reliant on a smaller number of crops, with maize, wheat, and rice in particular now representing over half of the global food supply from crops. Therefore, the importance of reliable production of these crops with the highest nutritional content possible cannot be overstated.

The exact mechanisms of how the HRW has positive effects on plant growth is not always clear but as far as an agricultural setting is concerned, as long as the HRW treatments have a cost benefit and are safe – to use and for consumption – it does not matter too much if the molecular mechanisms remain obscure. What matters is that H_2 can enhance plant growth.

6.4 Yields Following Molecular Hydrogen Treatments

From the above, it is clear that H_2 treatments can improve plant growth, but this means little unless the useable yield of the plant – for industrial use or consumption – is increased. It may be that the plant crop has better quality, and some of these are mentioned elsewhere in this chapter (and listed in Table 6.2). For example, the treatment of tomato with H_2 to reduce nitrite levels is discussed below (Zhang *et al.*, 2019). Zhu *et al.* (2022) mentioned that brown rice had improved nutrition values after HRW treatment. In a similar way, irrigation of tomatoes with HRW improved the nutritional value of fruits in a study by Yao *et al.* (2024), while Li *et al.* (2021) reported better taste and aroma in strawberries. All these reports show that it may not be the mass yield of the crop which is important but the quality of that yield, and it appears that improvements can be seen following H_2 treatment.

In a comprehensive study of the quality and quantity of rice produced in the presence of ditch water or hydrogen nanobubble water, Cheng *et al.* (2021) showed that the yield of rice was improved by the presence of H_2, as well as the quality of the rice grains. The authors stated that application of HNW 'could increase the length, width, and thickness of brown/rough rice and white rice' and concluded that it would be an effective method to 'increase field and grain quality of rice.' The authors assayed the expression of several genes which could account for these changes, and also noted that the treated rice had less Cd accumulation.

Although not a crop plant, in *Arabidopsis*, seed size was enhanced by H_2 treatments (Cheng *et al.*, 2023), and as *Arabidopsis* is a model organism, it would be interesting to know how this translates to crop plants. However, the first papers are starting

Table 6.2. Examples of enhanced crop yields on treatment with molecular hydrogen.

Plant material treated	Method of treatment	Effect seen/reported	Reference(s)
Tomato	H_2 gas	Lower nitrite levels	Zhang *et al.* (2019)
Rice (brown)	Hydrogen-rich water (HRW)	Altered nutritional value	Zhu *et al.* (2022)
Rice	Hydrogen nanobubble water (HNW)	Better yield and quality	Cheng *et al.* (2021)
Tomato	HRW	Better nutritional value	Yao *et al.* (2024)
Tomato	HNW	26–40% increased yield, and better fruit quality	Li *et al.* (2024)
Arabidopsis		Altered seed size	Cheng *et al.* (2023)
Strawberry	HNW	Better flavor, colour, and aroma	Li *et al.* (2022a)

to appear on increased crop yields, for example, the very recent paper by Li *et al.* (2024) which reported that there was not only better quality of tomatoes on H_2 treatment, but an increase in yield of almost 40%, which is very significant when discussing food security for the future.

Although there is a limited number of reports (as summarized in Table 6.2) which focus on crop yields following H_2 treatments, some suggest that H_2 may enhance both the quantity and quality of crops used for human consumption. Certainly, more work needs to be carried out in this area.

6.5 Use of Molecular Hydrogen Post Harvest

One key issue with food security is getting the plant material to market and then to the consumer. Considerable loss is incurred between the field and the market, and then kitchen, estimated by Alexander *et al.* (2017) as 44% of crop dry matter being lost between crop harvest and human consumption. Taking into account losses at all stages of the process, it is estimated that approximately 30% of all food worldwide is wasted (Iannotti *et al.*, 2024). Therefore, anything that can be done to reduce this loss will be beneficial for future food security. In this vein, it has been suggested that incorporating H_2 into the food chain would have significant benefits. Relatively recently this has been reviewed (Hancock *et al.*, 2022b), but a summary and update of the literature will be given here.

It is not only the food industry but other markets which may benefit, such as the floriculture industry. For example, HRW has been shown to increase the vase life and improve the antioxidant content in a variety of flowers, including lily and rose (Ren *et al.*, 2017). This has been reviewed elsewhere, and as this discussion is not relevant to food security, further details can be found in the literature (Hancock *et al.*, 2022b).

6.5.1 Hydrogen treatment of fresh material

As mentioned above, plant materials can be treated with H_2 as a gas or in an enriched solution. Both these methods have been used to attempt to preserve food materials post harvest, as summarized in Table 6.3.

Using kiwifruit, Hu *et al.* (2018) found that fumigating with H_2 gas delayed softening of the fruit and reduced cell wall degradation. Interestingly, the generation of ethylene (C_2H_4), a gas involved in fruit ripening, was reduced. The enzymes for ethylene metabolism were less active and their transcripts downregulated. Therefore, ripening of the fruit was delayed by H_2 gas, an effect perhaps mediated by altered ethylene metabolism. Similar work with kiwifruit (fresh cut) was also reported in a paper by Zhao *et al.* (2021). Fruits were treated with HRW and the effect was a delay in the decrease of chlorophyll content, and concentration of ascorbic acid, phenolics, and flavonoids. An increase in MDA content was reduced, as was electrolyte leakage. The fruit were firmer and looked greener. Overall, there appeared to be a delay in ripening and the fruit had a longer shelf-life. If combined with treatment with slightly acidic electrolyzed water (SAEW), the effects were enhanced, so the authors suggested using both treatments.

Dong *et al.* (2022) treated *Rosa sterilis* fruit with HRW and found enhanced antioxidant capacity in the tissues. Hydrogen peroxide (H_2O_2) and superoxide ($O_2^{\cdot-}$) levels were reduced, and enzymes used to remove ROS were elevated, such as SOD, CAT, and glutathione reductase (GR). Glutathione and ascorbic acid levels were elevated. Interestingly, enzymes involved in energy metabolism were also increased, such as succinate dehydrogenase and cytochrome oxidase. In okras, HRW delayed fruit ripening. Higher levels of Na_2CO_3-soluble pectin, hemicellulose, and cellulose were measured following HRW treatment, and cell wall structures were better maintained. In litchi fruit there was delayed pericarp browning following HRW treatment (Yun *et al.*, 2021). Oxidants and the capacity to remove them were assessed in the pericarps of the fruits before and after HRW treatment. In general, antioxidant capacity seemed to increase and was further enhanced after day one after treatment. Such effects could account for the decrease in pericarp changes seen when HRW was used. Li *et al.* (2022a) used HRW with Chinese water chestnuts, finding that HRW delayed yellowing of the chestnuts, and there was less oxidative damage. There was also a low flavonoid accumulation and less

Table 6.3. Examples of the use of molecular hydrogen post harvest.

Material treated	Method of treatment	Effect seen/reported	Reference(s)
Post harvest			
Kiwifruit	H_2 gas	Delayed ripening	Hu *et al*. (2014, 2018)
Kiwifruit (fresh cut)	Hydrogen-rich water (HRW)	Delayed senescence	Zhao *et al*. (2021)
Rosa sterilis fruit	HRW	Increased antioxidants and better energy status	Dong *et al*. (2022)
Okras	HRW	Delayed fruit softening	Dong *et al*. (2023)
Litchi fruit	HRW	Less pericarp browning and changes in antioxidants	Yun *et al*. (2021)
Chinese water chestnuts	HRW	Delayed yellowing and less oxidative damage	Li *et al*. (2022b)
Tomato	H_2 gas	Delayed senescence and lower nitrite levels	Zhang *et al*. (2019)
Tomato	HRW	Increased resistance to *Botrytis cinerea*	Lu *et al*. (2017)
Bananas	HRW	Delayed ripening, repressed ethylene	Yun *et al*. (2022)
Hypsizygus marmoreus	HRW	Better nutrient retention, increased antioxidants	Chen *et al*. (2017a)
Rice (*Oryza sativa*)	Irrigation with hydrogen nanobubble water (HNW)	Increased storage, increased amino acids after storage	Cai *et al*. (2022)
Strawberry	HNW	Increased storage, firmness and increased lignin and cellulose	Jin *et al*. (2023)
Production processes			
Orange juice	N_2/H_2 gassing	Better ascorbic acid and colour	Alwazeer *et al*. (2003)
Strawberry	H_2 in packaging	Better firmness and antioxidants	Alwazeer and Özkan (2022)
Apricot	H_2 in drying gas	Better colour and improved flavonoid content	Alwazeer and Örs (2019)
Butter	HRW, and H_2/Mg	Reduced amines	Bulut *et al*. (2022)
Butter	HRW (H_2/Mg)	Reduced heavy metal uptake	Alwazeer *et al*. (2022)
Butter	HRW	Improved acidity and colour	Ceylan *et al*. (2022)
Cheese	H_2 in packaging	Lower microbial counts and freshness	Alwazeer *et al*. (2020)
Various plant materials	HRW	Extraction of phytochemicals	Alwazeer and Elnasanelkasim (2023); Alwazeer (2024)

expression of genes involved in the phenylpropanoid pathway.

Hydrogen gas was used by Zhang *et al.* (2019) for the treatment of tomato fruits. They found that exogenously applied H_2 reduced H_2 generation by the fruits and decreased senescence. Important for human health, they showed that H_2 application blocked the accumulation of nitrite in the fruit, and H_2 appeared to have an effect on the enzymes involved in nitrite metabolism. They suggested that H_2 treatment would be good for the 'preservation of fruit and vegetable products.' Lu *et al.* (2017) also worked with tomato and reported that treatment of the fruits with HRW increased the resistance of the fruit to *Botrytis cinerea*. Additionally, in the fruit which had been HRW treated there was an increase in polyphenol oxidase (PPO) activity, as well as an increase in NO accumulation. The authors concluded that H_2 may be acting as a stress signal and would improve plant pathogen defence.

In a report by Jin *et al.* (2023), strawberries were treated with HNW, resulting in increased storage time, better firmness, and an increase in hemicellulose, cellulose, and pectin content.

In an intriguing study, Cai *et al.* (2022) compared the irrigation of rice with ditch water or water enhanced with hydrogen nanobubbles. The rice produced was stored and then analysed. After 1 year in storage, the H_2-treated rice had increased levels of essential amino acids, especially lysine. What the authors describe as 'off-flavors' were reduced, and the authors say that there were 'lower levels of volatile substances, including pentanal, hexanal, heptanal, octanal, 1-octen-3-ol, and 2-heptanone.' The rice showed higher antioxidants and lower lipid peroxidation. Here, there is clearly a long-term effect of the use of H_2-based water treatments.

Other plant and nonanimal materials have also been studied (Table 6.3). Yun *et al.* (2022) showed that HRW delayed the ripening of bananas. As with the work on kiwifruits above (Hu *et al.*, 2018), Yun *et al.* (2022) found that HRW was repressing ethylene signaling, which would account for the delayed ripening in banana. There was also delayed degeneration of cell walls and starch in the bananas. All in all, the HRW was keeping the bananas fresher, as seen with the other fruits reported on.

Although not a plant species, work with HRW treatment has been carried out on the fungus *Hypsizygus marmoreus* (Chen *et al.*, 2017a). Here, the authors found similar data to that reported in plants. What the authors describe as 25% HRW was found to be the best application to retain most nutrients in the fungus, while electrolyte leakage and MDA content were also reduced by this treatment. The antioxidant capacity of the tissues was increased, both as enzyme activity and gene expression, and overall the authors concluded that this was a 'simple method to maintain the quality and extend the shelf-life of mushrooms.'

It can be seen, therefore, that using either H_2 fumigation or treatment with HRW can prolong the postharvest storage of a range of fruits. It is relatively cheap and easy to carry out, and several papers in the literature conclude that it is a technology that could easily be adopted for the agri-food industry in the future, with tangible benefits.

6.5.2 Hydrogen treatment of processed food products

Even after plant or animal material has been processed, there are still benefits reported from the use of H_2 (Alwazeer and Engin, 2022), as summarized in Table 6.3. As stated by Bulut *et al.* (2022): "In the food industry, H_2 is considered a food additive (E 949) with a non-toxic effect. In the EU it is classified in part C Group I of regulation 1129/2011, food additives permitted in all categories of foods at *quantum satis*." Therefore, there has been increasing interest in using H_2 in food production.

Following storage of butter, previous treatment by washing with HRW was shown to reduce the formation of amines, with a decrease in tryptamine, 2-phenylethylamine, spermidine, and spermine. Treatment with H_2 and Mg also reduced the level of histamine. The same group looked at the presence of heavy metals in butter, with and without H_2 treatment. Cultured-cream butter was washed with HRW or with ordinary water, and the levels of several heavy metals were determined. Arsenic (As), cadmium (Cd), antimony (Sb), and mercury (Hg) were all shown to decrease after HRW treatment, and it was suggested that this was an eco-friendly way to reduce the uptake of heavy metals by food during processing

(Alwazeer *et al.*, 2022). Also looking at the effect of H_2 on butter, Alwazeer's group showed that H_2 treatment decreased the acidity of the product, lowered peroxide levels, and enhanced the colour and appearance, which would make the product potentially more attractive for sale. They went on to say that this treatment leaves nothing harmful either in the product or the environment (Ceylan *et al.*, 2022).

With plant materials, the use of H_2 has also been shown to enhance the food product. For example, gassing orange juice with N_2 and H_2 increased ascorbic acid stability and colour retention (Alwazeer *et al.*, 2003). Similarly, including H_2 in the drying atmosphere, in the case of using apricot, showed benefits in, for example, the total flavonoid content (Alwazeer and Örs, 2019).

The incorporation of H_2 into food packaging has also been suggested. For example, using H_2 in strawberry packaging increased several desirable qualities of the fruit, including total soluble solids (TSS), firmness, and antioxidant content (Alwazeer and Özkan, 2022). Although not a plant product, similar data were obtained when H_2 was incorporated into the packaging of cheese (Alwazeer *et al.*, 2020).

Finally, HRW has been used to extract phytochemicals from a range of plant materials. These include the peels from tomato (*Lycopersicon esculentum*), apple (*Malus communis* L.), and lemon (*Citrus lemon* L.). Red cabbage leaves (*Brassica oleracea* var. *Capitata* F.) and carrot (*Daucus carota* subsp. *stavis*) were also used, and it was concluded that the 'hydrogen-rich solvent method is a green and cost-effective method for extracting phytochemicals from different plant materials' (Alwazeer and Elnasanelkasim, 2023; Alwazeer, 2024).

There is therefore a considerable body of evidence, albeit mainly from one research group, that shows that the use of H_2 in various aspects of food processing is a beneficial and eco-friendly thing to do.

6.6 Conclusions and Future Perspectives

There is little doubt that application of hydrogen gas-based treatments to plants has benefits. H_2 can be applied by a variety of methods, and has even been suggested to be the mediator of improvements in crops following legume crop rotation (Dong *et al.*, 2003). Often, H_2 is useful to alleviate stress responses, leading to better plant growth, as exemplified by several of the examples in Table 6.1. There are several molecular events downstream of H_2 treatments, and these often involve a change in cellular redox, alterations in ROS or NO accumulation, and changes in the antioxidant capacity of tissues, again exemplified by examples above.

However, the exact molecular targets of H_2 in most cases remain somewhat elusive (Hancock *et al.*, 2022a). This matters little to the agricultural industry which needs better growth, healthier plants, and higher crop yields. For the first two characteristics, there is a body of evidence, as seen in Table 6.1. 'Are there better yields?' is a more difficult question to address, with limited reports in the literature, as seen in Table 6.2. Admittedly, some crops are thought to be better, more nutritious, and healthier but many questions still remain.

- What is the cost/benefit analysis? The cost of obtaining H_2, storing, and applying it has to be outweighed by the increase in profit from the yield, either in quantity or quality. The price of H_2 is likely to come down as it is used for other things, such as transport, and in the future H_2 may be stored on farms for a variety of purposes.
- What is the best way to apply H_2 to crop plants? H_2 gas is difficult; HRW may not give a lasting effect, while some donors may release unwanted by-products. How is this going to be scaled up to meet agricultural need? To answer the last question, there appears to be a lack of field trials. Should there be more investment in research on a larger scale?
- Not all plants will respond the same, and H_2 treatments may need to be tailored to the crop plant, both in the concentration of H_2 given and the duration and repetition of treatments. More work will be required to optimize H_2-based treatment for different crops and different environmental conditions – for example, treating with HRW in the rain may not be sensible if the leaf surface needs to be treated.

One of the most likely uses of H_2 for agriculture is in postharvest and food processing. Here, the application of H_2 can be more easily controlled and there appear to be substantial benefits. This may apply to food production but also the floriculture industry.

In summary, the use of H_2 in the human food chain, and to improve food security, has great promise. There are still many questions to be answered, but it seems as though future work in this area is worth pursuing and one day H_2-based agricultural treatments may become a commonplace occurrence.

References

Alexander, P., Brown, C., Arneth, A., Finnigan, J., Moran, D. *et al.* (2017) Losses, inefficiencies and waste in the global food system. *Agricultural Systems* 153, 190–200.

Alwazeer, D. (2024) Hydrogen-rich solvent method in phytochemical extraction: Potential mechanisms and perspectives. *Phytochemical Analysis* 35(2), 203–219.

Alwazeer, D. and Örs, B. (2019) Reducing atmosphere drying as a novel drying technique for preserving the sensorial and nutritional notes of foods. *Journal of Food Science and Technology* 56(8), 3790–3800.

Alwazeer, D. and Çiğdem, A. (2022) Use of the molecular hydrogen in agriculture field. *Turkish Journal of Agriculture-Food Science and Technology* 10, 14–20.

Alwazeer, D. and Engin, T. (2022) Use of molecular hydrogen in food technologies. *Turkish Journal of Agriculture-Food Science and Technology* 10, 1205–1213.

Alwazeer, D. and Özkan, N. (2022) Incorporation of hydrogen into the packaging atmosphere protects the nutritional, textural and sensorial freshness notes of strawberries and extends shelf life. *Journal of Food Science and Technology* 59(10), 3951–3964.

Alwazeer, D. and Elnasanelkasim, M.A. (2023) Hydrogen-rich water as a green solvent for the extraction of phytochemicals from agri-food wastes. *Sustainable Chemistry and Pharmacy* 33, 101035.

Alwazeer, D., Tan, K. and Örs, B. (2020) Reducing atmosphere packaging as a novel alternative technique for extending shelf life of fresh cheese. *Journal of Food Science and Technology* 57(8), 3013–3023.

Alwazeer, D., Delbeau, C., Divies, C. and Cachon, R. (2003) Use of redox potential modification by gas improves microbial quality, color retention, and ascorbic acid stability of pasteurized orange juice. *International Journal of Food Microbiology* 89(1), 21–29.

Alwazeer, D., Ceylan, M.M., Bulut, M. and Koyuncu, M. (2022) Evaluation of the impact of hydrogen-rich water on the deaccumulation of heavy metals in butter. *Journal of Food Safety* 42(6), e13005.

Bjurstedt, H. and Severin, G. (1948) The prevention of decompression sickness and nitrogen narcosis by the use of hydrogen as a substitute for nitrogen, the Arne Zetterstrom method for deep-sea diving. *Military Surgeon* 103, 107–116.

Bose, J., Babourina, O. and Rengel, Z. (2011) Role of magnesium in alleviation of aluminium toxicity in plants. *Journal of Experimental Botany* 62, 2251–2264.

Bulut, M., Çelebi Sezer, Y., Ceylan, M.M., Alwazeer, D. and Koyuncu, M. (2022) Hydrogen-rich water can reduce the formation of biogenic amines in butter. *Food Chemistry* 384, 132613.

Cai, C., Zhao, Z., Zhang, Y., Li, M., Li, L. *et al.* (2022) Molecular hydrogen improves rice storage quality via alleviating lipid deterioration and maintaining nutritional values. *Plants* 11(19), 2588.

Ceylan, M.M., Bulut, M., Alwazeer, D. and Koyuncu, M. (2022) Evaluation of the impact of hydrogen-rich water on the quality attribute notes of butter. *Journal of Dairy Research* 89, 431–439.

Chen, Y. and Liao, W. (2017) Carbon monoxide is involved in hydrogen gas-induced adventitious root development in cucumber under simulated drought stress. *Frontiers in Plant Science* 8, 128.

Chen, M., Cui, W., Zhu, K., Xie, Y., Zhang, C. *et al.* (2014) Hydrogen-rich water alleviates aluminum-induced inhibition of root elongation in alfalfa via decreasing nitric oxide production. *Journal of Hazardous Materials* 267, 40–47.

Chen, H., Zhang, J., Hao, H., Feng, Z., Chen, M. *et al.* (2017a) Hydrogen-rich water increases postharvest quality by enhancing antioxidant capacity in *Hypsizygus marmoreus*. *AMB Express* 7(1), 1–10.

Chen, Q., Zhao, X., Lei, D., Hu, S., Shen, Z. *et al.* (2017b) Hydrogen-rich water pretreatment alters photosynthetic gas exchange, chlorophyll fluorescence, and antioxidant activities in heat-stressed cucumber leaves. *Plant Growth Regulation* 83(1), 69–82.

Cheng, P., Wang, J., Zhao, Z., Kong, L., Lou, W. *et al.* (2021) Molecular hydrogen increases quantitative and qualitative traits of rice grain in field trials. *Plants* 10(11), 2331.

Cheng, P., Wang, Y., Cai, C., Li, L., Zeng, Y. *et al.* (2023) Molecular hydrogen positively regulates nitrate uptake and seed size by targeting nitrate reductase. *Plant Physiology* 193(4), 2734–2749.

Cui, W., Fang, P., Zhu, K., Mao, Y., Gao, C. *et al.* (2014) Hydrogen-rich water confers plant tolerance to mercury toxicity in alfalfa seedlings. *Ecotoxicology and Environmental Safety* 105, 103–111.

Cui, W., Yao, P., Pan, J., Dai, C., Cao, H. *et al.* (2020) Transcriptome analysis reveals insight into molecular hydrogen-induced cadmium tolerance in alfalfa: The prominent role of sulfur and (homo) glutathione metabolism. *BMC Plant Biology* 20(1), 58.

DiLisi, G.A. (2017) The *Hindenburg* disaster: Combining physics and history in the laboratory. *Physics Teacher* 55(5), 268–273.

Dixon, B.J., Tang, J. and Zhang, J.H. (2013) The evolution of molecular hydrogen: A noteworthy potential therapy with clinical significance. *Medical Gas Research* 3(1), 1–12.

Dong, B., Zhu, D., Yao, Q., Tang, H. and Ding, X. (2022) Hydrogen-rich water treatment maintains the quality of *Rosa sterilis* fruit by regulating antioxidant capacity and energy metabolism. *LWT* 161, 113361.

Dong, Z., Wu, L., Kettlewell, B., Caldwell, C.D. and Layzell, D.B. (2003) Hydrogen fertilization of soils – is this a benefit of legumes in rotation? *Plant Cell & Environment* 26(11), 1875–1879.

Dong, W., Shi, L., Li, S., Xu, F., Yang, Z. *et al.* (2023) Hydrogen-rich water delays fruit softening and prolongs shelf life of postharvest okras. *Food Chemistry* 399, 133997.

FAO, IFAD, UNICEF, WFP, and WHO (2023) Urbanization, agrifood systems transformation and healthy diets across the rural–urban continuum. In: *The State of Food Security and Nutrition in the World 2023*. FAO, Rome. Available at: https://openknowledge.fao.org/server/api/core/bitstreams/8865ca80-bd21-4d71-bae2-e4ed12786cba/content

Fu, X., Ma, L., Gui, R., Li, Y., Yang, X. *et al.* (2020) Hydrogen rich water (HRW) induces plant growth and physiological attributes in fragrant rice varieties under salt stress. Available at: www.researchsquare.com/article/rs-21074/v1

Fu, X., Ma, L., Gui, R., Ashraf, U., Li, Y. *et al.* (2021) Differential response of fragrant rice cultivars to salinity and hydrogen rich water in relation to growth and antioxidative defense mechanisms. *International Journal of Phytoremediation* 23(11), 1203–1211.

Ghori, N.H., Ghori, T., Hayat, M.Q., Imadi, S.R., Gul, A. *et al.* (2019) Heavy metal stress and responses in plants. *International Journal of Environmental Science and Technology* 16(3), 1807–1828.

Guo, W., Chen, S., Hussain, N., Cong, Y., Liang, Z. *et al.* (2015) Magnesium stress signaling in plant: Just a beginning. *Plant Signaling & Behavior* 10(3), e992287.

Han, A.L., Park, S.H. and Park, M.S. (2017) Hydrogen treatment protects against cell death and senescence induced by oxidative damage. *Journal of Microbiology and Biotechnology* 27(2), 365–371.

Hancock, J.T., Russell, G., Craig, T.J., May, J., Morse, H.R. *et al.* (2022a) Understanding hydrogen: Lessons to be learned from physical interactions between the inert gases and the globin superfamily. *Oxygen* 2(4), 578–590.

Hancock, J.T., Russell, G. and Stratakos, A.C. (2022b) Molecular hydrogen: The postharvest use in fruits, vegetables and the floriculture industry. *Applied Sciences* 12(20), 10448.

Hancock, J.T., LeBaron, T.W., May, J., Thomas, A. and Russell, G. (2021) Molecular hydrogen: Is this a viable new treatment for plants in the UK? *Plants* 10(11), 2270.

Hu, H., Li, P., Wang, Y. and Gu, R. (2014) Hydrogen-rich water delays postharvest ripening and senescence of kiwifruit. *Food Chemistry* 156, 100–109.

Hu, H., Zhao, S., Li, P. and Shen, W. (2018) Hydrogen gas prolongs the shelf life of kiwifruit by decreasing ethylene biosynthesis. *Postharvest Biology and Technology* 135, 123–130.

Huang, P., Li, C., Liu, H., Zhao, Z. and Liao, W. (2021) Hydrogen gas improves seed germination in cucumber by regulating sugar and starch metabolisms. *Horticulturae* 7(11), 456.

Huang, P., Li, C., Che, P., Liu, H., Zhao, Z. *et al.* (2023) Hydrogen gas enhanced seed germination by increasing trehalose biosynthesis in cucumber. *Journal of Plant Growth Regulation* 42(6), 3908–3922.

Iannotti, L., Kleban, E., Fracassi, P., Oenema, S. and Lutter, C. (2024) Evidence for policies and practices to address global food insecurity. *Annual Review of Public Health* 45(1), 375–400.

Islam, M.A., Shorna, M.N.A., Islam, S., Biswas, S., Biswas, J. *et al.* (2023) Hydrogen-rich water: A key player in boosting wheat (*Triticum aestivum* L.) seedling growth and drought resilience. *Scientific Reports* 13(1), 22521.

Jiang, Y., Ye, Q., Ma, L., Yang, X., Zhang, J. *et al.* (2023) Regulation of growth and physiological properties of fragrant rice seedlings by hydrogen-rich water (HRW) under nitrogen-deficient conditions. *Journal of Plant Growth Regulation* 42(4), 2221–2231.

Jin, Q., Zhu, K., Cui, W., Xie, Y., Han, B.I.N. *et al.* (2013) Hydrogen gas acts as a novel bioactive molecule in enhancing plant tolerance to paraquat-induced oxidative stress via the modulation of heme oxygenase-1 signalling system. *Plant, Cell & Environment* 36(5), 956–969.

Jin, Z., Liu, Z., Chen, G., Li, L., Zeng, Y. *et al.* (2023) Molecular hydrogen-based irrigation extends strawberry shelf life by improving the synthesis of cell wall components in fruit. *Postharvest Biology and Technology* 206, 112551.

Kawamura, T., Higashida, K. and Muraoka, I. (2020) Application of molecular hydrogen as a novel antioxidant in sports science. *Oxidative Medicine and Cellular Longevity* 2328768.

Köktürk, M., Yıldırım, S., Eser, G., Bulut, M. and Alwazeer, D. (2022) Hydrogen-rich water alleviates the nickel-induced toxic responses (inflammatory responses, oxidative stress, DNA damage) and ameliorates cocoon production in earthworm. *Biological Trace Element Research* 1–11.

Li, L., Liu, Y., Wang, S., Zou, J., Ding, W. *et al.* (2020) Magnesium hydride-mediated sustainable hydrogen supply prolongs the vase life of cut carnation flowers via hydrogen sulfide. *Frontiers in Plant Science* 11, 595376.

Li, L., Lou, W., Kong, L. and Shen, W. (2021) Hydrogen commonly applicable from medicine to agriculture: From molecular mechanisms to the field. *Current Pharmaceutical Design* 27(5), 747–759.

Li, F., Hu, Y., Shan, Y., Liu, J., Ding, X. *et al.* (2022a) Hydrogen-rich water maintains the color quality of fresh-cut Chinese water chestnut. *Postharvest Biology and Technology* 183, 111743.

Li, L., Wang, J., Jiang, K., Kuang, Y., Zeng, Y. *et al.* (2022b) Preharvest application of hydrogen nanobubble water enhances strawberry flavor and consumer preferences. *Food Chemistry* 377, 131953.

Li, M., Zhu, G., Liu, Z., Li, L., Wang, S. *et al.* (2024) Hydrogen fertilization with hydrogen nanobubble water improves yield and quality of cherry tomatoes compared to the conventional fertilizers. *Plants* 13(3), 443.

Liu, Y., Pan, J., Ni, S., Xing, B., Cheng, K. *et al.* (2022) Transcriptome and metabonomics combined analysis revealed the defense mechanism involved in hydrogen-rich water-regulated cold stress response of *Tetrastigma hemsleyanum*. *Frontiers in Plant Science* 13, 889726.

Liu, B., Lv, P., Zhang, X., Xia, C., Liu, X. *et al.* (2024) Zn-fe primary battery-enabled controlled hydrogen release in stomach for improving insulin resistance in obesity-associated type 2 diabetes. *Bioactive Materials* 33, 242–250.

Lu, H., Wu, B., Wang, Y., Liu, N., Meng, F. *et al.* (2017) Effects of hydrogen-rich water treatment on defense responses of postharvest tomato fruit to botrytis cinerea. *Journal of Henan Agricultural Sciences* 46, 64–68.

Ma, L., Kong, L., Gui, R., Yang, X., Zhang, J. *et al.* (2021) Application of hydrogen-rich water modulates physio-biochemical functions and early growth of fragrant rice under Cd and Pb stress. *Environmental Science and Pollution Research International* 28(41), 58558–58569.

Mansory, S., Bahreini, M. and Tadi, S.H. (2023) Comparison between the effect of activated waters on lentil seed germination using various plasma reactors and hydrogen injection system.

Negrão, S., Schmöckel, S.M. and Tester, M.J.A.O.B. (2017) Evaluating physiological responses of plants to salinity stress. *Annals of Botany* 119(1), 1–11.

Nguyen, T.K. and Lim, J.H. (2022) Is it a challenge to use molecular hydrogen for extending flower vase life? *Plants* 11(10), 1277.

Noor, M.N.Z., Alauddin, A.S., Wong, Y.H., Looi, C.Y., Wong, E.H. *et al.* (2023) A systematic review of molecular hydrogen therapy in cancer management. *Asian Pacific Journal of Cancer Prevention* 24(1), 37–47.

Nor A'azizam, N.M., Chopra, S., Guleria, P., Kumar, V. and Abd Rahim, M.H. (2024) Harnessing the potential of mutation breeding, CRISPR genome editing, and beyond for sustainable agriculture. *Functional & Integrative Genomics* 24(2), 44.

Ohsawa, I., Ishikawa, M., Takahashi, K., Watanabe, M., Nishimaki, K. *et al.* (2007) Hydrogen acts as a therapeutic antioxidant by selectively reducing cytotoxic oxygen radicals. *Nature Medicine* 13(6), 688–694.

Ohta, S. (2014) Molecular hydrogen as a preventive and therapeutic medical gas: Initiation, development and potential of hydrogen medicine. *Pharmacology & Therapeutics* 144(1), 1–11.

Rahman, M.H., Bajgai, J., Fadriquela, A., Sharma, S., Trinh Thi, T. *et al.* (2021) Redox effects of molecular hydrogen and its therapeutic efficacy in the treatment of neurodegenerative diseases. *Processes* 9(2), 308.

Ren, P.J., Jin, X., Liao, W.B., Wang, M., Niu, L.J. *et al.* (2017) Effect of hydrogen-rich water on vase life and quality in cut lily and rose flowers. *Horticulture, Environment, and Biotechnology* 58(6), 576–584.

Russell, G., Nenov, A., Kisher, H. and Hancock, J.T. (2021) Molecular hydrogen as medicine: An assessment of administration methods. *Hydrogen* 2(4), 444–460.

Russell, G., May, J. and Hancock, J.T. (2024) An interplay of gases: Oxygen and hydrogen in biological systems. *Oxygen* 4(1), 37–52.

Shin, D., Cho, E.S.R., Bang, H.T. and Shim, K.S. (2016) Effects of oxygenated or hydrogenated water on growth performance, blood parameters, and antioxidant enzyme activity of broiler chickens. *Poultry Science* 95(11), 2679–2684.

Song, R., Zhang, X., Feng, C., Zhang, S., Song, L. *et al.* (2023) Exogenous hydrogen promotes germination and seedling establishment of barley under drought stress by mediating the ASA-GSH cycle and sugar metabolism. *Journal of Plant Growth Regulation* 42(5), 2749–2762.

Su, N., Wu, Q., Chen, H., Huang, Y., Zhu, Z. *et al.* (2019) Hydrogen gas alleviates toxic effects of cadmium in brassica campestris seedlings through up-regulation of the antioxidant capacities: Possible involvement of nitric oxide. *Environmental Pollution* 251, 45–55.

Vaidya, S., Deng, C., Wang, Y., Zuverza-Mena, N., Dimkpa, C. *et al.* (2024) Nanotechnology in agriculture: A solution to global food insecurity in a changing climate? *NanoImpact* 34, 100502.

Von Elbe, G. and Lewis, B. (1942) Mechanism of the thermal reaction between hydrogen and oxygen. *Journal of Chemical Physics* 10(6), 366–393.

Wang, Y., Lv, P., Kong, L., Shen, W. and He, Q. (2021) Nanomaterial-mediated sustainable hydrogen supply induces lateral root formation via nitrate reductase-dependent nitric oxide. *Chemical Engineering Journal* 405, 126905.

Wang, X., An, Z., Liao, J., Ran, N., Zhu, Y. *et al.* (2023) The role and mechanism of hydrogen-rich water in the *Cucumis sativus* response to chilling stress. *International Journal of Molecular Sciences* 24(7), 6702.

Wang, R., Yang, X., Chen, X., Zhang, X., Chi, Y. *et al.* (2024) A critical review for hydrogen application in agriculture: Recent advances and perspectives. *Critical Reviews in Environmental Science and Technology* 54(3), 222–238.

Wu, Q., Su, N., Cai, J., Shen, Z. and Cui, J. (2015) Hydrogen-rich water enhances cadmium tolerance in Chinese cabbage by reducing cadmium uptake and increasing antioxidant capacities. *Journal of Plant Physiology* 175, 174–182.

Wu, Q., Su, N., Huang, X., Ling, X., Yu, M. *et al.* (2020) Hydrogen-rich water promotes elongation of hypocotyls and roots in plants through mediating the level of endogenous gibberellin and auxin. *Functional Plant Biology* 47(9), 771–778.

Xie, Y., Mao, Y., Zhang, W., Lai, D., Wang, Q. *et al.* (2014) Reactive oxygen species-dependent nitric oxide production contributes to hydrogen-promoted stomatal closure in Arabidopsis. *Plant Physiology* 165(2), 759–773.

Xu, S., Zhu, S., Jiang, Y., Wang, N., Wang, R. *et al.* (2013) Hydrogen-rich water alleviates salt stress in rice during seed germination. *Plant and Soil* 370(1–2), 47–57.

Yang, L., Tian, J., Zhu, M., Yu, B. and Sun, Y.I. (2023) Hydrogen-rich water improvement in root growth in maize exposed to saline stress. *Cereal Research Communications* 52(2), 581–590.

Yao, Y., Zhao, Z., Ding, Z., Yao, K., Yang, Y. *et al.* (2024) Hydrogen-rich water irrigation promotes fruit ripening and nutritional composition in tomato. *Postharvest Biology and Technology* 213, 112920.

Yun, Z., Gao, H., Chen, X., Chen, Z., Zhang, Z. *et al.* (2021) Effects of hydrogen water treatment on antioxidant system of litchi fruit during the pericarp browning. *Food Chemistry* 336, 127618.

Yun, Z., Gao, H., Chen, X., Duan, X. and Jiang, Y. (2022) The role of hydrogen water in delaying ripening of banana fruit during postharvest storage. *Food Chemistry* 373, 131590.

Zeng, J., Ye, Z. and Sun, X. (2014) Progress in the study of biological effects of hydrogen on higher plants and its promising application in agriculture. *Medical Gas Research* 4, 1–4.

Zhang, J., Hao, H., Chen, M., Wang, H., Feng, Z. *et al.* (2017) Hydrogen-rich water alleviates the toxicities of different stresses to mycelial growth in *Hypsizygus marmoreus*. *AMB Express* 7(1), 107.

Zhang, Y., Zhao, G., Cheng, P., Yan, X., Li, Y. *et al*. (2019) Nitrite accumulation during storage of tomato fruit as prevented by hydrogen gas. *International Journal of Food Properties* 22(1), 1425–1438.

Zhao, X., Meng, X., Li, W., Cheng, R., Wu, H. *et al*. (2021) Effect of hydrogen-rich water and slightly acidic electrolyzed water treatments on storage and preservation of fresh-cut kiwifruit. *Journal of Food Measurement and Characterization* 15(6), 5203–5210.

Zheng, M., Yu, H., Xue, Y., Yang, T., Tu, Q. *et al*. (2021) The protective effect of hydrogen-rich water on rats with type 2 diabetes mellitus. *Molecular and Cellular Biochemistry* 476(8), 3089–3097.

Zhou, L., Wang, S., Zhang, L. and Hu, J. (2021) Generation and stability of bulk nanobubbles: A review and perspective. *Current Opinion in Colloid & Interface Science* 53, 101439.

Zhu, C., Yang, L., Nie, P., Zhong, L., Wu, Y. *et al*. (2022) Effects of hydrogen-rich water on the nutritional properties, volatile profile and texture of germinated brown rice. *International Journal of Food Science & Technology* 57(12), 7666–7680.

Zhu, Y., Liao, W., Wang, M., Niu, L., Xu, Q. *et al*. (2016) Nitric oxide is required for hydrogen gas-induced adventitious root formation in cucumber. *Journal of Plant Physiology* 195, 50–58.

Zulfiqar, F., Russell, G. and Hancock, J.T. (2021) Molecular hydrogen in agriculture. *Planta* 254(3), 56.

7 Application of Biotechnologies for Crop Improvement and Food Production: Case Studies From the BIOALI Network

Sandra Sharry[1,2*], Lorenzo Burgos[3], Daniela Dalzotto[2,4], Patricia Boeri[2,4], Lucrecia Piñuel[2,4] and Iselen Trujillo[5]

[1]National Research System (SNI), Panama; [2]Universidad Nacional de Río Negro, Viedma, Argentina; [3]Fruit Tree Biotechnology Group - Plant Breeding Department, CEBAS-CSIC,El Puntal, Spain; [4]CIT-RIO NEGRO, Universidad Nacional de Río Negro, Viedma, Argentina; [5]Centro de Estudios para el Desarrollo Agroecológico Tropical (CEDAT), Instituto de Estudios Científicos y Tecnológicos (IDECYT), Universidad Nacional Experimental Simón Rodríguez (UNESR), Caracas, Venezuela

Abstract

This chapter highlights the work conducted by the BIOALI Network and presents various case studies that demonstrate the application of biotechnologies in crop improvement and food production. The objective of the BIOALI Network was to establish an interdisciplinary platform for exchanging scientific knowledge and experiences in order to enhance breeding programs for food crops. The network members, comprising different actors involved in food production, utilized various biotechnological approaches to achieve their specific objectives. These approaches included breeding plants for virus resistance, characterizing fruit flours from diverse native fruit species, and propagating trees and pastures in agroforestry systems for animal feed. By working collaboratively on the production of food crops using a multidisciplinary approach, the network members were able to synergize their efforts and increase applicability across Latin American and Caribbean countries. This is particularly important given the challenges posed by population growth, climate change, and increasing demand for food resources.

Keywords: agroforestry systems, food production, plant breeding, climate change

7.1 Introduction

The challenge for the world in the next 25 years will be to produce the same amount of food that has been generated in the last 8000 years, as a global population of 9 billion people is expected by 2050. This requires not only an increase in food production but also ensuring its quality and safety. It is important to debate what type of food we consume and how we produce it. Agricultural production faces significant challenges such as higher input costs, efficient pest

*Corresponding author: sesharry@unrn.edu.ar

 Plant Biotechnology and Sustainable Agriculture (ed. Tariq Aftab)
DOI: 10.1079/9781800627659.0007

and disease control, and the global climate crisis increasing pressure on farmers. Conventional and family farming are affected by increased demand for safe and high-quality products, price volatility, and competition in an increasingly global market (Izquierdo and Sharry, 2019). Furthermore, climate change demands improvement in cultivated varieties to adapt to biotic (pests) and abiotic (droughts, floods) stress conditions.

Approximately 70% of changes in natural habitats and significant biodiversity loss are a direct or indirect result of the current food system. Unsustainable diets are having a negative impact on our limited natural resources. To address these challenges, it is necessary to apply sustainable technologies and innovations to improve our current food systems and maximize efficient use of agricultural land. This includes more effective methods for crop production with lower energy and water consumption, as well as biotechnological advancements focused on enhancing desirable traits such as pest resistance or drought tolerance.

But it is not just about producing food; it is about adequately feeding the global population, which implies producing certain types of foods that are currently lacking to ensure nutritional security. Additionally, we must consider that one of the global goals is to reduce loss and waste of food. This challenge is addressed in Sustainable Development Goal (SDG) 12 on sustainable consumption and production, specifically target 12.3 which aims to halve per capita global food waste at retail and consumer levels by 2030.

We face major challenges related to sustainable food production amid global population growth, climate change, and other environmental challenges. According to the High-Level Expert Forum (FAO), adoption of sustainable and effective production methods is crucial along with adaptation to climate change. There is no one-size-fits-all solution, and part of the answer lies in science, advanced technology, and innovation. Biotechnology offers one of the options proposed by experts when combined with healthy and transparent public policies fostering sustainable practices within the agri-food sector. It is also important to educate and inform consumers about the environmental and social implications of their food choices in order to drive positive changes towards more sustainable patterns.

This chapter briefly describes the application of biotechnologies for food security and shows how the biotechnology technical cooperation network (BIOALI) makes it possible to coordinate actions and research to advance towards improved food production through biotechnologies. It also presents three cases of diverse applications for the improvement, characterization, and propagation of food resources.

7.2 Biotechnology Applied to the Sustainable Production of Safe and Healthy Food

Biotechnology is a field that utilizes biological systems and living organisms, or their derivatives, to modify or create specific products. In the context of food production, biotechnology plays a crucial role in optimizing processes, enhancing safety and quality, and improving nutritional value. This discipline encompasses various areas such as genetic improvement of crops and animals, development of preservation and processing methods, and efficient utilization of natural resources. By employing biotechnological techniques and innovations, it is possible to achieve innovative solutions in areas like food safety, human and animal nutrition, sustainable production practices, and environmental conservation. Ultimately, food biotechnology offers tools to address current challenges related to sustainable and safe food production while ensuring an adequate supply for a growing population. Figure 7.1 illustrates the applications of biotechnology in achieving environmentally friendly, economically viable, socially responsible food production practices.

The Food and Agriculture Organization of the United Nations (FAO) is very receptive to the use of biotechnology in agriculture, livestock, forestry, fisheries, aquaculture, and agro-industry to combat hunger and poverty. It recognizes that it can play a decisive role in meeting the needs of an increasingly urbanized and growing population, thus contributing to the sustainability of the planet.

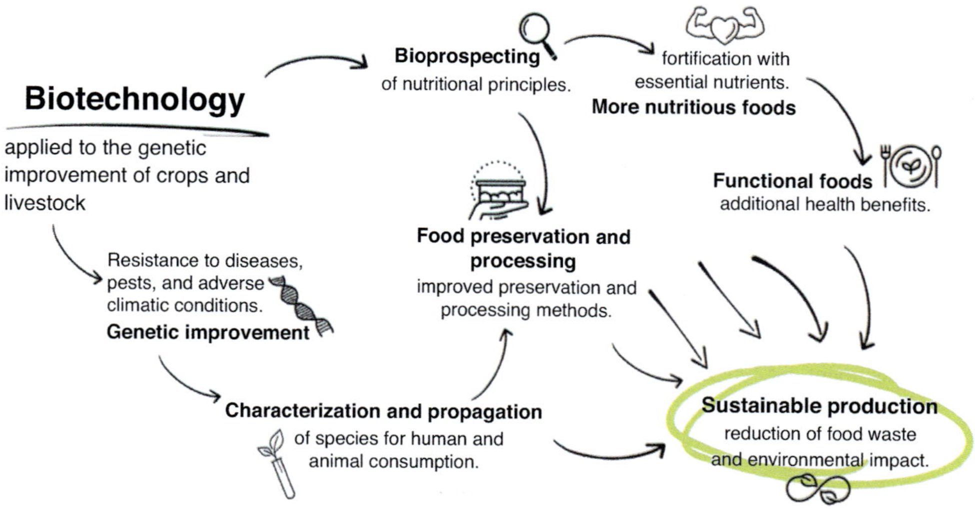

Fig. 7.1. Biotechnology applied to the sustainable production of safe and healthy food.

7.2.1 Biotechnology to strengthen programs for the improvement of species of socio-economic interest (BIOALI network)

Facilitating collaboration and knowledge sharing in Ibero-America

In the context of today's knowledge society, research networks play a key role in collaboration and knowledge sharing. To operate effectively in this globalized world, it is necessary to seek, forge, and strengthen strategic alliances and partnerships with other entities. These include academia, research centers, health and other sector institutions, nongovernmental organizations, bilateral and multilateral cooperation agencies, and international financial institutions, among others. The resulting partnerships create opportunities for the advancement of human resources, while the collection and exchange of knowledge and information contribute to improving food production. Research networks are transforming modes of knowledge and technology production by revaluing collaboration, enabling transdisciplinarity, contributing to the internationalization of scientific communities and allowing broader R&D objectives to be addressed.

BIOALI is a network of scientific and technological cooperation formed by institutions and experts in the field of food biotechnology from Ibero-American countries. Its main objective is to promote collaboration, knowledge exchange, and capacity building in food biotechnology among member countries, with the aim of promoting sustainable development and food security in the region. This network operates under the auspices of the Ibero-American Program for Science and Technology for Development (CYTED; www.cyted.org/BIOALI). The BIOALI Network carries out various activities such as collaborative research projects, academic training, organization of international seminars and conferences, student exchange programs, and dissemination of knowledge through scientific publications. The work carried out by this network contributes to scientific-technological advancement in food biotechnology as well as knowledge transfer among member countries. Furthermore, it promotes linkages between academia and industry to drive innovation and competitiveness in the agri-food sector.

Experiences have been developed and shared to increase the productivity of fruits including *Musa* spp., *Theobroma cacao*, and *Prunus* sp. Improvement programs in quantitative and qualitative terms have used new biotechnological tools (especially - omics) to meet the challenge of population growth and food demand in the current climate change scenario. This improvement will be carried out through the exploration of genomes, the fortification of products, the development of methods for early diagnosis of main pathogens by means of phenotyping, physiological and metabolomics techniques, and the obtaining of indicators of resistance to the main biotic and abiotic stresses derived from climate change (see Fig. 7.1). Working on the improvement of species of food interest in parallel and even considering the establishment of combined agricultural systems can produce synergy in the achievement of results and increase applicability in Ibero-American countries. The participation of producer companies and associations in the proposal ensures the transfer of results as well as the development of projects that meet their real needs. This cooperation network allows the countries and companies involved to optimize their resources and access complex techniques and infrastructures, ensuring that developments in the scientific and technological sector reach society.

Members of the network are not only working on crop improvement, but also in different areas (Fig. 7.2). For example, the BIOALI Group at the University of Antioquia, Colombia, is an interdisciplinary group which conducts applied research in packaging development, by-product utilization, food safety, effect of processing on food quality, product development, and prebiotic bacteria.

In conclusion, the BIOALI Network plays an important role in strengthening research capabilities and increasing synergies among Ibero-American countries in food biotechnology with a view to improving sustainable production of safe and nutritious foods.

7.2.2 Biotechnological approaches to produce virus and viroid-free apricot plants

Fruit Tree Biotechnology Group, Plant Breeding Department, CEBAS-CSIC, Spain

Cultivated species of *Prunus* are affected by viruses and viroids associated with vegetative propagation techniques. Rubio *et al.* (2017) indicated that 44 viruses and three viroids have been described in the main cultivated *Prunus* species

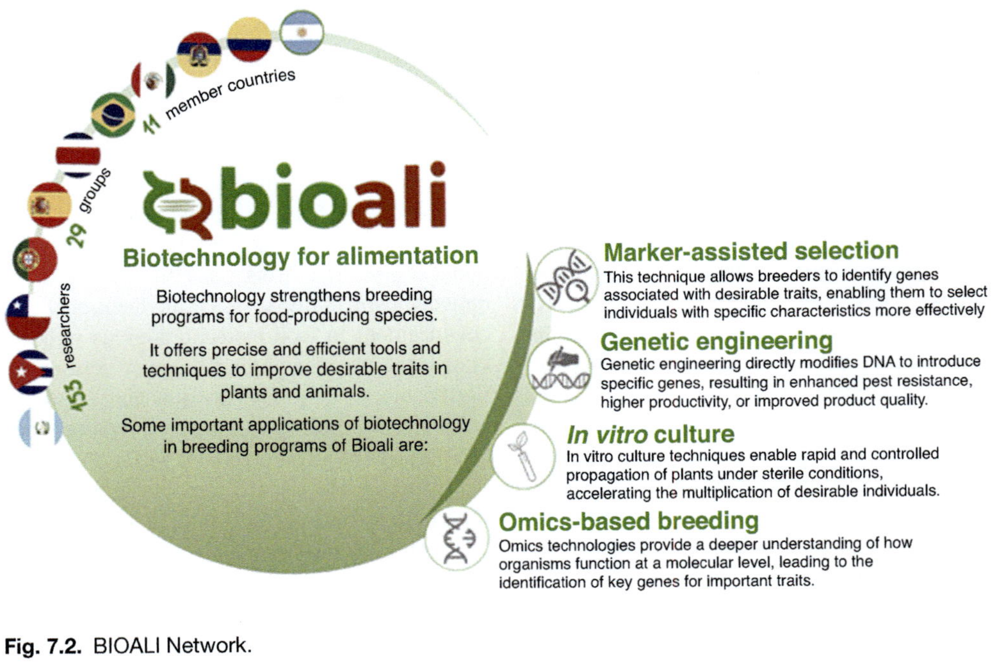

Fig. 7.2. BIOALI Network.

such as peach, plum, apricot, almond, sweet and sour cherry. *Prunus* yields and fruit quality can be negatively affected by these pathogens. Unfortunately, the health status of traditional apricot orchards endangers the future of apricot production. Orchards in Spanish productive areas have high infection rates with apple chlorotic leaf spot virus (García-Ibarra *et al.*, 2010) and plum pox virus (PPV) (Egea *et al.*, 1999), which causes sharka disease. Other viruses found with lower frequency include *Prunus* necrotic ring spot virus and apple mosaic virus (Rodriguez Bonilla and Cieniewicz, 2022). Finally, higher levels of the hop stunt viroid (HSVd) were also detected in many of the analyzed samples (Pallás *et al.*, 2003; Rubio *et al.*, 2013) and in most apricot cultivars in Spain (Amari *et al.*, 2007).

All these results indicate the need for efficient methodologies to produce virus- and viroid-free *Prunus* trees for new plantations after removing infected trees from traditional plantings. Production of virus-free plants is necessary to control viral diseases, import novel cultivars from other countries, exchange breeding materials between countries or regions, and preserve plant germplasm. To fulfill this aim, it would be paramount to enforce a certification program which ensures the production of clean material in nurseries as is compulsory in citrus (BOE-A-1995–14422) or grapes (BOE-A-2003–3835) in Spain. Virus-free peaches and almonds as well as Japanese plums have been successfully produced by micrografting (Juarez *et al.*, 1988), although the efficiency was very low or zero for viroid-free plants (Conejero *et al.*, 2013). On the other hand, micrografting has been very difficult and presented an extremely low efficiency in apricots (personal communication, J. Juarez and A. Conejero). *In vitro* techniques represent the most successful approaches for virus eradication.

Biotechnological approaches have been used in different species to produce virus- and viroid-free plants. *In vitro* culture combined with chemotherapy, thermotherapy or cryotherapy has produced clean plants in different species (Barba *et al.*, 2017; Wang *et al.*, 2018). Thermotherapy consists of exposing plants to high temperatures with the objective of slowing down or inhibiting replication of pathogens and infection of meristematic cells. Chemotherapy is based on antiviral compounds, such as ribavirin, which are added to culture media to inhibit plant virus replication (Szabó *et al.*, 2024). Cryopreservation techniques can be adapted as a reliable methodology to eliminate viruses. Meristematic cells, with a lower probability of being infected, also survive better in ultra-low temperatures than differentiated cells with a less dense cytoplasm, higher vacuoles, and water content (Brison *et al.*, 1997; Wang *et al.*, 2006).

Plants have been sprayed with silver nanoparticles (AgNPs) as a preventive treatment to reduce the accumulation of viral particles and symptoms of bean yellow mosaic virus (Elbeshehy *et al.*, 2015) and tomato mosaic virus (El-Dougdoug *et al.*, 2018). In different studies AgNPs were sprayed on leaves after virus inoculation and, in these cases, they reduced or completely eliminated symptoms of tomato spotted wilt virus in potatoes (Shafie *et al.*, 2018), banana bunchy top virus (Mahfouze *et al.*, 2020) or sunhemp rosette virus in beans (Jain and Kothari, 2014).

It is of vital importance to produce virus-free plants in order to effectively manage viral diseases, introduce new germplasm, share breeding resources across countries or regions, and preserve plant genetic diversity (Wang *et al.*, 2018). *In vitro* techniques represent the most successful approaches for virus eradication. Meristem culture or rescue was the methodology traditionally used to produce virus-free plants since meristems are plant tissues where viruses are not found or, if they are, it is at very low concentrations (Hosokawa, 2008; Kumar and Reddy, 2011). However, some viruses and viroids are able to invade meristematic regions quickly and therefore decrease or eliminate the efficiency of the meristem culture methodology (Polák *et al.*, 2008). Shoot apical meristems are usually free of pathogen infection but some pathogens (such as PPV and HSVd) are able to infect meristematic tissues (Bradamante *et al.*, 2021) and nontreated meristems are often infected. This phenomenon could be attributed to pathogens spreading cell to cell at a higher rate than meristematic cells divide (Gosalvez-Bernal *et al.*, 2006). Therefore, limiting virus spread would be helpful to obtain virus-free shoot apical meristems.

For instance, thermotherapy followed by meristem culture produced PPV elimination with 70.9% efficiency in 'Redtop' and 62.8% in

'Dixired' peach cultivars (Zarghami and Ahmadi, 2023) and 82.4% in nectarine (Manganaris *et al.*, 2003). Exposing apricot plants to 37°C for 20 days and isolating 5 mm-length meristems achieved an 82.4% efficiency (Koubouris *et al.*, 2007).

We have found that rescuing meristems from *in vitro* shoots cultured at 28°C for 1 day and 30°C the following 2 days (acclimatization) followed by a temperature treatment consisting in alternating between 38°C and 32°C every 4 hours for 30, 35, 40, or 45 days produced similar efficiencies. From 18 to 75% of surviving 'Canino' meristems developed in virus-free plants. Additionally, from 12 to 22% of 'Mirlo Rojo' surviving meristems developed in viroid-free plants (unpublished results).

Extended exposure to high temperatures causes damage to the plants, such as growth inhibition, wilting, and necrosis, resulting in fewer buds. Thermotherapy also requires specific equipment and energy consumption to generate and maintain the necessary temperatures.

Cryotherapy protocols often achieve high efficiency rates for virus elimination but, in contrast, during the freezing and thawing processes plant cells are damaged, which leads to low survival rates. Nevertheless, 50% of PPV-free plum (Brison *et al.*, 1997) and 100% of apricot (Seker *et al.*, 2015) plants were produced by conditioning them with PVS2 and a two-step (first –40°C and then liquid nitrogen) freezing protocol. Also, 100% PPV-free plums were obtained by applying both PVS2 and PVS3 (solutions used in vitrification-based shoot tip cryopreservation procedures) for 20 and 60 min respectively before freezing them in liquid nitrogen (Jevremović *et al.*, 2023). Unfortunately, cryotherapy requires skilled personnel and protocols are species or even cultivar specific (Wang *et al.*, 2018).

In contrast, the application of chemotherapy is straightforward. Chemotherapy has been used to produce PPV-free plums (Vescan *et al.*, 2011). Treatments with 50 mg l^{-1} ribavirin for 4 weeks followed by rescue of 5–10 mm-length meristem produced 'Delia' and 'Tita' healthy plums with 12.9 and 33.3% efficiency, respectively. Additionally, the application of 40 and 60 mg l^{-1} of ribavirin to the culture medium for 6 weeks, even without meristem rescue, was able to produce PPV-free plum shoots with 15.4 and 16.7% efficiency rates (Paunovic *et al.*, 2007). Application of chemical treatments, as well as AgNPs, is straightforward but expensive. Moreover, extended exposure to antivirals affects shoot viability and can cause growth inhibition or necrosis, even after removing the antiviral from the medium (Paunovic *et al.*, 2007; Vescan *et al.*, 2011).

Therefore, it is important to explore different biotechnological approaches to produce virus- and viroid-free plants in order to have a toolbox available so that the most appropriate technique is used for each pathogen and cultivar or species.

7.2.3 Exploring the nutritional potential of carob trees and chañar fruits in Argentine Patagonia

CIT-RIO NEGRO, Sede Viedma, Universidad Nacional de Río Negro (UNRN-CONICET)

The United Nations Development Programme (UNDP) indicates that one of its focuses is on recovering and promoting indigenous knowledge about the types and varieties of food crops. Valuing native biodiversity products such as little-known fruits and flours maintains biodiversity and is part of food sovereignty.

Carob trees (*Neltuma* spp.) represent a significant resource for the production of food and other human consumption products, many of which are linked to traditional knowledge. Currently, the nutritional composition of flours from various carob tree fruits is being investigated to create new foods. These flours are a nutritionally attractive option compared to conventional flours, as they improve the protein profile and bioactive components of foods. Despite the benefits of these flours, the Argentine Food Code (CAA) only permits the use of five *Neltuma* species present in Argentina, possibly due to the lack of complete nutritional characterization.

The alpataco (*Neltuma alpataco* (Phill.) C.E. Hughes and G.P. Lewis) and flexuosa (*Neltuma flexuosa* (D.C.) C.E. Hughes and G.P. Lewis) are native species of the arid zone of northern Patagonia, Argentina, and both are considered by the CAA. The whole flour of *N. alpataco* has a nutritional value of 317.15 kcal/100 g, as well as a high percentage of proteins and carbohydrates (10 and 62% of dry weight, respectively), low lipid content (3.23%), and an unsaturated/saturated fat ratio of 4:1, making it a potential food resource (Boeri *et al.*, 2017). In *N. alpataco* and *N. flexuosa*, a polyphenol content of between 0.45g and 0.50g of gallic acid equivalents/100g of flour has been identified. Additionally, compounds such as (−)-epigallocatechin gallate, trans-picein, and V-viniferin were identified for the first time in the genus (Hoffmann *et al.*, 2024). An *in vivo* study revealed that extracts from *N. alpataco* and *N. flexuosa* pods had an antioxidant effect in zebrafish larvae. Furthermore, *N. alpataco* extract was able to inhibit enzymes associated with metabolic syndrome: α-amylase, α-glucosidase, and lipase. The results demonstrate the potential of this nonconventional flour as a promising ingredient for functional foods.

On the other hand, the caldén (*Neltuma caldenia* (Burkart) C.E. Hughes and G.P. Lewis), an endemic carob tree of Argentina whose fruits are traditionally consumed and recognized for their medicinal value, has not yet been incorporated into the CAA. Dalzotto *et al.* (2022) identified that the whole flour from the fruits of this species is mainly composed of carbohydrates, followed by proteins, ash, and lipids, with average values of 84, 11%, 3%, and 2%, respectively, providing a nutritional value of 4 kcal/g. This indicates that the flour from *N. caldenia* fruits maintains the healthy nutritional characteristics of the algarrobo flours introduced in the CAA, demonstrating the industrial potential of caldén.

Chañar (*Geoffroea decorticans* (Gill. ex Hook. et Arm.) Burk) is another native plant distributed in arid and semiarid regions of South America, whose fruits are a food resource for rural and native communities. It has been indicated that whole flours from this species in the northern and southern regions of Argentina have a total polyphenol content of between 87 and 124 mg GAE/100g of dry flour, demonstrating *in vitro* antioxidant activity (Maschio *et al.*, 2021). However, the composition and activity of these compounds may vary depending on the geographical location due to the different environmental pressures or stimuli in each region. Figure 7.3 show plants and integral fruit flours from carob tree and chañar.

In conclusion, the genetic biodiversity of regional resources offers the possibility of obtaining new foods of high nutritional value for human consumption, which promote diverse and local diets. In crisis contexts, such as pandemics or war conflicts, eating locally and biodiverse is an option to ensure food security. In order to safely use these foods derived from biodiversity, biotechnologies can be applied to characterize them, determine their nutritional properties (bioprospection), and propagate and conserve them.

7.2.4 *In vitro* propagation of pastures and trees to support agroforestry systems

Centro de Estudios para el Desarrollo Agroecológico Tropical (CEDAT), Instituto de Estudios Científicos y Tecnológicos (IDECYT), Universidad Nacional Experimental Simón Rodríguez (UNESR), Caracas, Venezuela

Animal production in dual-purpose systems is an important activity for smallholders in most tropical regions of Latin America, and is an alternative for food security within the framework of sustainability (Riesco *et al.*, 1982; Orantes-Zebadúa *et al.*, 2014; Rae Simms *et al.*, 2021). However, traditional systems are often marginal in the economic sense and use land use practices that are not considered sustainable in the long term, in particular those applied in dual-purpose cattle production in hillside areas and tropical forest margins (Toledo and Formoso, 1993). The main limitations to increased productivity in this system are the limited supply and quality of feed, the low production potential of the animals, and the management systems used.

Several research studies have identified forage legumes with great potential to increase animal productivity per unit area, which would allow alternative land use with livestock in the most fragile areas (CIAT, 1992; Argel, 2006). These forage crops can be integrated with annual or tree crops (Veiga and Serrão, 1990) as components of sustainable land use systems.

Fig. 7.3. Plant and integral fruit flour of (A) *Neltuma alpataco*, (B) *Neltuma flexuosa*, (C) *Neltuma caldenia*, and (D) *Geoffroea decorticans*.

The economic and ecophysiological potential of the species that make up the floristic inventory of neotropical savannahs must be developed, due to the relatively low natural fertility of the soils associated with this type of ecosystem. It is necessary to increase the development of plant species adapted to these edaphoclimatic conditions, and the study of forage legumes requires the efficient generation of new technologies that can be implemented in sustainable development programs, respecting the criteria for the maintenance of biodiversity, seeking to generate partial responses to the problems of desertification, environmental degradation and climatic changes in this type of ecosystem (Rubio and Molina, 2016; Caicedo-Aldaz *et al.*, 2020; Guerrero-Rodríguez *et al.*, 2021).

The major constraints to widespread adoption are the lack of information on their use and integration into existing feeding systems, and a compelling demonstration of their economic viability. Moreover, because smallholders are often involved in mixed livestock-crop systems, decision making about resource use is a complex process. Legume-based pastures contribute to more sustainable land use through N fixation (Ara *et al.*, 1990), phosphorus (P) recycling

(Oberson *et al.*, 1999), and increased biological activity in the soil (Decaëns *et al.*, 1994); however, their usefulness at the farm level is still limited (Rubio and Molina, 2016).

It can be said, then, that the key to increasing productivity is developing production strategies that allow combining improved forages with those existing on farms to optimize their use and overcome nutritional deficiencies, and a strategy that takes into account animal genotype options, land use, and the farmer's capacity to implement the new technology. To achieve this, it is necessary to integrate forage utilization alternatives, development of system models, diagnosis and characterization, and applied research at the farm level with participatory methods (Alatorre-Hernández *et al.*, 2018: Ruiz *et al.*, 2012). In Latin America there is a wide range of plants with productive potential that have not been studied, and that have an immense potential for use. These could be the starting point for various investigations to find new natural components to generate various benefits (Nájera-Garduño *et al.*, 2016).

Despite the importance of legumes, their adoption by producers, especially in the case of forage species, has been scarce. Because their regeneration rate in natural environments is extremely low or their wide distribution is mostly due to seasonal reasons, there is a need to implement methodologies for rapid '*in vitro*' propagation and genetic improvement, and to develop techniques to take advantage of these resources and guarantee high productivity without causing destruction of stable and diverse ecosystems (Ashmore and Engelman, 1997; Altman and Loberant, 1998; Pérez Ponce, 1998; Marinucci *et al.*, 2004; Pérez *et al.*, 2006; Trujillo, 2008; Maiquetía Mendoza *et al.*, 2020). Their use in silvopastoral systems, whether arboreal or annual and perennial grasses, constitutes an economic alternative by reducing animal feed costs, in the case of herbaceous legumes, and as a timber source, as in the case of alcornoque and cují.

The development of highly efficient technologies from the biological, quantitative, and economic points of view, which facilitate a greater degree of automation in the processes of storage and multiplication of valuable germplasm, will in turn reduce costs in the production of *in vitro* plants of these species.

For micropropagation through organogenesis, various explants have been tested – buds, leaf sections, meristems, etc. – and different culture media have been used (Murashige and Skoog, 1962; Schenk and Hildebrandt, 1972). Through this technique, successful results have been obtained in canafistola (*Cassia moschata*) by using TDZ (thidiazuron) as a source of cytokinins, since less oxidation occurs, which is a typical problem in the cultivation of woody legumes, generated due to the production of phenolic compounds and browning of the explants and the medium (Trujillo *et al.*, 2002). The process of '*in vitro*' propagation through indirect organogenesis has been developed for species such as *Centrosema macrocarpum*, *Centrosema venosum*, and *Stylosanthes humilis* (Pérez *et al.*, 2006; Trujillo, 2008; Maiquetía Mendoza *et al.*, 2020).

In the case of *Bowdichia virgiloides* (called the cork oak in Venezuela), the cultivated explants had a positive response to all media after 21 days of culture, finding that the medium with a benzyl amino purine (BA) concentration of 0.5 mg/l was the most indicated for induction of cork oak micropropagation. For the multiplication stage, the best results were obtained with BA at a concentration of 1 mg/l with an approximate number of three shoots per explant (Fig. 7.4). Subsequently, the shoots stopped growing, so it was necessary to decrease the concentration of BA to 0.1 mg/l, which resulted

Fig. 7.4. *In vitro* multiplication stage in cork oak (*Bowchidia virgilioides*).

in elongation of the shoots. In the rooting phase, the concentration of cytokinins (BA) was eliminated and the Murashige & Skoog (MS) medium used in the previous phases was supplemented with naphthalene acetic acid (NAA) (0.01 mg/l) (Trujillo, 2008; Trujillo *et al.*, 2016).

The importance of establishing an *in vitro* micropropagation process lies in the fact that it allows the conservation of forage species, as well as multiplying, renewing, preserving, and conserving genetic materials, thus helping to maintain the biodiversity of savannahs.

7.3 Conclusion

The need to promote bioeconomy and biotechnology to increase sustainable production and achieve food sovereignty is justified today. The use of modern biotechnological tools, such as genetic modification and editing, should be encouraged to produce more and better food. Biotechnologies play a crucial role in addressing current challenges in food production, from improving nutrition and product properties to promoting sustainable and efficient agricultural practices. However, the use of these technologies must also ensure equitable access for all stakeholders involved in food production.

Collaboration between research institutions within the BIOALI network is an example of development cooperation in biotechnology in Ibero-America. Overall, biotechnology within the framework of the new paradigm of bioeconomy holds part of the key to enabling a transformation of our economy and our food systems. While in the past we have mainly focused on providing farmers with sustainable alternatives to chemical pesticides and reducing food waste, for example by extending shelf-life, current innovative biotechnological solutions offer society as a whole the possibility to shift towards a more sustainable diet by valuing local foods and improving resistance and tolerance to biotic and abiotic stressors.

In conclusion, biotechnology holds great promise for driving sustainable agricultural practices that contribute to global food security while minimizing environmental impacts. However, it is essential that these technologies are used in an inclusive manner that ensures equitable access for all actors involved in food production.

References

Alatorre-Hernández, A., Guerrero-Rodríguez, J. de D., Olvera-Hernández, J.I., Aceves-Ruíz, E. and Vaquera-Huerta, H. (2018) Leguminosas forrajeras herbáceas de enredadera, recursos poco valorados en el trópico seco. *Revista Fitotecnia Mexicana* 41(4), 403–411.

Altman, A. and Loberant, B. (1998) Micropropagation: Clonal plant propagation in vitro. In: Altman, A. (ed.) *Agricultural Biotechnology*. Marcel Dekker, New York, pp. 19–42.

Amari, K., Ruiz, D., Gómez, G., Sánchez-Pina, M.A., Pallás, V. *et al.* (2007) An important new apricot disease in Spain is associated with hop stunt viroid infection. *European Journal of Plant Pathology* 118(2), 173–181.

Ara, M., Sanchez, P.A. and Y Vela, J.W. (1990) Contribución nitrogenada de *Desmodium ovalifolium* CIAT 350 asociada con *Bracharia decumbens*. In: *Red Internacional de Evaluación de Pastos Tropicales. RIEPT-Amazonía. Documento de Trabajo No. 75*. INIAA, Lima, pp. 737–744.

Argel, P.J. (2006) Contribución de los forrajes mejorados a la productividad ganadera en sistemas de doble propósito. *Archivos Latinoamericanos de Producción Animal* 14(2), 65–72.

Ashmore, S.E. and Engelman, E. (1997) *Status Report on the Development and Application of In Vitro Techniques for the Conservation and Use of Plant Genetics Resources*. IPGRI, Roma, p. 67.

Barba, M., Hosakawa, M., Wang, Q.C., Taglienti, A. and Zhang, Z. (2017) Viroid elimination by thermotherapy, cold therapy, tissue culture, in vitro micrografting, or cryotherapy. In: Hadidi, A., Flores, R., Randles, J. and Palukaitis, P. (eds) *Viroids and Satellites*. Academic Press, New York, pp. 425–435.

Boeri, P., Piñuel, L., Sharry, S. and Barrio, D. (2017) Caracterización nutricional de la harina integral de algarroba (Prosopis alpataco) de la norpatagonia Argentina. *Revista de La Facultad de Agronomía de La Universidad Nacional de La Plata* 116(1), 129–140.

Bradamante, G., Mittelsten Scheid, O. and Incarbone, M. (2021) Under siege: Virus control in plant meristems and progeny. *Plant Cell* 33(8), 2523–2537.

Brison, M., Boucaud, M.-T., Pierronnet, A. and Dosba, F. (1997) Effect of cryopreservation on the sanitary state of a cv prunus rootstock experimentally contaminated with plum pox potyvirus. *Plant Science* 123(1–2), 189–196.

Caicedo-Aldaz, J.C., Puyol-Cortez, J.L., López, M.C. and Ibáñez-Jacome, S.S. (2020) Adaptabilidad en el sistema de producción agrícola: Una mirada desde los productos alternativos sostenibles. *Revista de Ciencias Sociales* 26(4), 308–327.

CIAT (Centro Internacional de Agricultura Tropical) (1992) *Programa de Pastos Tropicales. Informe Anual 1992*. CIAT, Cali-Palmira.

Conejero, A., Romero, C., Cunill, M., Mestre, M.A., Martínez-Calvo, J. *et al*. (2013) *In vitro* shoot-tip grafting for safe prunus budwood exchange. *Scientia Horticulturae* 150, 365–370.

Dalzotto, D., Piñuel, L., Sharry, S. and Boeri, P. (2022) Alimentos nativos: Caracterización nutricional de harina de caldén (*Prosopis caldenia* Burkart) de dos provincias Argentinas. In: *Congreso Internacional de Ciencia y Tecnología de Alimentos*. Available at: http://rid.unrn.edu.ar:8080/bitstream/20.500.12049/9659/4/2022%20Calden%20CICYTAC%20.pdf

Decaëns, T., Lavelle, P., Jiménez Jaén, J.J., Escobar, G. and Rippstein, G. (1994) Impact of land management on soil macrofauna in the oriental llanos of Colombia. *European Journal of Soil Biology* 30, 157–168.

Egea, J., Burgos, L., Martínez-Gómez, P. and Dicenta, F. (1999) Apricot breeding for sharka resistance at CEBAS-CSIC, Murcia (Spain). *Acta Horticulturae* 488(488), 153–158.

Elbeshehy, E.K.F., Elazzazy, A.M. and Aggelis, G. (2015) Silver nanoparticles synthesis mediated by new isolates of bacillus spp., nanoparticle characterization and their activity against bean yellow mosaic virus and human pathogens. *Frontiers in Microbiology* 6, 1–13.

El-Dougdoug, N.K., Bondok, A.M. and El-Dougdoug, K.A. (2018) Evaluation of silver nanoparticles as antiviral agent against Tomv and PVY in tomato plants. *Middle East Journal of Applied Sciences* 8, 100–111.

García-Ibarra, A., Martínez-Gómez, P., Rubio, M., Dicenta, F., Soler, A. *et al*. (2010) First report of apricot latent virus and plum bark necrosis stem pitting-associated virus in apricot from Spain. *Plant Disease* 94(2), 275.

Gosalvez-Bernal, B., Garcia-Castillo, S., Pallas, V. and Sanchez-Pina, M.A. (2006) Distribution of carnation viruses in the shoot tip: Exclusion from the shoot apical meristem. *Physiological and Molecular Plant Pathology* 69(1–3), 43–51.

Guerrero-Rodríguez, J. de D., Bustamante-González, A., Aceves-Ruiz, E., Vargas-López, S., Calderón-Sánchez, F. *et al*. (2021) Abundancia, producción de materia seca y valor nutricional de leguminosas arbustivas forrajeras del trópico seco. *Ecosistemas y Recursos Agropecuarios* 8, II.

Hoffmann, E., Boeri, P., Monasterio, R., Fontana, A., Puppo, M.C. *et al*. (2024) Bioactivity of *Prosopis alpataco* and *Prosopis flexuosa* flours: Healthy alternatives as ingredients for nutritional foods. *Food Bioscience* 58, 103778.

Hosokawa, M. (2008) Leaf primordia-free shoot apical meristem culture: A new method for production of viroid-free plants. *Journal of the Japanese Society for Horticultural Science* 77(4), 341–349.

Izquierdo, J. and Sharry, S. (2019) Cambio climático, conservación de la biodiversidad y protección eficiente de loscultivos: el valor natural de los servicios ecosistémicos. In: Burger, C., Sharry, S., Crippa, J., Ambroso, F.D. and Lima, L. (eds) *Liber Amicorum En Homenaje a La Profesora Teodora Zamudio. Universidad Del Museo Social Argentino, Buenos Aires, Argentina*, Vol. 1. Buenos Aires, Argentina, pp. 155–179.

Jain, D. and Kothari, S.L. (2014) Green synthesis of silver nanoparticles and their application in plant virus inhibiton. *Journal of Mycology Plant Pathology* 44, 1816–1819.

Jevremović, D., Vasilijević, B., Anđelić, T. and Vujović, T. (2023) Effect of D and V cryo-plate methods for plum pox virus eradication from two plum cultivars. *Plant Cell, Tissue and Organ Culture* 152(3), 529–538.

Juarez, J., Arregui, J.M., Camarasa, E., Cambra, M., Llácer, G. *et al*. (1988) Recovery of virus-free peach trees from selected clones by shoot-tip grafting *In Vitro*. In: *Proceedings of the XIV International Symposium on Fruit Tree Virus Diseases*, Vol. 235. pp. 77–84.

Koubouris, G.C., Maliogka, V.I., Efthimiou, K., Katis, N.I. and Vasilakakis, M.D. (2007) Elimination of *Plum pox* virus through *in vitro* thermotherapy and shoot tip culture compared to conventional heat treatment in apricot cultivar Bebecou. *Journal of General Plant Pathology* 73(5), 370–373.

Kumar, N. and Reddy, M.P. (2011) *In vitro* plant propagation: A review. *Journal of Forest and Environmental Science* 27, 61–72.

Mahfouze, H.A., El-Dougdoug, N.K. and Mahfouze, S.A. (2020) Virucidal activity of silver nanoparticles against *Banana bunchy* top virus (BBTV) in banana plants. *Bulletin of the National Research Centre* 44(1), 1–11.

Maiquetía Mendoza, M.Y., Vargas Cedeño, T.E., Toro García, M. and García, E.C. (2020) Estudios en la germinación y propagación in vitro de tres especies de leguminosae: *Calopogonium sp.*, stylosantes capitata y cassia moschata. *Revista Científica Ecociencia* 7(6), 1–25.

Manganaris, G.A., Economou, A.S., Boubourakas, I.N. and Katis, N.I. (2003) Elimination of PPV and PNRSV through thermotherapy and meristem-tip culture in nectarine. *Plant Cell Reports* 22(3), 195–200.

Marinucci, L., Ruscitti, M. and Abedini, W. (2004) Morfogénesis in vitro de leguminosas forestales nativas de la República Argentina. *Revista de La Facultad de Agronomía, La Plata* 105(2).

Maschio, J.I., Boeri, P., Dalzotto, D., Piñuel, L. and Sharry, S. (2021) Caracterización química y nutricional de harina de chañar de diferente distribución geográfica. *Bionatura* 6(1), 1563–1568.

Murashige, T. and Skoog, F. (1962) A revised medium for rapid growth and bio assays with tobacco tissue cultures. *Physiologia Plantarum* 15(3), 473–497.

Nájera-Garduño, A., Piedra-Matias, R., Albarrán-Portillo, B. and García-Martínez, A. (2016) Cambios en la ganadería doble propósito en el trópico seco del Estado de México. *Agrociencia* 50(6), 701–710.

Oberson, A., Friesen, D.K., Tiessen, H., Morel, C. and Stahel, W. (1999) Phosphorus status and cycling in native savanna and improved on an acid low-P Colombian oxisol. *Nutrient Cycling in Agroecosystems* 55(1), 77–88.

Orantes-Zebadúa, M.A., Platas-Rosado, D., Córdova-Avalos, V., Santos-Lara, M.C. and Córdova-Avalos, A. (2014) Caracterización de la ganadería de doble propósito en una región de Chiapas, México. *Ecosistemas y Recursos Agropecuarios* 1(1), 49–58.

Pallás, V., Gomez, G., Amari, K., Cañizares, M.C., Candresse, T. *et al.* (2003) Hop stunt viroid in apricot and almond. In: *Viroids. CSIRO, Collinwood*. pp. 168–170.

Paunovic, S., Ruzic, D., Vujovic, T., Milenkovic, S. and Jevremovic, D. (2007) *In vitro* production of *plum pox virus* – free plums by chemotherapy with ribavirin. *Biotechnology & Biotechnological Equipment* 21(4), 417–421.

Pérez, A., Trujillo, I., Vidal, M.C. and Lima, N. (2006) Propagación *in vitro* de *Stylosanthes Capitata Vogel*: Una especie de gran potencial forrajero. *Acta Botánica Venezuelica* 29(2), 335–346.

Pérez Ponce, J.N. (1998) Propagación y mejora genética de plantas por biotecnología. In: *Instituto de Biotecnología de Las Plantas*. Santa Clara, p. 390.

Polák, J., Jokeš, M., Ducháčová, M., Hauptmanová, A. and Komínek, P. (2008) Electron microscopy of structures present in embryonic cells of plants infected with *Plum* pox virus. *Plant Protection Science* 44(3), 81–84.

Rae Simms, S., Barragán Fuentes, M.E. and Cinco Martínez, S. (2021) Modelo de innovación en ganadería sostenible para la península de Yucatan. In: *The Nature Conservancy*, Vol. 74. Universidad Autónoma de Yucatán, Mexico.

Riesco, A., Meini, G., Torre, M., Huamán, H., Reyes, C. *et al.* (1982) *Análisis exploratorio de los sistemas de fundos de pequeños productores en la Amazonía, Región de Pucallpa*. IVITA-CIID, Lima, Perú, p. 47.

Rodriguez Bonilla, F. and Cieniewicz, E. (2022) Distribution and diversity of prunus necrotic ringspot virus, prune dwarf virus, and peach latent mosaic viroid in wild Prunus spp in South Carolina and Georgia. *PhytoFrontiers* 2(4), 363–370.

Rubio, L.A. and Molina, E. (2016) Las leguminosas en alimentación animal. *Arbor* 192(779), a315.

Rubio, M., Garcia-Ibarra, A., Soler, A., Martínez-Gómez, P. and Dicenta, F. (2013) Relationship of viruses and viroids with apricot "viruela" disease. *Phytopathologia Mediterranea* 52, 148–153.

Rubio, M., Martínez-Gómez, P., Marais, A., Sánchez-Navarro, J.A., Pallás, V. *et al.* (2017) Recent advances and prospects in prunus virology. *Annals of Applied Biology* 171, 125–138.

Ruiz, S.C., González, U.J. and Juan, U.J. (2012) Estudio Propuesta de desarrollo del cultivo de leguminosas para la pequeña agricultura. In: *Instituto de Investigaciones Agropecuarias Centro Regional de Investigaciones Quilamapu*. Ministerio de Agricultura, Chile, p. 89.

Schenk, R.U. and Hildebrandt, A.C. (1972) Medium and techniques for induction and growth of monocotyledonous and dicotyledonous plant cell cultures. *Canadian Journal of Botany* 50(1), 199–204.

Seker, M.G., Süzerer, V., Elibuyuk, I.O. and Çiftçi, Y.Ö. (2015) In vitro elimination of PPV from infected apricot shoot tips via chemotherapy and cryotherapy. *International Journal of Agriculture and Biology* 17(5), 1066–1070.

Shafie, R.M., Salama, A.M. and Farroh, K.Y. (2018) Silver nanoparticles activity against tomato spotted wilt virus. *Middle East Journal of Agriculture Research* 7, 1251–1267.

Szabó, L.K., Desiderio, F., Kirilla, Z., Hegedűs, A., Várallyay, É. *et al.* (2024) A mini-review on *in vitro* methods for virus elimination from *Prunus* sp. fruit trees. *Plant Cell, Tissue and Organ Culture* 156(2).

Toledo, J.M. and Formoso, R. (1993) Sostenibilidad de pastos sembrados en el trópico y subtrópico. In: *Memorias Del XVII Congreso Internacional de Pradera*. Nueva Zelandia-Australia, pp. 1891–1896.

Trujillo, I., Ribón, I., Pérez, A. and Vidal, M. (2002) Propagación *in vitro* de cañafístola (*Cassia moschata*): Una especie de gran potencial para la suplementación de la alimentación de ganado bovino. In: *VIII Congreso Latinoamericano de Botánica*. Cartagena, Colombia.

Trujillo, I. (2008) Biotecnología aplicada al desarrollo sustentable de sabanas. *Acta Biologica Venezulica* 28, 45–56.

Trujillo, I., Pèrez, O., Brucato, M.G. and Sharry, S. (2016) Conservation in vitro of Alcornoque (*Bowchidia virgilioide*), a native leguminous savanna tree. In: *4th International Conference of the IUFRO Working Party*. Buenos Aires, Argentina.

Veiga, J.B. and Serrão, E.A.S. (1990) Sistemas silvapastoris e produção animal nos trópicos umidos: A experiência da amazônia brasileira. In: *Pastagens. Sociedade Brasileira de Zootecnica*. Piracicaba, pp. 37–68.

Vescan, L.A., Pamfil, D. and Zagrai, I. (2011) *In vitro* techniques for plum pox virus elimination from two infected Romanian plum cultivars. *Bulletin UASVM Animal Science and Biotechnologies* 68, 389–395.

Wang, M.-R., Cui, Z.-H., Li, J.-W., Hao, X.-Y., Zhao, L. *et al.* (2018) In vitro thermotherapy-based methods for plant virus eradication. *Plant Methods* 14, 87.

Wang, Q., Liu, Y., Xie, Y. and You, M. (2006) Cryotherapy of potato shoot tips for efficient elimination of *Potato Leafroll Virus* (PLRV) and *Potato Virus Y* (PVY). *Potato Research* 49(2), 119–129.

Zarghami, R. and Ahmadi, B. (2023) Production of plum pox virus-free and prunus necrotic ringspot virus-free regenerants using thermotherapy and meristem-tip culture in prunus persica l. *Erwerbs-Obstbau* 65(4), 719–727.

8 Development of Doubled Haploid Wheat Genotypes Through Intergeneric Hybridization with Maize

Md. Yeakub Khan[1], Sabrina Parvin Mita[1] and Akbar Hossain[2*]
[1]Biotechnology Division, Bangladesh Wheat and Maize Research Institute, Dinajpur, Bangladesh; [2]Soil Science Division, Bangladesh Wheat and Maize Research Institute, Dinajpur, Bangladesh

Abstract

The goal of current agricultural research is to develop crop species in a sustainable way by combining genomes from different genetic backgrounds through widespread hybridization. Breeders dislike it when preferential uniparental chromosome removal makes it difficult to recover stable hybrids in some intergeneric crosses. A crucial step in producing doubled haploid (DH) lines in plants is chromosomal elimination, which is the rapid loss of chromosomes from one parent during the zygote stage, resulting in haploids. Wheat haploid embryos are produced when emasculated wheat (*Triticum aestivum* L.) and maize (*Zea mays* L.) are crossed as pollen donors. This is due to the total loss of maize chromosomes. Pure lines with stable characteristics can be created rapidly via chromosomal doubling and embryo rescue procedures. The method is known as the "Wheat × Maize System." More creative chromosomal elimination technique protocols were created in order to increase the haploid induction efficiency manifolds. The formation of true breeding lines can be sped up by wheat breeders by producing haploid plants from hybrids and then doubling their chromosomes. Numerous uses for DH populations exist in plant breeding, including the production of cultivars and germplasm, the transfer of traits from wild types, the study of some aspects of quantitative genetics, and whole-genome mapping. This chapter examines the several haploid induction methods being used worldwide and rates them according to how well they induce haploid phenotype in wheat.

Keywords: abiotic and biotic stresses, doubled haploid (DH), quantitative genetics, genome mapping

8.1 Introduction

Wheat (*Triticum aestivum*; 2n = 6x = 42, AABBDD) is the most widely grown cereal crop in the world and one of the central pillars of global food security. It is a key staple meal that is consumed throughout the world. We must increase wheat yield with the meager resources at our disposal in order to feed the world's rapidly expanding population. To enhance wheat production sustainably, the most practical and environmentally benign approach is to apply genetic improvement to create varieties with high yield potential, resistance or tolerance to biotic and abiotic challenges, and acceptable end-use quality. Inbreeding along

*Corresponding author: akbarhossainwrc@gmail.com

DOI: 10.1079/9781800627659.0008

with selection, comprehensive testing of breeding materials at several sites to find suitable and stable genotypes with desired agronomic properties, and introduction and generation of genetic variation are all necessary for such crop improvement.

Sexual crosses—usually single, three-way, double, or backcrosses—and mutant breeding are two methods for generating variation. These variations have been fixed and homozygous genotypes developed by breeders using a variety of techniques. Single seed descent (SSD), backcrossing, selfing in the field, plus in some cases the use of off-season nurseries and shuttle breeding approaches—conventional inbreeding methods—allow for the isolation of homozygous and homogeneous genotypes. However, this process involves multiple cycles of inbreeding and selection, making it the most laborious, costly, and time-consuming part of any breeding program. Moreover, the majority of selections in traditional plant breeding contain some heterozygous loci, and fully homozygous lines are uncommon (Snape, 1989; Raina, 1997; Baenziger *et al.*, 2006). It takes several generations to produce homozygous wheat lines using the conventional breeding method. In a single generation, homozygous wheat lines might be produced using doubled haploid technology.

Instead of requiring 6–10 generations of inbreeding through selfing or sibcrossing, haploid induction and doubled haploid (DH) technologies can produce pure lines with genotypic homozygosity of 100% (Prigge *et al.*, 2012). This represents a major advancement in the creation of novel varieties (Dunwell, 2010). With this method, the breeding process will be significantly shortened and genetically homozygous lines can be obtained in just one generation (Tadesse *et al.*, 2012).

In 1922, jimson weed was found to be the first haploid plant to occur naturally (Blakeslee *et al.*, 1922). After that, haploid rice and jimson weed plants were produced, respectively, using anther and microspore culture procedures (Guha and Maheshwari, 1964; Niizeki and Oono, 1968). Breeders progressively started to understand DH technology, leading to the development of oilseed rape and barley cultivars (Ho and Jones, 1980). Doubled haploid technology has been used in many plant species, including *Arabidopsis*, *Avena sativa*, pearl millet, maize, tomato, cucumber, melon, and so forth, in addition to improved haploid induction and chromosome doubling (Kalinowska *et al.*, 2019).

Though they are uncommon in nature, DH plants can be produced by *in vivo* or *in vitro* procedures (Dwivedi *et al.*, 2015). Gynogenesis (bud, ovary, or unfertilized ovule culture) and androgenesis (anther or microspore culture) are examples of *in vitro* cultivation. *In vivo* techniques include intra- or interspecific hybridization, commonly referred to as the chromosomal elimination method, and inducing parthenogenesis by inactive pollen pollination (ionizing radiation, chemical agents, and high temperature) (Zargar *et al.*, 2022). Plant species determine which approaches are more or less effective (Kalinowska *et al.*, 2019).

Due to its genotype dependence, slow regeneration, unstable ploidy, and albino seedlings, *in vitro* anther or microspore culture is limited in wheat DH breeding (Liu *et al.*, 2002; Tawkaz, 2011). Ionizing radiation and chemical pollen treatments can lead to aneuploidy and hardly produce haploidy (Ding *et al.*, 2017). On the other hand, haploid seed production in wheat can be induced by mutations of the pollen-specific phospholipase gene *TaMTL* (Liu *et al.*, 2020). Tang *et al.* (2023) have developed a visual screening method to detect haploid seeds using the wheat Tamtl mutant. The Wheat × Maize System includes doubling of the maternal chromosome and removal of the foreign chromosome in both *in vitro* and *in vivo* settings. The Wheat × Maize System is a popular technique for creating DH wheat because it is 2–3 times more efficient in producing green plantlets than anther culture (Niroula and Bimb, 2009) exhibits little to no genotype dependence in wheat, and does not require haploid identification. In Yunnan, China, where maize and wheat can grow in the same season, the chromosome removal approach is widely applied and various wheat varieties, including Yunmai 110, have been created (Ding *et al.*, 2022).

8.2 Concept of Haploids

At the First International Symposium on Haploids in Higher Plants, which took place in Guelph, Canada (Kasha, 1974), the possibilities of haploidy breeding were first examined.

However, Blakeslee first reported this event in *Datura stramonium* in 1922, which led to the discovery of spontaneous development in haploid plants (Blakeslee *et al.*, 1922). Then, comparable findings in wheat (*Triticum aestivum*), tobacco (*Nicotiana tabacum*), and a few more species were reported (Forster *et al.*, 2007). Although haploids rarely emerge spontaneously, this phenomenon has little practical significance. After haploid embryos were successfully produced *in vitro* from *Datura* anthers in 1964 (Guha and Maheshwari, 1964, 1966), haploid creation in tobacco (Nitsch and Nitsch, 1969) garnered interest once more regarding the possibility of haploidy in plant breeding. Since that time, about 250 plant species from nearly every family in the plant world have had the phenomenon studied, leading to the publication of protocols (Maluszynski *et al.*, 2003).

Sporophytic plants with the gametophytic chromosomal number are known as haploids (Hs), whereas haploid plants that experienced spontaneous or induced chromosome duplication are known as doubled haploids (DHs). Most often, 'doubled haploid' is misinterpreted as 'dihaploid.' A potato (*Solanum tuberosum*, AAAA=48) is an example of an autotetraploid genotype (2n=4x), from which haploids (1n=2x) are really derived to become dihaploids. Breeding for crop development has been made much easier by the ability to achieve homozygosity in a single generation as opposed to multiple generations of selfing.

8.3 Origin and Generation of Doubled Haploids

The genus *Triticum*, which includes all cultivated wheat, was split into three main taxonomic groupings by Schultz in 1913: einkorn, emmer, and dinkel. Sakamura's ground-breaking cytological study from 1918 confirmed that Schultz's three wheat groups were classified according to different chromosome counts: dinkels are hexaploids (2n = 6x = 42), emmers are tetraploids (2n = 4x = 28), and einkorns are diploids (2n = 2x = 14), all of which have the same basic chromosome count of x = 7. The genome formulas for cultivated einkorn (*T. monococcum* L., 2n = 2x = 14), emmer (*T. turgidum* L., 2n = 4x = 28), and dinkel (*T. aestivum* L., 2n = 6x = 42) were soon after identified as AA, AABB, and AABBDD, respectively, based on cytogenetic study by Kihara in 1924.

Different haploids are formed by each group of wheat. Sporophytes with gametic chromosomal number (n) are known as haploids. With genomic constitutions of A, AB, and ABD, respectively, the haploids from einkorn, emmer, and dinkel have n = x = 7, n = 2 x = 14, and n = 3 x = 21 chromosomes (Heyne, 1987; Fehr, 1991; Folling, 2002). Haploids can arise through many methods of induction or spontaneously in the natural world. Though it is rare and hence of limited practical relevance, occurrence of spontaneous development of haploid plants has been recognized since Blakeslee *et al.* (1922) first described it in *Datura stramonium* L. The induction of haploids and chromosomal doubling result in the genotypes known as DHs.

The first instance of chromosome elimination was in tobacco (*Nicotiana tabacum*, *N. sylvestris*) during an interspecific hybridization, which produced haploid *N. tabacum* plants (Clausen and Mann, 1924). Later, an early study of the Wheat × Maize System was conducted by Laurie and Bennett (1986), who created an intergeneric system between wheat and maize in which the removal of maize chromosomes occurred independently of wheat Kr genes (Laurie and Bennett, 1989; Laurie and Snape, 1990). In order to create haploid plants from commercial wheat types, the Wheat × Maize System gained popularity and was frequently employed (Laurie and Reymondie, 1991; Campbell *et al.*, 2000; Jauhar *et al.*, 2009). Currently, one of the most useful techniques for wheat is the Wheat × Maize System (Niroula *et al.*, 2007; Moradi *et al.*, 2009; Li *et al.*, 2020).

The Wheat × Maize programmed haploid production was also tested by Gurtay *et al.* (2021), who also used wheat anther culture and spontaneous doubling in androgenesis—which involves using polyploid wheat varieties that are ancient, local, and contemporary—to acquire DH plants (4 and 6). Hexaploid wheat thus acquired more haploid embryos than tetraploid wheat, but the rate of plant regeneration remained the same. It appears that wheat anther culture and Wheat × Maize hybridization are impacted by the D genome.

Centromeres, particularly those found in the wheat D genome, are clearly crucial in the process of eradicating the maize chromosome and causing wheat to become haploid. To fully understand how the wheat D genome works to remove all the paternal chromosomes, more research is necessary.

8.4 Doubled Haploid Induction Techniques in Wheat

In 1964, haploid embryo formation from *Datura* anther *in vitro* culture was achieved (Guha and Maheshwari, 1964). This led to the successful *in vitro* haploid production of rice (*Oryza sativa* L.) (Niizeki and Oono, 1968) and tobacco (*Nicotiana tabacum* L.) (Nakata and Tanaka, 1968; Nitsch and Nitsch, 1969). These findings demonstrated the potential of haploidy for plant breeding. Following this, haploid bread wheat (*Triticum aestivum* L.) was also generated through anther culture (Ouyang *et al.*, 1973; Picard and De Buyser, 1973), isolated microspore culture (Wei, 1982), and wide hybridization with wild barley (*Hordeum bulbosum* L.) and maize (*Zea mays* L.) (Barclay, 1975; Laurie and Bennett, 1986, 1988; Inagaki and Tahir, 1990). The two most widely used systems are anther culture and wheat × maize cross.

8.4.1 Anther culture

Early in the 1970s, anther culture was used to successfully regenerate bread wheat plants for the first time (Ouyang *et al.*, 1973; Picard and De Buyser, 1973). Anther culture takes advantage of the fact that some *in situ* pollen grains are embryogenic. Only when these pollen grains are placed on artificial media can they grow into embryos. Selection of primary wheat spikes containing anthers with pollen in the mid-late uninucleate stage of development, as shown in Fig. 8.1, is the first step in the anther cultivation process (Ouyang *et al.*, 1973; Picard and De Buyser, 1973; Liu *et al.*, 2002).

Many cereal breeding efforts employ anther culture, which produces DHs more affordably than intergeneric crosses (Snape *et al.*, 1986; Tadesse *et al.*, 2012). In contrast, anther culture depends on genotype (Khiabani *et al.*, 2008;

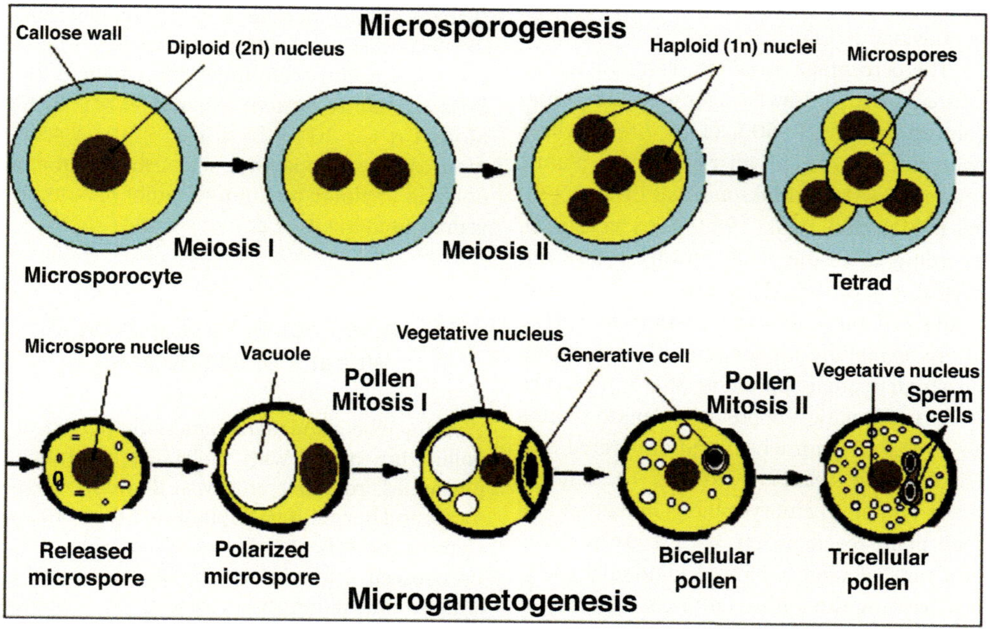

Fig. 8.1. Stages of pollen development. Source: Tadesse *et al.* (2012).

Grauda *et al.*, 2010; Tawkaz, 2011; El-Hennawy *et al.*, 2011; Tadesse *et al.*, 2012). A number of system components have been studied in an effort to counteract the genetic effect, including the donor plants' growth environment, the medium's composition, the medium's physical state (liquid or solid), cold pretreating anthers prior to culturing, raising the incubation temperature during the first few days of culture, and exposing anther explants to γ irradiation (Karsai *et al.*, 1994; Karimzadeh *et al.*, 1995; Xynias *et al.*, 2001; Zamani *et al.*, 2003; Shirdelmoghanloo *et al.*, 2009; Tawkaz, 2011).

The stage of pollen development is critical since small little variations might result in significant output reductions. Following aseptic dissection, calli are transferred to a solid plant regeneration medium and incubated in a culture environment at 25°C and 16 hours of daylight duration for approximately 30 days. Anthers are then cultivated on induction medium and incubated in darkness at 26–28°C for 4–6 weeks. Transferring green plantlets to culture tubes with 20 ml of modified plantlet regeneration medium (PLL) is done. Vigorous seedlings are transplanted into pots with a 2:1:1 soil:sand:peat-moss combination and housed in a plastic container after 1–2 months of hardening (Tawkaz, 2011).

The percentage of albino plants in anther culture is another drawback for wheat breeding. Albinism affects 20–50% of all bread wheat genotypes, or 30% of all regenerated plants, on average (Abd El-Maksoud and Bedő, 1992; Abd El-Maksoud *et al.*, 1993; Tawkaz, 2011). According to Cistué *et al.* (2006, 2009) and Labbani *et al.* (2007), durum wheat (*T. turgidum* L. subsp. *durum*) is very resistant to anther culture, exhibiting a poor rate of regeneration and a high frequency of albino plants. Ayed *et al.* (2010) tried a variety of pretreatments to address this issue in durum wheat and discovered that a 5 week cold treatment (4°C) of the anthers greatly enhanced embryogenesis induction and green plant regeneration, with a green:albino plant ratio of almost 75%. Additionally useful for increasing the anther culture's efficiency in durum wheat is mannitol pretreatment (Labbani *et al.*, 2007; Ayed *et al.*, 2010).

8.4.2 Wheat × Maize System: origin, principle, and benefits

The first instance of chromosome elimination was in tobacco (*N. tabacum*, *N. sylvestris*) during an interspecific hybridization, which produced haploid *N. tabacum* plants (Clausen and Mann, 1924). Similarly, Gaines and Aase (1926) used an intergeneric hybridization (*T. compactum* Host *Aegilops cylindrica* L.) to create haploid wheat. Moreover, Barclay (1975) produced haploid wheat plants by employing the bulbosum technique—that is, *Hordeum bulbosum*—as pollen donors. The dominant crossability inhibitor genes Kr1 and Kr2, which are found on chromosomes 5B and 5A, are sensitive to *H. bulbosum*. Nevertheless, wheat genotypes lacking dominant Kr1 and/or Kr2 cannot use the bulbosum method.

Later, an early study of the Wheat × Maize System was conducted by Laurie and Bennett (1986), who created an intergeneric system between wheat and maize in which the removal of maize chromosomes occurred independently of wheat Kr genes (Laurie and Bennett, 1989; Laurie and Snape, 1990). In order to create haploid plants from commercial wheat types, the Wheat × Maize System gained popularity and was frequently employed (Laurie and Reymondie, 1991; Campbell *et al.*, 2000; Jauhar *et al.*, 2009). Currently, one of the most useful techniques for wheat is the Wheat × Maize System (Niroula *et al.*, 2007; Moradi *et al.*, 2009; Li *et al.*, 2020). A schematic of the comparison of conventional and doubled haploidy techniques (embryo culture) is presented in Fig. 8.2.

8.5 Advances in Research on the Wheat × Maize System

Planting wheat and maize, emasculating wheat, pollinating maize with pollen, treating with hormones, rescuing embryos, doubling treatment, and harvesting DH plants are the primary steps in the Wheat × Maize System (Fig. 8.3). Pseudoseed formation frequency (PFF), embryo formation frequency (EFF), haploid regeneration frequency (HRF), haploid formation frequency (HFF), and haploid doubling

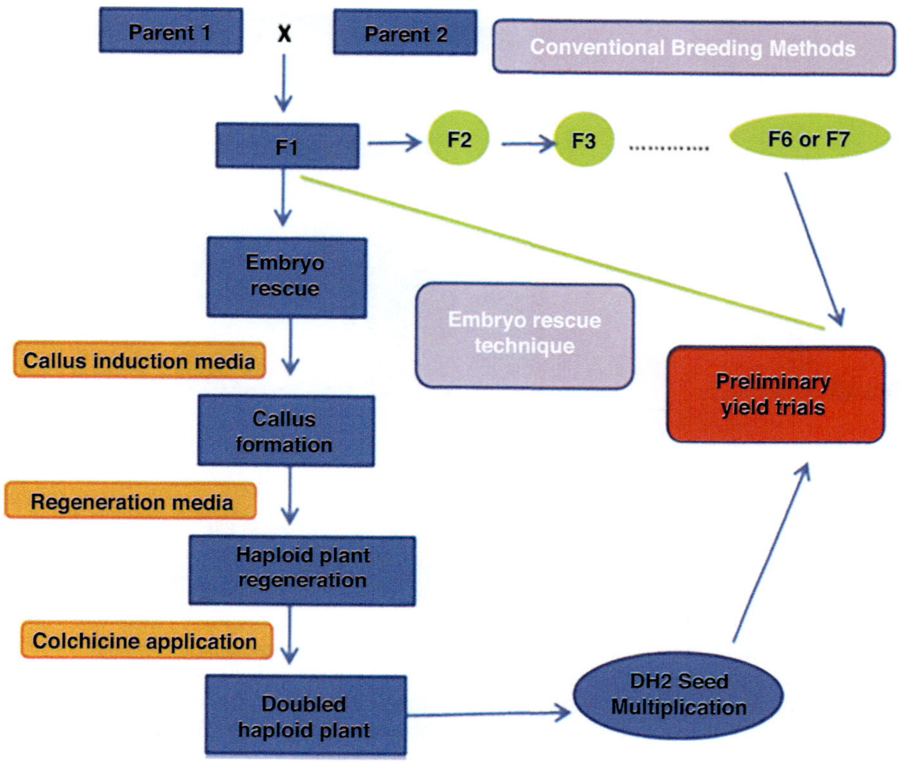

Fig. 8.2. Schematic description of comparison of conventional and doubled haploidy technique (embryo culture). Source: Patial *et al.* (2019).

frequency are among the many parameters that regulate the final wheat DH rate during the process (Verma *et al.*, 1999; Brazauskas *et al.*, 2005; Mayel *et al.*, 2016; Jeberson *et al.*, 2021; Kapoor *et al.*, 2021).

The embryo rate (obtained wheat haploid embryos after crossing and hormone treatment), the plantlet rate (obtained wheat haploid plantlets after haploid embryo rescue), and the doubling rate (obtained wheat DH plantlets after chromosome doubling) ultimately reflect the inheritance of PFF, EFF, and HRF, which is independent (Sharma *et al.*, 2005; Dhiman *et al.*, 2012). The Wheat × Maize System was progressively improved over time, and is currently a potent tool for studying wheat genetics and breeding. This covers uses such as QTL mapping (Tehseen *et al.*, 2022), thermophotosensitive genic male sterile line creation (Li *et al.*, 2020), molecular cytogenetic characterization (Singh *et al.*, 2019), and so forth.

8.6 Impact of Genotypes on the Wheat × Maize System

In the ongoing search for a more effective haploid inducer pollen parent, Zenkteler and Nitzsche (1984) revealed that pollinating hexaploid wheat with maize resulted in the induction of haploid embryos. The crossability inhibitor genes did not affect the maize pollen, which made the system successful (Laurie and Bennett, 1989). It was also applicable to a variety of wheat genotypes and maize genotypes that were resistant to androgenesis (Cherkaoui *et al.*, 2000; Chaudhary *et al.*, 2002; Singh *et al.*, 2004; Tadesse *et al.*, 2012).

Moreover, cytological data later provided proof that maize pollen regularly germinated and developed into the wheat embryo sac, which is where the maize pollen fertilizes the wheat egg (Laurie and Bennett, 1986). When maize

Fig. 8.3. Main procedures of the Wheat × Maize System. (A) Manual emasculation of wheat. (B) Collecting maize pollen grains. (C) Manual pollination (applying maize pollen on wheat pistils). (D) Auxin treatment. (E) Immature seeds harvested. (F) Embryo isolation. (G) Embryo rescue. (H) Embryo regeneration. (I) Haploid plantlets. (J) Chromosome doubling by colchicine. (K) Transplanted DH plantlets. (L) DH plants. Source: Guan *et al.* (2024).

and wheat hybridized, a hybrid zygote with 10 maize chromosomes and 21 wheat chromosomes was created (Laurie and Bennett, 1988). Maize chromosomes do not migrate toward the spindle poles in the karyotypically unstable hybrid zygotes during cell divisions. It is possible that after three or four mitotic cell divisions, their centromeres lose their ability to bind to spindle microtubules as a result of a progressive loss of centromere activity, which leads to the eradication of maize chromosomes (Laurie and Bennett, 1988). Following the initial reports on the technique, numerous further studies on the creation of haploid wheat via wide crossing have been published (Matzk and Mahn, 1994; Suenaga, 1994; Morshedi and Darvey, 1995).

Many genotypes of maize and wheat react to the system, but their effectiveness varies (Suenaga *et al.*, 1998). The percentage of pollinated florets that make haploid embryos has an impact on the efficiency of haploid production. Haploid embryo production has been observed to vary widely based on a range of factors, ranging from as low as 1% (Suenaga and Nakajima, 1989) to as high as 53% (Morshedi and Darvey, 1995). Genetic variations among individual wheat and maize lines (Chaudhary *et al.*, 2002Chaudhary *et al.*, 2015; Sharma *et al.*, 2005; Dhiman *et al.*, 2012) and the timing and application of exogenous growth substances to stimulate ovule development (Subrahmanyam and Kasha, 1973) are among the factors influencing the yield of haploid embryos.

8.7 Environmental Factors Involved in the Wheat × Maize System

When cut plants were cultured *in vitro* in a nutrient solution (40 g/L sucrose, 10 mg/L silver nitrate, 3 g/L calcium phosphate, and 8 ml/L sulfurous acid) under controlled conditions of 22–23°C in light, 16–17°C in dark, and an ambient humidity of 70% (Gu *et al.*, 2008), this achieved a haploid embryo rate of 31.6% in the Wheat × Maize System. However, when field-grown plants were used, the haploid embryo rate was just 9.6%.

Khan and Ahmad (2011) went on to cultivate 25 hexaploid wheat genotypes from fields *in vitro* using a tillering solution that contained 8 ml/L sulfurous acid, 40 g/L sucrose, and 100 mg/L 2,4-D. They examined the regulated elements that affect the development of haploid seeds, including temperature during pollen collection, pollination timing, light intensity, and relative humidity. Therefore, at 21–26°C, pollination at 24 hour post emasculation, 10,000 Lux (140 μmol/m2/s) of light intensity, and 60–65% relative humidity at 20–22°C are the ideal conditions for maize pollen.

The haploid induction rate between wheat F1s and *Z. mays/I. cylindrica* under various circumstances was examined by Khan *et al.* (2017). The F1s in the greenhouse produced DH at a significantly higher rate than the F1s in the field. Therefore, ideal environmental conditions can be identified for an enhanced Wheat × Maize System, as the growth conditions of both wheat and maize are crucial to the System. Khan and Ahmad's (2011) proposed environmental parameters can be used as a guide for technical advancement.

Environmental factors, particularly temperature, play a significant role during and after pollination. According to Pratap *et al.* (2005), auxin analogues are also essential for the induction and upkeep of haploid wheat embryos. Research conducted on wheat × maize hybrids and self-pollinated wheat plants using cytology techniques demonstrated that the development of haploid embryos does not result in the formation of an endosperm. Instead, the endosperm nuclei stay free within the cytoplasm, fail to progress into the cellular stage, and ultimately degenerate (Zhang *et al.*, 1996). The hybridization technique of wheat × maize is more efficient than anther culture and bulbosum for producing haploids in bread wheat due to its reduced genotypic specificity, lack of albinism, and practicability (Wang *et al.*, 1991). Furthermore, this system's higher frequency of haploids makes it better suited for breeding (Kasha *et al.*, 1995; Pratap *et al.*, 2005).

8.8 Pollination of Wheat Using Maize Pollen

For the Wheat × Maize System to be conducted, a regulated atmosphere, or growth state, is ideal. However, immature wheat spikes were removed during heading and subsequently treated to *in vitro* cultivation due to the restricted size of an environmentally controlled facility (Gu *et al.*, 2008; Khan and Ahmad, 2011; Hussain *et al.*, 2012). These days, it is easy to use large, contemporary greenhouses that can hold an adequate number of wheat and maize plants year round. Consequently, the *in vitro* cultivation of immature wheat spikes is not required. Any responsible pollination, according to Laurie and Bennett (1989), is dependent on the wheat floret state, with the best florets having a feathery stigma. In their investigation of the impact of spikelet location on wheat–maize compatibility, Martins-Lopes *et al.* (2001) discovered that central spikelets produced better results. Therefore, in order to produce more haploid embryos under regulated circumstances, maize pollen should be administered to middle spikelets with a feathery stigma.

8.9 Hormone Treatment for the Wheat × Maize System

For haploid production, phytohormone therapy after wheat × maize pollination is essential. Hormone therapy increases haploid embryo survival rates and ovarian development, making haploid embryo rescue on media more feasible and successful (Laurie and Bennett, 1988; Laurie and Reymondie, 1991; Mahato and Chaudhary, 2019). Numerous hormones, including 2,4-dichlorophenoxyacetic acid (2,4-D), dicamba, picloram, indole-3-acetic acid (IAA), phenylacetic acid (PAA), silver nitrate, 1-naphthaleneacetic acid (NAA), kinetin, 6-benzyladenine (BA), and zearalenone, were studied in an effort to optimize the Wheat ×Maize System (Shubham *et al.*, 2023). Among these, callus induction and organ regeneration are two common uses for 2,4-D. Additionally, 2,4-D controls both somatic and zygotic embryogenesis during early and postembryogenic plant development (Juzon *et al.*, 2022).

The amount, timing, and technique of hormone application in the Wheat × Maize System should be established. In hexaploid wheat, 2,4-D efficiently generates haploid embryos at 100 ppm (Suenaga and Nakajima, 1989; Kaushik *et al.*, 2004; Pant and Khanna,

2017). In tetraploid wheat, 2,4-D efficiently stimulates haploid formation at 250 ppm (Mahato and Chaudhary, 2019). Various techniques for applying 2,4-D were examined by Kaushik *et al.* (2004) including spray, tiller injection, dipping, and spikelet culture. Out of these, only the spikelet culture approach exhibited good behavior in recovering embryos. In spite of this, we chose the spraying approach due to its ease of use and great efficacy, and as a result, we have achieved an average embryo rate of 12.9%.

8.9.1 Chromosome doubling

Until their fertility is restored to allow for the development of fully formed plants with a doubled chromosomal complement, haploids are not economically valuable. The term 'doubled haploids' refers to these changed plants technically since they have the same copy of the DNA. The chromosomes can double naturally or through external stimulation.

8.9.2 Induced doubling

Since haploids seldom experience natural chromosomal complement doubling, it must be induced by using antimitotic drugs to stop mitotic division at a specific moment. Chemicals such as colchicine, oryzalin, (S)-N-(5, 6, 7, 9-tetrahydro-1, 2, 3, 10-tetramethoxy-9oxobenzo[a]heptalen-7-yl) acetamide are among the many substances that researchers worldwide have worked with. According to research by Eigsti and Dustin (1955) and Hassawi and Liang (1991), colchicine was the most effective chromosomal doubling agent when administered to bread wheat callus or plants. However, each lab uses a different technique for applying colchicine. It is often applicable to haploid plants both *in vitro* and *in vivo*.

8.9.3 *In vitro* application of colchicines

Almost all laboratories employ the *in vitro* colchicine application methodology, which was created by Inagaki (1985). The protocol has been improved and several studies have been conducted on different aspects of the technique, such as the stage of haploid plants (Navarro-Alvarez *et al.*, 1994; Inagaki, 1997; Xynias and Roupakias, 2000), the concentration of the colchicine (Matzk and Mahn, 1994; Mohammadi *et al.*, 2007), supplementary agents (Stefanowaska, 1989), and physical factors (Hua *et al.*, 2005a, b), etc.

Using a 0.1% colchicine solution enhanced with 2% dimethyl sulfoxide and around 0.05% Tween 20, haploid plants with roots clipped to 2–3 cm are immersed in the solution for 5 hours at 20°C in the basic procedure. The method remained efficient even with a little temperature variation (Jauhar *et al.*, 2000; Maluszynski *et al.*, 2003).

A high rate of death in green plants following colchicine treatment prompted the researchers to hunt for earlier stages of chromosomal doubling. Hansen and Andersen (1998) examined the impact of colchicine in MS media on haploid embryos and found that the number of doubled haploids increased the most when the embryos were cultivated for 24 hours. In comparison to the conventional method, Chen *et al.* (2001) found improved doubling in colchicine containing MS media.

By cultivating the embryos in the presence of colchicine, Jobet *et al.* (2003) created a procedure for the generation of doubled haploid plants from crosses between wheat and maize. After comparing the effects of colchicine-containing MS medium at different stages of embryonic development and regeneration, Daniel *et al.* (2005) came to the conclusion that treating the plants with colchicine after regeneration resulted in the maximum survival of doubled haploid plants. When compared to plants treated at the 2–3 tillers stage, Ahmad and Chowdhry (2005) observed that plants at the 5–7 tillers stage exhibited greater plant fertility and decreased plant mortality.

8.10 Application of Doubled Haploids

In advanced breeding programs of many agricultural species, the induction and regeneration

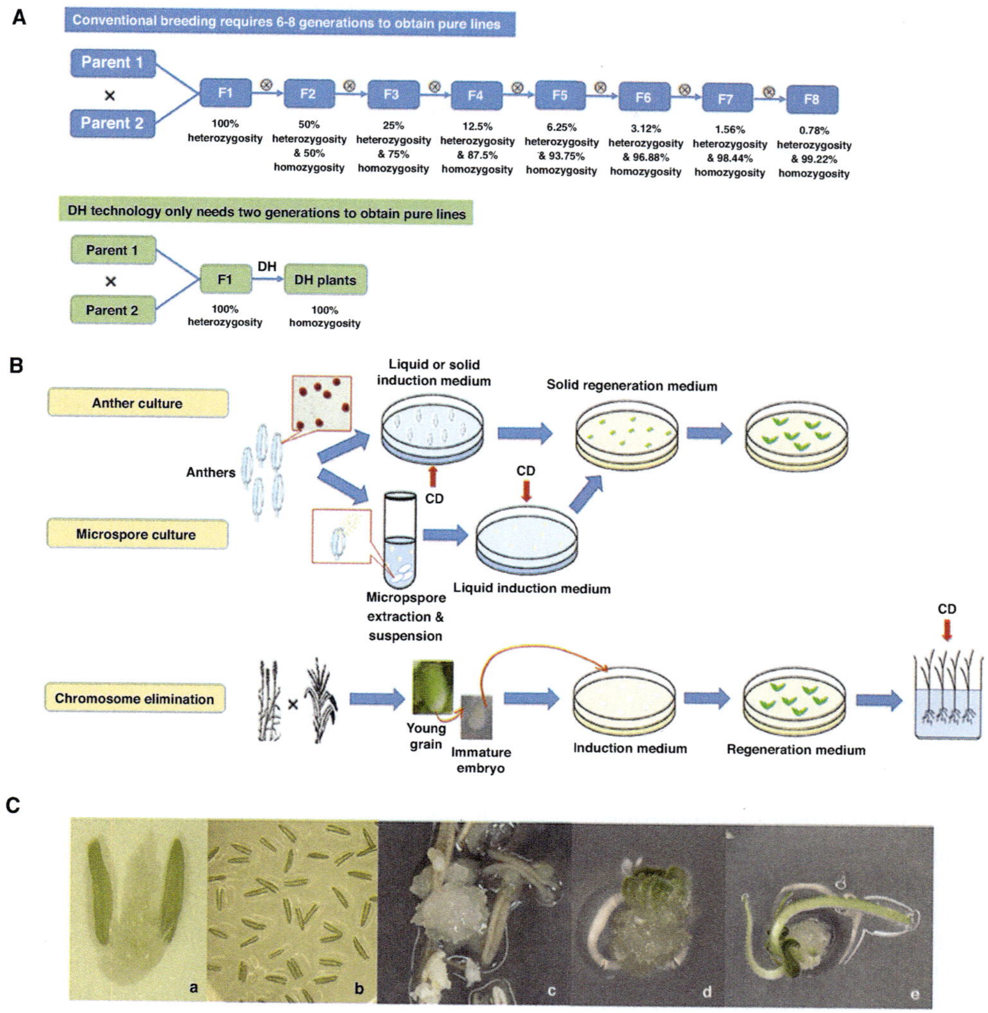

Fig. 8.4. Comparison of conventional and anther culture breeding methods. Conventional methods take at least 8–9 years for production of homozygous lines. Anther culture takes only 3 years. This doubled haploid (DH) method enables production of a homozygous plant in one generation, one additional year for seed increase, and the possibility to undertake screening and preliminary yield trials in the third year. Source: Tadesse *et al.* (2012).

of haploids, followed by the spontaneous or induced doubling of chromosomes, are often employed. When it comes to traditional plant breeding, it usually takes six generations or more after crossing to get a homozygous population that is suitable for screening and initial yield experiments in the F7 generation. With one more year for seed increase (DH2), the DH method allows for the creation of a homozygous plant in one generation. In the third year (DH3), as shown in Fig. 8.4, screening and preliminary yield experiments can be conducted. As a result, DH technology significantly accelerates the inbred developmental processes by eliminating multiple generations of labor-intensive inbreeding and improving the predictiveness of genotyping and phenotyping. In contrast to spring wheat, where two generations may be advanced annually by the use of SSD or a shuttle breeding strategy, winter wheat only experiences one cycle of generation advancement each year, making the time gained through the DH system particularly significant in this crop.

8.11 Applications of Doubled Haploids in Fundamental and Applied Plant Breeding Research

Doubled haploids are useful for the following reasons.

- Doubled haploid technology shortens the number of selection cycles required to achieve the required homozygosity, cutting the varietal development period from 12 to 13 to 6–7 years.
- DH lines can be used to create a valuable genetic stock or they can be issued straight as variety.
- It is possible to generate DH lines from hybrid crossings that have yield potential comparable to that of hybrids (Bentolila *et al.*, 1992). In wheat, for instance, the creation of hybrid DH lines derived from heterotic F1 crossings fared well when compared to their respective parental lines and control, with very few considerably outperforming the higher yielding parent and the control cultivar (El-Hennawy *et al.*, 2011).
- DH lines are 100% homozygous, which makes them the ideal material for quantitative genetic investigations.
- Landraces can use *in vitro* DH technology to overcome the loss of variability brought on by greater homozygosity in outbreeding species at later generations (Reif *et al.*, 2005). The cultigens gene pool can be expanded by using DH technologies to unleash the genetic diversity from extremely varied landrace populations.
- The elements of quantitative genetics, including as gene linkage, the number of genes regulating the trait, interactions, additive variances, and gene locations on chromosomes, are studied using DH lines.
- The resistance to foliar diseases of wheat, important agronomic traits, variation in grain yield in response to varying nitrogen application, grain quality, common bunt resistance and plant height, leaf rust (LR) resistance, stem rust resistance, stripe rust resistance, and adult plant yellow rust and stem rust resistance have all been identified and mapped using DH lines in wheat.
- Using a DH mapping population, the QTL governing ergot sclerotia size in hexaploid wheat has been found (Gordon *et al.*, 2015).
- The homozygous character of DH populations makes them perfect for genomic mapping.
- In order to increase wheat's host plant resistance, DH and MAS technologies have been effectively used. Fusarium head blight (*Fusarium graminearum*) (Yang *et al.*, 2003), stem rust (*Puccinia graminis* f. sp. *tritici*) (Hiebert *et al.*, 2016), and stripe rust (*Puccinia striiformis* f. sp. *tritici*) (Bakhtiar *et al.*, 2014), for instance, are all resisted by wheat DH.

8.12 Conclusion

The most effective way to modernize wheat breeding and generate fully homozygous lines for the production and use of hybrids is without doubt the DH technology. However, DH line production calls for new abilities for its implementation. The first and most important step in ensuring the success of DH production is giving the workforce proper training in the many aspects of DH production, such as chromosome doubling, haploid identification, haploid induction, and DH line recovery. Second, the process for producing haploid plants needs to be standardized because the plants are frequently fragile and susceptible to different environmental conditions. Third, DH technology will produce faster hybrid development by enhancing genetic gains and breeding efficiency by integration with other biotechnological methods such as marker-assisted selection, induced mutagenesis, and transgenic development.

A lot of work has been put into producing DH using various methods, but the decreased quantity of DH plants produced has limited its widespread application. The addition of DH technology to wheat genetics and breeding programs is very helpful in shortening the breeding cycle, producing novel cultivars, and enhancing our comprehension of genetic features that are crucial for agronomy. Therefore, to design the most effective plans for implementing the DH system for wheat breeding, industry members, researchers from many disciplines, and breeders must collaborate.

References

Abd El-Maksoud, M.M. and Bedő, Z. (1992) Half-diallel analysis of different characters in wheat anther culture. *Acta Agronomica Hungarica* 41, 235–242.

Abd El-Maksoud, M.M., Karsai, I. and Bedo, Z. (1993) Agronomic traits of wheat lines developed by the doubled haploid, single seed descent and pedigree methods after three cycles of selection. *Acta Agronomica Hungarica* 42, 377–382.

Ahmad, J. and Chowdhry, M.A. (2005) The efficiency of chromosome doubling in haploids derived from wheat × maize crossing. *Pakistan Journal of Biological Sciences* 8(12), 1707–1711.

Ayed, O.S., Buyser, J., Picard, E., Trifa, Y. and Amara, H.S. (2010) Effect of pre-treatment on isolated microspores culture ability in durum wheat (*Triticum turgidum* subsp. durum Desf.). *Journal of Plant Breeding and Crop Science* 2(2), 030–038.

Baenziger, P.S., Russell, W.K., Graef, G.L. and Campbell, B.T. (2006) Improving lives: 50 years of crop breeding, genetics, and cytology (C-1). *Crop Science* 46(5), 2230–2244.

Bakhtiar, F., Afshari, F., Najafian, G. and Mohammadi, M. (2014) Backcross-breeding and doubled-haploid facilitated introgression of stripe rust resistance in bread wheat. *Archives of Phytopathology and Plant Protection* 47(14), 1675–1685.

Barclay, I.R. (1975) High frequencies of haploid production in wheat (*Triticum aestivum*) by chromosome elimination. *Nature* 256(5516), 410–411.

Bentolila, S., Hardy, T., Guitton, C. and Freyssinet, G. (1992) Comparative genetic analyses of F2 plants and anther culture derived plants of maize. *Genome* 35(4), 575–582.

Blakeslee, A.F., Belling, J., Farnham, M.E. and Bergner, A.D. (1922) A haploid mutant in the jimson weed, "*Datura Stramonium*". *Science* 55(1433), 646–647.

Brazauskas, G., Pašakinskienė, I. and Ruzgas, V. (2005) Improved approaches in wheat × maize crossing for wheat doubled haploid production. *Biologija* 51(4).

Campbell, A.W., Griffin, W.B., Burritt, D.J. and Conner, A.J. (2000) Production of wheat doubled haploids via wide crosses in New Zealand wheat. *New Zealand Journal of Crop and Horticultural Science* 28(3), 185–194.

Chaudhary, H.K., Singh, S. and Sethi, G.S. (2002) Interactive influence of wheat and maize genotypes on the induction of haploids in winter × spring hexaploid wheat hybrids. *Journal of Genetics & Breeding* 56(3), 259–266.

Chaudhary, H.K., Badiyal, A. and Jamwal, N.S. (2015) New frontiers in doubled haploidy breeding in wheat. *Agricultural Research Journal* 52(4), 1–12.

Chen, Q.F., Wang, C.L., Lu, Y.M., Shen, M., Afza, R. *et al.* (2001) Anther culture in connection with induced mutations for rice improvement. *Euphytica* 120, 401–408.

Cherkaoui, S., Lamsaouri, O., Chlyah, A. and Chlyah, H. (2000) Durum wheat × maize crosses for haploid wheat production: Influence of parental genotypes and various experimental factors. *Plant Breeding* 119(1), 31–36.

Cistué, L., Soriano, M., Castillo, A.M., Vallés, M.P., Sanz, J.M. *et al.* (2006) Production of doubled haploids in durum wheat (*Triticum turgidum* L.) through isolated microspore culture. *Plant Cell Reports* 25(4), 257–264.

Cistué, L., Romagosa, I., Batlle, F. and Echávarri, B. (2009) Improvements in the production of doubled haploids in durum wheat (*Triticum turgidum* L.) through isolated microspore culture. *Plant Cell Reports* 28(5), 727–735.

Clausen, R.E. and Mann, M.C. (1924) Inheritance in *Nicotiana tabacum*: V. the occurrence of haploid plants in interspecific progenies. *Proceedings of the National Academy of Sciences* 10(4), 121–124.

Daniel, G., Baumann, A. and Schmucker, S. (2005) Production of wheat doubled haploids (*Triticum aestivum* L.) by wheat x maize crosses using colchicine enriched medium for embryo regeneration. *Cereal Research Communications* 33, 461–468.

Dhiman, R., Rana, V. and Chaudhary, H. (2012) Himalayan maize – potential pollen source for maize mediated system of chromosome elimination approach in DH breeding of bread wheat. *Cereal Research Communications* 40(2), 246–255.

Ding, M., Zhao, H., Gu, J., Li, H., Liu, K. *et al.* (2017) Research and breeding application progress of the technique of producing double haploid of wheat by wide hybridization between wheat and maize. *Agricultural Science & Technology* 18(12), 2202–2208.

Ding, M., Yang, Z., Cui, Y., Li, H., Liu, K. *et al.* (2022) A high-yield and multi-resistant new wheat variety bred using diploid technology—Yunmai 110. *Journal of Triticeae Crops* 42, 1589.

Dunwell, J.M. (2010) Haploids in flowering plants: Origins and exploitation. *Plant Biotechnology Journal* 8(4), 377–424.

Dwivedi, S.L., Britt, A.B., Tripathi, L., Sharma, S., Upadhyaya, H.D. *et al.* (2015) Haploids: Constraints and opportunities in plant breeding. *Biotechnology Advances* 33(6), 812–829.

Eigsti, O.J. and Dustin, P. (1955) *Colchicine in Agriculture, Medicine, Biology and Chemistry*. Iowa State College Press, Ames.

El-Hennawy, M.A., Abdalla, A.F., Shafey, S.A. and Al-Ashkar, I.M. (2011) Production of doubled haploid wheat lines (*Triticum aestivum* L.) using anther culture technique. *Annals of Agricultural Sciences* 56(2), 63–72.

Fehr, W. (1991) *Principles of Cultivar Development: Theory and Technique*. Macmillan, New York.

Folling, L. (2002) Erratum to: Transformation of wheat (*Triticum aestivum* L.) microspore-derived callus and microspores by particle bombardment. *Plant Cell Reports* 20, 1098–1105.

Forster, B.P., Heberle-Bors, E., Kasha, K.J. and Touraev, A. (2007) The resurgence of haploids in higher plants. *Trends in Plant Science* 12(8), 368–375.

Gaines, E.F. and Aase, H.C. (1926) A haploid wheat plant. *American Journal of Botany* 13(6), 373–385.

Gordon, A., Basler, R., Bansept-Basler, P., Fanstone, V., Harinarayan, L. *et al.* (2015) The identification of QTL controlling ergot sclerotia size in hexaploid wheat implicates a role for the Rht dwarfing alleles. *Theoretical and Applied Genetics* 128, 2447–2460.

Grauda, D., Lepse, N., Strazdiņa, V., Kokina, I., Lapiņa, L. *et al.* (2010) Obtaining of doubled haploid lines by anther culture method for the Latvian wheat breeding. *Agronomy Research* 8, 545–552.

Gu, J., Liu, K., Li, S., Tian, Y., Yang, H. *et al.* (2008) Study on the in vitro culture of cut plants in wheat haploid embryo induction by a wheat × maize cross. *Frontiers of Agriculture in China* 2(4), 391–395.

Guan, X., Peng, J. and Fu, D. (2024) Technology for production of wheat doubled haploid via maize pollen induction—updated review. *Agronomy* 14(2), 375.

Guha, S. and Maheshwari, S.C. (1964) In vitro production of embryos from anthers of Datura. *Nature* 204(4957), 497–497.

Guha, S. and Maheshwari, S.C. (1966) Cell division and differentiation of embryos in the pollen grains of Datura in vitro. *Nature* 212(5057), 97–98.

Gurtay, G., Kutlu, I. and Avci, S. (2021) Production of haploids in ancient, local and modern wheat by anther culture and maize pollination. *Acta Biologica Cracoviensia. Series Botanica* 63(1), 43–53.

Hansen, N.J.P. and Andersen, S.B. (1998) Efficient production of doubled haploid wheat plants by in vitro treatment of microspores with trifluralin or APM. *Plant Breeding* 117(5), 401–405.

Hassawi, D.S. and Liang, G.H. (1991) Antimitotic agents: Effects on double haploid production in wheat. *Crop Science* 31(3), 723–726.

Heyne, E.G. (1987) *Wheat and Wheat Improvement*, Vol. 2. American Society of Agronomy, Madison.

Hiebert, C.W., Kassa, M.T., McCartney, C.A., You, F.M., Rouse, M.N. *et al.* (2016) Genetics and mapping of seedling resistance to Ug99 stem rust in winter wheat cultivar Triumph 64 and differentiation of Srtmp, Srcad, and Sr42. *Theoretical and Applied Genetics* 129(11), 2171–2177.

Ho, K.M. and Jones, G.E. (1980) Mingo barley. *Canadian Journal of Plant Science* 60(1), 279–280.

Hua, C., Chuan, M.X. and Zhong, L.W. (2005a) Chromosome doubling in wheat haploids from intergeneric cross between wheat and maize. *Journal of Triticeae Crops* 25, 20–22.

Hua, C., Chuan, M.X., Hong, S.Q., Qiang, Q.Y. and Zhong, L.W. (2005b) Approaches to increasing of the frequency of haploid embryo production in wheat x maize crosses. *Journal of Triticeae Crops* 25, 20–24.

Hussain, M., Niaz, M., Iqbal, M., Iftikhar, T. and Ahmad, J. (2012) Emasculation techniques and detached tiller culture in wheat x maize crosses. *Journal of Agricultural Research* 50(1), 1–19.

Inagaki, M. (1985) Chromosome doubling of wheat haploids obtained from crosses with *Hordeum bulbosum* L. *Japanese Journal of Breeding* 35, 193–195.

Inagaki, M. (1997) Technical advances in wheat haploid production using ultra-wide crosses. *JIRCAS Journal* 4, 51–62.

Inagaki, M. and Tahir, M. (1990) Comparison of haploid production frequencies in wheat varieties crossed with *Hordeum bulbosum* L. and maize. *Japanese Journal of Breeding* 40(2), 209–216.

Jauhar, P.P., Dogramaci-Altuntepe, M., Peterson, T.S. and Almouslem, A.B. (2000) Seedset on synthetic haploids of durum wheat: Cytological and molecular investigations. *Crop Science* 40(6), 1742–1749.

Jauhar, P.P., Xu, S.S. and Baenziger, P.S. (2009) Haploidy in cultivated wheats: Induction and utility in basic and applied research. *Crop Science* 49(3), 737–755.

Jeberson, M.S., Chaudhary, H.K., Chahota, R.K. and Wani, S.H. (2021) Doubled haploid production in advanced back cross generations and molecular cytogenetic characterization of rye chromatin in triticale × wheat derived doubled haploid lines. *Biocell* 45(6), 1651–1659.

Jobet, C., Zuniga, J. and Hugo, C.Q. (2003) Plantas doble haploides generadas por cruza intergenerica de trigo x maiz. *Agricultura Técnica* 63, 323–328.

Juzoń, K., Warchoł, M., Dziurka, K., Czyczyło-Mysza, I.M., Marcińska, I. *et al.* (2022) The effect of 2,4-dichlorophenoxyacetic acid on the production of oat (*Avena sativa* L.) doubled haploid lines through wide hybridization. *PeerJ* 10, e12854.

Kalinowska, K., Chamas, S., Unkel, K., Demidov, D., Lermontova, I. *et al.* (2019) State-of-the-art and novel developments of *in vivo* haploid technologies. *Theoretical and Applied Genetics* 132(3), 593–605.

Kapoor, C., Chaudhary, H.K., Sharma, P., Relan, A., Manoj, N.V. *et al.* (2021) *In vivo* haploid induction potential of himalayan maize (*Zea mays*) and cogon grass (*Imperata cylindrica*) gene pools in different segregational cycles of intra and inter-generic crosses of wheat. *Plant Genetic Resources* 19(6), 522–529.

Karimzadeh, G., Kovács, G. and Barnabás, B. (1995) Effects of cold treatment and different culture media on the androgenic capacity of two winter wheat genotypes. *Cereal Research Communications* 23, 223–227.

Karsai, I., Bedo, Z. and Hayes, P.M. (1994) Effect of induction medium pH and maltose concentration on *in vitro* androgenesis of hexaploid winter triticale and wheat. *Plant Cell, Tissue and Organ Culture* 39(1), 49–53.

Kasha, K.J. (1974) Haploids in higher plants: Advances and potential. In: *Proceedings of the 1st International Symposium*. University of Guelph.

Kasha, K.J., Yao, Q., Simion, E., Hu, T. and Oro, R. (1995) *Production and Application of Doubled Haploids in Crops*. IAEA, Vienna.

Kaushik, N., Sirohi, M. and Khanna, V.K. (2004) Influence of age of the embryo and method of hormone application on haploid embryo formation in wheat x maize crosses. In: *Proceedings of the 4th International Crop Science Congress*, Vol. 26. Brisbane.

Khan, M.A. and Ahmad, J. (2011) In vitro wheat haploid embryo production by wheat x maize cross system under different environmental conditions. *Pakistan Journal of Agricultural Sciences* 48(1), 49–53.

Khan, H., Bhardwaj, S.C., Gangwar, O.P., Prasad, P. and Rathore, R. (2017) Efficiency of double haploid production in wheat through wide hybridization and embryo rescue. *Indian Journal of Genetics and Plant Breeding* 77(3), 428–430.

Khiabani, B.N., Vedadi, C., Rahmani, E. and Shalmani, M.A.M. (2008) Response of some Iranian wheat genotypes to another culture system. *Indian Journal of Biotechnology* 7, 531–535.

Labbani, Z., Buyser, J. and Picard, E. (2007) Effect of mannitol pretreatment to improve green plant regeneration on isolated microspore culture in *Triticum turgidum* ssp. *durum* cv. *Jennah Khetifa'*. *Plant Breeding* 126(6), 565–568.

Laurie, D.A. and Bennett, M.D. (1986) Wheat × maize hybridization. *Canadian Journal of Genetics and Cytology* 28(2), 313–316.

Laurie, D.A. and Bennett, M.D. (1988) The production of haploid wheat plants from wheat x maize crosses. *Theoretical and Applied Genetics* 76(3), 393–397.

Laurie, D.A. and Bennett, M.D. (1989) The timing of chromosome elimination in hexaploid wheat × maize crosses. *Genome* 32(6), 953–961.

Laurie, D.A. and Snape, J.W. (1990) The agronomic performance of wheat doubled haploid lines derived from wheat x maize crosses. *Theoretical and Applied Genetics* 79(6), 813–816.

Laurie, D.A. and Reymondie, S. (1991) High frequencies of fertilization and haploid seedling production in crosses between commercial hexaploid wheat varieties and maize. *Plant Breeding* 106(3), 182–189.

Li, H., Li, S., Abdelkhalik, S., Shahzad, A., Gu, J. *et al.* (2020) Development of thermo-photo sensitive genic male sterile lines in wheat using doubled haploid breeding. *BMC Plant Biology* 20, 1–10.

Liu, H., Wang, K., Jia, Z., Gong, Q., Lin, Z. *et al.* (2020) Efficient induction of haploid plants in wheat by editing of TaMTL using an optimized Agrobacterium-mediated CRISPR system. *Journal of Experimental Botany* 71(4), 1337–1349.

Liu, W., Zheng, M.Y., Polle, E.A. and Konzak, C.F. (2002) Highly efficient doubled-haploid production in wheat (*Triticum aestivum* L.) via induced microspore embryogenesis. *Crop Science* 42(3), 686–692.

Mahato, A. and Chaudhary, H.K. (2019) Auxin induced haploid induction in wide crosses of durum wheat. *Cereal Research Communications* 47(3), 552–565.

Maluszynski, M., Kasha, K.J. and Szarejko, I. (2003) Published doubled haploid protocols in plant species. In: Maluszynski, M., Kasha, K.J., Forster, B.P. and Szarejko, I. (eds) *Doubled Haploid Production in Crop Plants: A Manual*. Springer Netherlands, Dordrecht, pp. 309–335.

Martins-Lopes, P., Guedes-Pinto, H., Pinto-Carnide, O. and Snape, J. (2001) The effect of spikelet position on the success frequencies of wheat haploid production using the maize cross system. *Euphytica* 121, 265–271.

Matzk, F. and Mahn, A. (1994) Improved techniques for haploid production in wheat using chromosome elimination. *Plant Breeding* 113(2), 125–129.

Mayel, A., Chaudhary, H.K., Badiyal, A. and Jamwal, N.S. (2016) Comparative pollination efficiency of freshly harvested pollen of *Imperata cylindrica* and *Zea mays* for haploid induction in bread wheat. *Cereal Research Communications* 44(1), 162–171.

Mohammadi, P.P., Moieni, A. and Jalali-Javaran, M. (2007) Colchicine induced embryogenesis and doubled haploid production in maize (*Zea mays* L.) anther culture. *Iranian Journal of Biotechnology* 5(3), 140–146.

Moradi, P., Hagh, N.A., Bozorgipour, R. and Sharma, B. (2009) Development of yellow rust resistant doubled haploid lines of wheat through wheat × maize crosses. *International Journal of Plant Production* 3, 1735–6814.

Morshedi, A.R. and Darvey, N.L. (1995) High frequency of embryos in wheat × maize crosses. *SABRAO Journal* 27(1/2), 17–22.

Nakata, K. and Tanaka, M. (1968) Differentiation of embryoids from developing germ cells in another culture of tobacco. *Japanese Journal of Genetics* 43(1), 65–71.

Navarro-Alvarez, W., Baenziger, P.S., Eskridge, K.M., Hugo, M. and Gustafson, V.D. (1994) Addition of colchicine to wheat another culture media to increase doubled haploid plant production. *Plant Breeding* 112(3), 192–198.

Niizeki, H. and Oono, K. (1968) Induction of haploid rice plant from another culture. *Proceedings of the Japan Academy* 44(6), 554–557.

Niroula, R.K. and Bimb, H.P. (2009) Overview of wheat x maize system of crosses for dihaploid induction in wheat. *World Applied Sciences Journal* 7(8), 1037–1045.

Niroula, R.K., Bimb, H.P., Thapa, D.B., Sah, B.P. and Nayak, S. (2007) Production of haploid wheat plants from wheat (*Triticum aestivum* L.) x maize (*Zea mays* L.) cross system. *Himalayan Journal of Sciences* 4(6), 65–69.

Nitsch, J.P. and Nitsch, C. (1969) Haploid plants from pollen grains. *Science* 163(3862), 85–87.

Ouyang, T.W., Hu, H., Chuang, C.C. and Tseng, C.C. (1973) Induction of pollen plants from anthers of *Triticum aestivum* L cultured *in vitro*. *Scientia Sinica* 16, 79–90.

Pant, U. and Khanna, V.K. (2017) Effect of hormonal treatments on haploid formation and in vitro haploid regeneration in wheat x maize system. *International Journal of Plant Sciences* 12(2), 234–239.

Patial, M., Pal, D., Thakur, A., Bana, R.S. and Patial, S. (2019) Doubled haploidy techniques in wheat (*Triticum aestivum* L.): An overview. In: *Proceedings of the National Academy of Sciences, India Section B: Biological Sciences*, pp. 27–41 (Vol. 89).

Picard, E. and De Buyser, J. (1973) Obtention de plantlets haploides de *Triticum aestivum* L. A partir de cultures d'antheres in vitro. *CR Academie Des Sciences* 277, 1463–1466.

Pratap, A., Sethi, G.S. and Chaudhary, H.K. (2005) Relative efficiency of different gramineae genera for haploid induction in triticale and triticale x wheat hybrids through the chromosome elimination technique. *Plant Breeding* 124(2), 147–153.

Prigge, V., Xu, X., Li, L., Babu, R., Chen, S. *et al.* (2012) New insights into the genetics of *in vivo* induction of maternal haploids, the backbone of doubled haploid technology in maize. *Genetics* 190(2), 781–793.

Raina, S. (1997) Doubled haploid breeding in cereals. *Plant Breeding Reviews* 15, 141–186.

Reif, J.C., Hamrit, S., Heckenberger, M., Schipprack, W., Peter Maurer, H. *et al.* (2005) Genetic structure and diversity of European flint maize populations determined with SSR analyses of individuals and bulks. *Theoretical and Applied Genetics* 111(5), 906–913.

Sharma, S., Sethi, G.S. and Chaudhary, H.K. (2005) Influence of winter and spring wheat genetic backgrounds on haploid induction parameters and trait correlations in the wheat × maize system. *Euphytica* 144(1), 199–205.

Shirdelmoghanloo, H., Moieni, A. and Mousavi, A. (2009) Effects of embryo induction media and pre-treatments in isolated microspore culture of hexaploid wheat (*Triticum aestivum* L. cv. *Falat*). *Falat). African Journal of Biotechnology* 8(22), 6134–6140.

Shubham, G., Sandal, S.S., Walia, P. and Upadhyay, V. (2023) Application of auxins in haploid embryo induction in hexaploidy wheat. *International Journal of Environment and Climate Change* 13, 251–255.

Singh, A.K., Zhang, P., Dong, C., Li, J., Trethowan, R. *et al.* (2019) Molecular cytogenetic characterization of stem rust and stripe rust resistance in wheat-*Thinopyrum bessarabicum*-derived doubled haploid lines. *Molecular Breeding* 39, 1–11.

Singh, S., Sethi, G.S. and Chaudhary, H.K. (2004) Differential responsiveness of winter and spring wheat genotypes to maize-mediated production of haploids. *Cereal Research Communications* 32(2), 201–207.

Snape, J.W. (1989) Doubled haploid breeding: Theoretical basis and practical applications. *Review of Advances in Plant Biotechnology* 2, 19–31.

Snape, J.W., Simpson, E., Parker, B.B., Friedt, W. and Foroughi-Wehr, B. (1986) Criteria for the selection and use of doubled haploid systems in cereal breeding programmes. In: Horn, W., Jensen, C., Odenbach, W. and Schieder, O. (eds) *Genetic Manipulation in Plant Breeding*. De Gruyter, Berlin.

Stefanowaska, G. (1989) The effect of colchicine action depending on the genotype of F1 hybrids *Triticum aestivum* x *Secale cereale* and on the methods of colchicines treatment. *Genetica Polonica* 30, 19–25.

Subrahmanyam, N.C. and Kasha, K.J. (1973) Selective chromosomal elimination during haploid formation in barley following interspecific hybridization. *Chromosoma* 42(2), 111–125.

Suenaga, K. (1994) Doubled haploid system using the intergeneric crosses between wheat (*Triticum aestivum*) and maize (*Zea mays*). *Bulletin of the National Institute of Agrobiology Resources* 9, 83–139.

Suenaga, K. and Nakajima, K. (1989) Efficient production of haploid wheat (*Triticum aestivum*) through crosses between Japanese wheat and maize (*Zea mays*). *Plant Cell Reports* 8(5), 263–266.

Suenaga, K., Morshedi, A.R. and Darvey, N.L. (1998) Evaluation of teosinte lines as pollen parents for wheat haploid production. *Cereal Research Communications* 26(2), 119–125.

Tadesse, W., Inagaki, M., Tawkaz, S., Baum, M. and Ginkel, M. (2012) Recent advances and application of doubled haploids in wheat breeding. *African Journal of Biotechnology* 11(89), 15484–15492.

Tang, H., Wang, K., Zhang, S., Han, Z., Chang, Y. *et al.* (2023) A fast technique for visual screening of wheat haploids generated from TaMTL-edited mutants carrying anthocyanin markers. *Plant Communications* 4(3), 100569.

Tawkaz, S. (2011) Response of some wheat genotypes to anther culture technique for doubled haploid production. Doctoral dissertation, Sudan Academy of Science, Kartum, Sudan.

Tehseen, M.M., Tonk, F.A., Tosun, M., Randhawa, H.S., Kurtulus, E. *et al.* (2022) QTL mapping of adult plant resistance to stripe rust in a doubled haploid wheat population. *Frontiers in Genetics* 13, 900558.

Verma, V., Bains, N.S., Mangat, G.S., Nanda, G.S., Gosal, S.S. *et al.* (1999) Maize genotypes show striking differences for induction and regeneration of haploid wheat embryos in the wheat × maize system. *Crop Science* 39(6), 1722–1727.

Wang, J.L., Sun, J.S., Lu, T.G., Fang, R., Cui, H.R. *et al.* (1991) Fertilization and embryo development in wheat × maize crosses. *Acta Botanica Sinica* 33, 674–679.

Wei, Z.M. (1982) Pollen callus culture in *Triticum aestivum*. *Theoretical and Applied Genetics* 63(1), 71–73.

Xynias, I.N. and Roupakias, D.G. (2000) Primary triticale production from indigenous genetic material with desirable agronomic traits. *Journal of Genetics and Breeding* 54, 277–282.

Xynias, I.N., Zamani, I.A., Gouli-Vavdinoudi, E. and Roupakias, D.G. (2001) Effect of cold pretreatment and incubation temperature on bread wheat (*Triticum aestivum* L.) anther culture. *Cereal Research Communications* 29, 331–338.

Yang, Z.P., Gilbert, J., Somers, D.J., Fedak, G., Procunier, J.D. *et al.* (2003) Marker assisted selection of Fusarium head blight resistance genes in two doubled haploid populations of wheat. *Molecular Breeding* 12(4), 309–317.

Zamani, I., Gouli-Vavdinoudi, E., Kovacs, G., Xynias, I., Roupakias, D. *et al.* (2003) Effect of parental genotypes and colchicine treatment on the androgenic response of wheat F_1 hybrids. *Plant Breeding* 122(4), 314–317.

Zargar, M., Zavarykina, T., Voronov, S., Pronina, I. and Bayat, M. (2022) The recent development in technologies for attaining doubled haploid plants *in vivo*. *Agriculture* 12(10), 1595.

Zenkteler, M. and Nitzsche, W. (1984) Wide hybridization experiments in cereals. *Theoretical and Applied Genetics* 68(4), 311–315.

Zhang, J., Friebe, B., Raupp, W.J., Harrison, S.A. and Gill, B.S. (1996) Wheat embryogenesis and haploid production in wheat × maize hybrids. *Euphytica* 90(3), 315–324.

9 Bioeconomy and Biotechnology: Innovative Strategies in the Utilization of Resources in the Argentine North Patagonia

Daniela Dalzotto[1,2], Miguel José Morón Rivera[1,2], Luciana Di Sario[1,2], Agustina Cabrera[1,2], Mariana Dalponte[1,2], Sandra Sharry[1,2,3] and Patricia Boeri[1,2*]

[1]Universidad Nacional de Río Negro (UNRN), Sede Atlántica, Viedma, Argentina; [2]CIT-RIO NEGRO, Sede Viedma, Universidad Nacional de Río Negro (UNRN-CONICET), Viedma, Argentina; [3]Universidad Nacional de La Plata (UNLP), Facultad de Ciencias Agrarias y Forestales, La Plata, Argentina

Abstract

With a growing global population, ensuring food security will be a crucial challenge due to climate change, natural resource degradation, and socio-economic inequalities. The 2030 Agenda of the United Nations, with its 17 Sustainable Development Goals (SDGs), addresses these challenges, with SDG 2 specifically aiming to ensure food security, improve nutrition, and promote sustainable agriculture. In the arid and semiarid North Patagonia, agriculture is vital for regional development but faces challenges including soil degradation and biodiversity loss. The bioeconomy, through biotechnology and bioprospecting, emerges as an innovative solution, using biological resources to enhance sustainable agricultural productivity and adapt to climate change. Implementing the 2030 Agenda and its SDGs in this region is crucial for ensuring food security, promoting sustainable economic growth, and strengthening agricultural resilience against current challenges. This chapter explores the contributions of the bioeconomy in North Patagonia, local initiatives for food security, and the importance of agriculture in the region.

Keywords: agricultural resilience, climate change, Sustainable Development Goals (SDGs), natural resource degradation

9.1 Introduction

The global population is continually expanding, with estimates suggesting it will reach 9.7 billion by 2050, leading to a greater demand for food. This challenge is compounded by climate change (CC), the degradation of natural resources, and economic and social inequalities (PNUD, 2015). In this context, ensuring that everyone has access to sufficient, safe, and nutritious food becomes a fundamental priority for human well-being and sustainable development, but it also represents a significant challenge.

*Corresponding author: pboeri@unrn.edu.ar

(ed. Tariq Aftab)
DOI: 10.1079/9781800627659.0009

The 2030 Agenda of the United Nations offers a comprehensive response to these challenges by addressing the social, environmental, and economic dimensions of sustainable development, through the adoption of 17 Sustainable Development Goals (SDGs). These objectives have specific and measurable targets to reduce poverty and inequality, and combat CC and environmental degradation. In the context of food security, SDG 2 'Zero Hunger' stands out, promoting sustainable agriculture and improving nutrition through eight specific goals adopted and adapted by each country and region according to its particular circumstances. The four pillars of food security (availability, access, utilization, and stability) are integral to achieving these goals.

The agricultural and livestock sectors are directly linked to food security, as they are the primary source of income and employment for 80% of the rural poor (FAO, n.d.). It is crucial to adopt sustainable agricultural practices that increase productivity, preserve natural resources, and reduce environmental impact. The bioeconomy, defined as an economy that sustainably utilizes biological resources, emerges as a key instrument to address the interrelated challenges of food security, resource depletion, and CC (Rodríguez *et al.*, 2017). At the same time, it offers the opportunity to foster sustainable economic growth through the development of new activities and value chains based on bioproducts. In this context, the integration of biotechnology and bioprospecting is a fundamental tool for the realization of the bioeconomy. Plant biotechnology, by enabling the genetic manipulation of crops and enhancing their resistance to adverse conditions, offers the capability to increase agricultural productivity and adapt to CC challenges. Meanwhile, bioprospecting explores and uses biodiversity to discover new resources and compounds with potential applications in food, medicinal, or industrial applications (Jaramillo *et al.*, 2019).

Agricultural practices have been crucial for the development of economically marginal regions like North Patagonia (Argentina). However, the diverse geography and climate of the region profoundly influence its farming techniques and productivity. For these reasons, adopting management systems that combine mitigation of their activities, impacts and adaptation to CC, alongside implementing biotechnology and bioprospecting, are crucial tools of interest for regional improvement.

In this context, this chapter explores the importance of agriculture in North Patagonia, highlighting local initiatives aimed at achieving food security, as well as the contributions of the bioeconomy in the region. Through a multidisciplinary approach, the aim is to understand how technological innovations, sustainable management of natural resources, and effective public policies can converge to ensure a prosperous and resilient future for agriculture in this marginal region.

9.2 Argentine North Patagonia: An Agricultural Perspective of the Region

Agriculture plays a strategic role in the Argentine economy, and the growing demand for agricultural products presents a significant opportunity for the equitable development of the regions of the country through primary production, value addition, and agroindustry. Argentina bears great responsibility for future food security, being one of the countries with the highest potential for food production (Andrade *et al.*, 2017). This agricultural prowess is underpinned by the remarkable diversity of climates and geographical features of this country. This is attributed to its extensive latitudinal range that extends over 33° latitude from north to south and almost 20° longitude from east to west, with a significant altitudinal difference from 0 to nearly 7000 meters above sea level and a vast maritime coastline stretching 4725 km. It encompasses a climate ranging from tropical and subtropical conditions in the northwest and northeast to extreme cold in the mountainous regions and the south. This leads to the existence of diverse environments such as salt flats and deserts, temperate forests, subtropical jungles, and high-altitude grasslands.

Seventy percent of Argentine territory can be classified as arid or semiarid (MAyDS, 2020), with a significant portion located in the southern region, known as Patagonia. This region is characterized by extensive plateaus and mountain ranges of Precambrian origin,

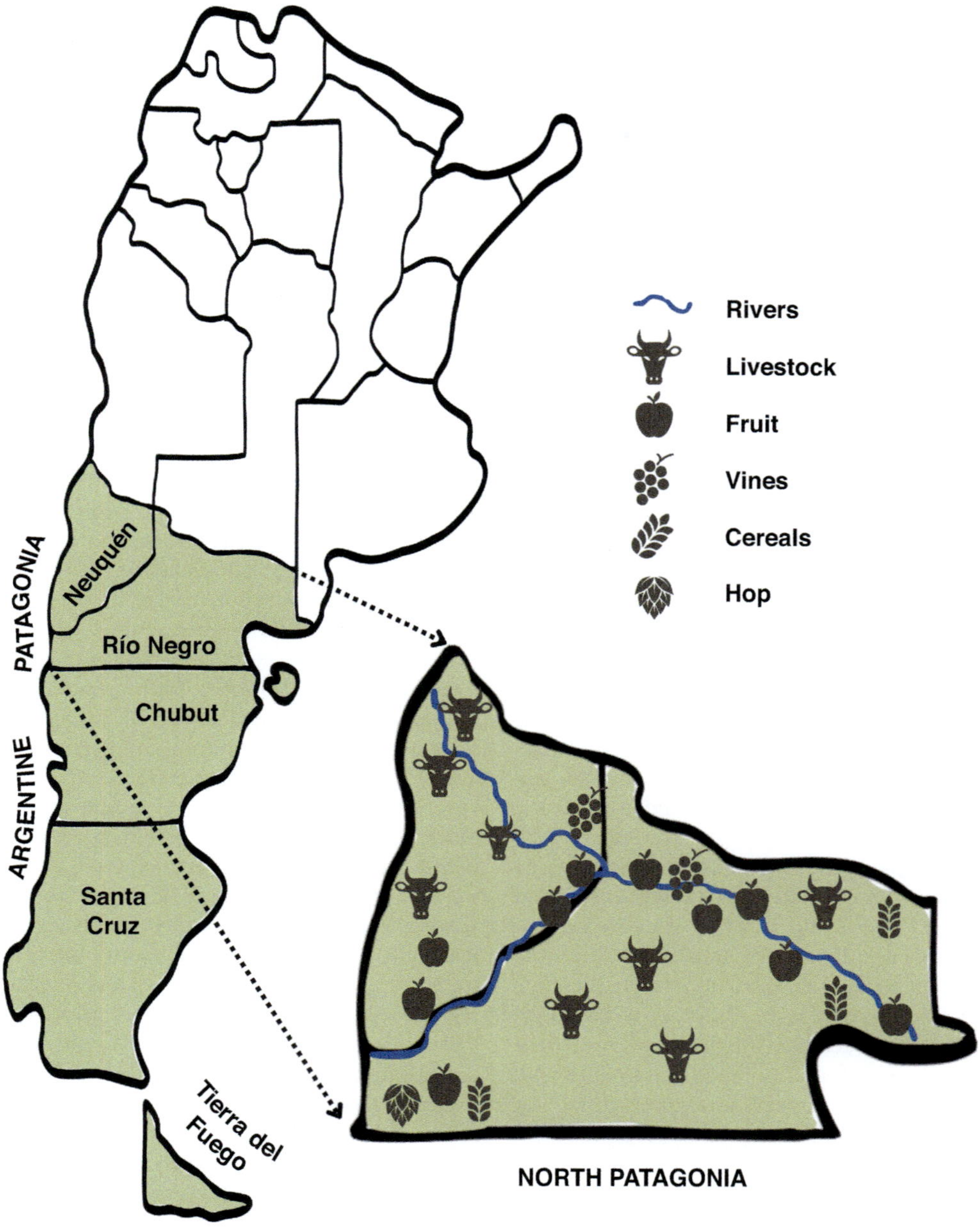

Fig. 9.1. Location and provinces comprising Argentine Patagonia and distribution of the main economic activities of North Patagonia related to livestock farming and agriculture.

spanning across the provinces of Río Negro, Neuquén, Chubut, Santa Cruz, and Tierra del Fuego (Fig. 9.1). Patagonia experiences strong winds and variable rainfall; the Andean zone receives over 600 mm, the central and coastal areas around 200 mm, and Tierra del Fuego approximately 300 mm. Arid and semiarid ecosystems face ecological constraints such as discontinuous resource availability, poor soil organic matter and nutrient content, persistent water scarcity, desertification, land degradation, and biodiversity loss, all of which impact the livelihoods of local communities (Gaur and Squires, 2018). Ecosystems in these drylands are highly fragile and particularly sensitive to anthropogenic changes in the climate system (Huang *et al.*, 2016; Ramarao *et al.*, 2019; IPCC, 2022). In the Patagonian region, desertification processes affect 84% of the surface area (Del Valle *et al.*, 1998). Specifically, assessments in Río Negro Province show that out of 4.25 million hectares surveyed, 56% have lost their productive capacity. Although complete reversal is not possible, there is potential for recovery through appropriate management. An additional 11% is in a severe state (land converted to unproductive desert), while the remainder shows varying degrees of moderate to light degradation, where recovery is possible through suitable management practices (Bran *et al.*, 2000; Oliva *et al.*, 2016; Gaitán *et al.*, 2020).

North Patagonia, represented by the provinces of Río Negro and Neuquén, features a varied geography from west to east (from the mountains to the sea) and from north to south (from the large valleys to the plateau). North Patagonia has four ecoregions: the Monte de Llanuras y Mesetas, the Espinal, the Patagonian Forests, and the Patagonian Steppe, all presenting different ecosystems and agricultural activities (Morello *et al.*, 2012). Historically, agriculture has been fundamental and strategic in the development of the North Patagonian region (Blanco and Tozzini, 2017), where economic and social growth have been closely linked to the expansion of export fruit farming (Bendini, 2002). According to the Ministerio de Economía (2023) of Argentina, the primary agricultural activities in this area are mostly associated with the production of pears and apples (90.2% of the national total), with production of fine fruits and other crops of agronomic interest, like hops (see Fig. 9.1). Another significant activity in the region is wine production, as 89.5% of the hectares dedicated to vineyards in Patagonia are located in Río Negro and Neuquén (INV, 2023). The regional production system is based on traditional sectors that supply the domestic market and are part of the main export chains of the country. Given the arid and semiarid conditions, these production systems have been sustained by the rivers that cross the region.

Agriculture remains vital for the livelihoods of the rural population and the local economy in this territory. Agricultural practices are predominantly family run, enhancing social cohesion and providing a stable income source for many communities (Michael and Núñez, 2020). However, there are also corporate/industrial producers, specialized and diversified, with varying degrees of mechanization and access to land, resulting in different challenges for each subsector (Jocou and Rodríguez, 2024).

The strategic significance of agriculture in this area lies in its ability to adapt and adopt sustainable practices. The implementation of advanced technologies and innovative agricultural methods, such as precision agriculture and agroecology, in recent years has enabled tangible contributions (seeds, food, and fodder) and intangible ones (local knowledge) to maximize productive potential, improve external inputs efficiency, and minimize environmental impact (Romitelli *et al.*, 2017; Sheridan and Vásquez, 2023). These practices not only ensure the long-term sustainability of natural resources, but also enhance the resilience of agricultural systems to climate change. However, it is necessary that these practices are implemented effectively and made accessible to all producers, especially small family farmers who may face greater obstacles.

On the other hand, CC poses a significant threat to this potential agroclimatic, as Patagonia has been the region of the country with the greatest temperature increase in recent years (IPCC, 2021). In particular, northeastern Patagonia comprises an area of South American transition, between the Neotropical and Andean regions (Arana *et al.*, 2017). This transitional location accentuates the effects of CC, where an increase in temperature could induce these areas to adopt characteristics typical of Neotropical regions. This transformation could significantly alter thermal and water patterns, directly affecting crop growth and ecosystem development (Garcia *et al.*, 2014).

The agricultural aptitude of North Patagonia presents both opportunities and limitations due to these climatic trends. Increased heat availability could improve crop growth and development, yet it may also pose challenges, such as reduced winter chill necessary for high-chill cultivars and species. Additionally, extreme thermal events may affect crop phenology and yield, while decreasing water availability could impact the viability of irrigated agriculture in certain valleys and the degradation of natural grasslands (Brendel *et al.*, 2020).

The ecosystems of North Patagonia provide essential natural resources that play a crucial role in sustaining agriculture and local livelihoods (Laterra *et al.*, 2021), including regulating water balance, stabilizing the microclimate, and pollinating crops. However, economic activities such as extensive livestock farming and agriculture in the region have led to desertification, soil erosion, biodiversity loss, and the invasion of exotic species, exacerbated by natural and CC disturbances such as temperature and precipitation alterations, fires, and volcanic ash (Garcia *et al.*, 2014; Franzese and Ghermandi, 2014; Bidinost *et al.*, 2020; Valenta *et al.*, 2020). Thus, agriculture contributes to the loss of natural resources and climate change, but it is also part of the solution. Sustainable practices allow for balancing the maintenance of productivity with environmental, economic, and social costs (Garcia Martinez *et al.*, 2017). This way, the ecological value is maintained, contributing to ensuring food security.

9.3 Food Security in the Framework of the SDG: The Role of North Patagonia

The adoption of appropriate public policies is crucial for achieving long-term SDGs and addressing the challenges of CC in agriculture, which directly impacts global food demand. SDG 2 focuses on eradicating hunger, improving nutrition, promoting sustainable agriculture, and achieving food self-sufficiency. However, various factors including climate change, deepening inequalities, and the COVID-19 pandemic, along with their combined effects, have significantly exacerbated the ongoing rise in global hunger and food insecurity (UN, 2023). Therefore, the adoption, adaptation, and implementation of proposed initiatives are critical for addressing current hunger-related challenges.

Aligned with the reality of North Patagonia, the provinces of Río Negro and Neuquén have embraced and tailored some of the objectives outlined in the UN 2030 Agenda to progress towards achieving SDG 2. Both provinces are committed to these goals through various governmental bodies (provincial ministries and secretariats) implementing actions laid out to achieve food security. However, they have adopted a differentiated and adapted approach to the specific characteristics and requirements of each region. Notably, the province of Río Negro has given priority to targets 2.2 and 2.3 and established their respective indicators (Table 9.1) (Constanzo Caso, 2020). Target 2.2 in this province focuses on eliminating all forms of malnutrition by 2030, with specific objectives set for 2025. The chosen indicators reflect a direct concern for the most vulnerable groups, such as pregnant women, nursing mothers, and children. Furthermore, target 2.3 aims to enhance agricultural productivity and the incomes of small-scale producers, with a particular emphasis on women, indigenous peoples, and family farmers. This approach illustrates a comprehensive strategy addressing not only nutrition but also the economic and social development of the agricultural sector.

However, the province of Neuquén has adapted and prioritized targets 2.3, 2.4, 2.4*, and 2.5, and established their corresponding indicators (see Table 9.1) (COPADE, 2020). This province presents a more diverse and technical approach. Target 2.3 aims to increase horticultural and forage productivity and improve the incomes of small-scale producers, demonstrating a quantitative and structural approach to enhancing efficiency and sustainability in the agricultural sector. Target 2.4 introduces sustainability in food production systems, promoting adaptation to CC. Additionally, target 2.5 underscores the importance of preserving the genetic diversity of plants and animals, with indicators measuring genetic enhancements and the survey of native varieties. This set of targets and indicators reflects a strategy not only seeking to boost production and income but also ensuring long-term sustainability and agrobiodiversity.

In comparison, it can be observed that Río Negro concentrates on eradicating malnutrition and the socioeconomic development of its agricultural population, whereas Neuquén adopts a more technical and sustainable approach, centered on productive efficiency and the conservation of genetic resources. Both approaches are complementary and reflect the socioeconomic and environmental peculiarities of each

Table 9.1. Goals adopted and adapted by the provinces of North Patagonia and their respective indicators.

Province	Prioritized adapted targets	Indicators
Río Negro	2.2. By 2030, end all forms of malnutrition, including achieving, by 2025 at the latest, internationally agreed targets on stunting and wasting in children under 6 years of age, and address the nutritional needs of infants, pregnant and lactating women, and older persons.	2.2.1. Nutritional deficit in children under six years of age. 2.2.2. Nutritional deficit in pregnant and lactating women.
	2.3. By 2030, double the agricultural productivity and incomes of small-scale food producers, particularly women, indigenous peoples, family farmers, and pastoralists, including through secure and equal access to land, other productive resources and inputs, knowledge, financial services, markets, and opportunities for value addition and nonfarm employment.	2.3.1. Growth rate of value added by the agricultural sector at constant values. 2.3.2. Growth rate of registered employment in the agricultural sector. 2.3.3. Productivity of the agricultural sector according to main branches (agriculture and livestock).
Neuquén	2.3. By 2030, increase horticultural and forage productivity by 35%, sustain fruit farming, and improve the incomes of small-scale food producers.	2.3.1. Volume of production per unit of work according to the size of the producer in the strategic sectors at the provincial level (agricultural/livestock) and subsectors. 2.3.2. Annual variation in the number of registered family farmers expressed as a percentage. 2.3.3. Average incomes of small-scale food producers.
	2.4. By 2030, strive for the sustainability of food production systems through good agricultural practices that increase productivity and production, strengthen the capacity to adapt to climate change and variability, reduce agricultural risk, and progressively improve the quality of agroecosystems.	
	2.4*. Increase by 40% the area under differentiated production through certification and/or adherence to provincial programs of good agricultural practices or organic farming.	2.4.1. Percentage of agricultural and livestock areas developed following sustainable agricultural practices. 2.4.2. Amount of budget allocated to plans, programs, or projects related to sustainable agricultural production.
	2.5. By 2030, maintain the genetic diversity of seeds, cultivated plants, farm animals, and domesticated and wild related species, and promote access to the benefits derived from the use of genetic resources, including the application of scientific and technological knowledge.	2.5.1. Number of genetic improvements made using scientific and technological techniques. 2.5.2. Number of native varieties surveyed in the province of Neuquén.

province. However, integrating these efforts can offer a comprehensive model for implementing the 'Zero Hunger' target in regions with diverse needs and characteristics. The integration of specific policies, both in improving nutrition and in sustainable agricultural development, is crucial to effectively addressing hunger and malnutrition in Argentina.

9.4 Contributions From the Bioeconomy

In the current context, achieving food security goals involves addressing a series of future challenges. The complexity of these demands a multidisciplinary approach encompassing technological innovation, effective public policies, sustainable natural resource management, and agricultural education (Magrin *et al.*, 2014; Villagra *et al.*, 2023). From the bioeconomic perspective, Argentina can contribute to addressing these challenges through various aspects, such as biological resources, nature tourism, science and technology, biotechnology, the productive sector and human resource development, although these contributions vary depending on the region (MINCyT, 2023). Nationally, biological resources and the productive sector predominate, highlighting the importance of agriculture, livestock, and other natural resource-based activities to the bioeconomy of the country. Patagonia shows a strong contribution from the productive sector, similar to other regions, underscoring the significance of primary production.

However, according to MINCyT (2023), contributions from science and technology, and biotechnology in Patagonia are moderate compared to other parts of the country, indicating a notable although not predominant presence of scientific and biotechnological activities. In this context, harnessing contributions to the bioeconomic potential can foster sustainable development and innovation across various sectors. A viable option is to strengthen these sectors with more sustainable and technologically advanced practices. Therefore, integrating propagation strategies supported by biotechnologies, bioprospecting, and the utilization of agroindustrial waste would enhance the value of the region's resources. These advancements are crucial for achieving SDG 2 goals and promoting the socioeconomic development and resilience of local communities.

In North Patagonia, contributions in this regard include the bioprospecting of native flora, which has allowed the identification of biocompounds with potential use in the regional bioeconomy. This has led to chemical characterizations of extracts or fractions with biological activity of interest for industry (Quintin *et al.*, 2019). The incorporation of native plants representative of the North Patagonia region, such as those from the genus *Neltuma* and *Geoffroea*, can be considered as an alternative to diversify food matrices, improve food security, and reduce environmental impact. In this regard, Boeri *et al.* (2017) and Hoffmann *et al.* (2024) demonstrated that flours elaborated through the milling of the fruits of *Neltuma alpataco* (Phill. C.E. Hughes and G.P. Lewis) and *N. flexuosa* (D.C. C.E. Hughes and G.P. Lewis) (Fig. 9.2A,B), respectively are rich in fiber and polyphenols, which confer antioxidant properties and the ability to inhibit enzymes associated with metabolic syndrome. Flour derived from these fruits emerges as a promising ingredient for functional foods. Additionally, it has been reported that whole flour from *Geoffroea decorticans* (Gill. ex Hook. et. Arm.) Burk (chañar) fruits (Fig. 9.2C) could be used as a source of natural antioxidants for food use (Maschio *et al.*, 2021).

Additionally, Boeri *et al.* (2020) successfully extracted phenolic compounds from *Berberis microphylla* G. Forst (calafate), considered a superfruit (Fig. 9.2E). These authors identified around 14 phenolic compounds, primarily quercetin and caffeic acid, which demonstrated high antioxidant capacity through *in vitro* and *in vivo* methods using the zebrafish model. This highlights the essential role of bioprospecting native species, especially in vulnerable areas such as arid and semiarid regions, as it not only diversifies the food supply but also promotes health. Therefore, future challenges should focus on developing healthy and safe alimentary products, based on native species such as those of the *Neltuma*, *Geoffroea*, and *Berberis* genera to promote access to safe and nutritious food that could help reduce forms of malnutrition.

Many plant compounds investigated for their potential use in various industries are

Fig. 9.2. (A) *N. alpataco* fruits. (B) *N. flexuosa* fruits. (C) *G. decorticans* flour. (D) *L. divaricata* Cav. fruits. (E) *B. microphylla* fruit and seeds. (F) *H. lupulus in vitro* plants. (G) Grape pomace.

secondary metabolites (SMs). SMs include a wide variety of compounds necessary for the environmental interaction of plants, such as survival, adaptation, and competitiveness. SMs are synthesized in very low concentrations under natural conditions but their levels can vary depending on different environmental stimuli. Factors such as cultivation difficulties, low productivity, phytogeographic and seasonal variations, and economic costs limit the industrial production of these phytochemicals (Halder *et al.*, 2019).

Therefore, different plant biotechnology techniques, such as *in vitro* plant tissue culture (PTC), have been developed over more than three decades as an attractive alternative for the production of economically important SMs. Elicitation is one of the most effective and widely used biotechnological tools for inducing new biocompounds. Elicitors are biotic or abiotic stimuli that can induce or enhance the biosynthesis of specific SMs of interest. In this regard, the *Larrea* genus has been studied (Fig. 9.2D) due to its traditional use in various preparations, such as tea, for its antiinflammatory and antioxidant properties. These are attributed to the presence of flavonoids, tannins, and essential oils. Consequently, these species have been introduced into tissue culture both to obtain bioactive compounds like nordihydroguaiaretic acid (NDGA) and quercetin, and for *in vitro* elicitation (Palacio *et al.*, 2012; Cabrera, 2023).

Another strategy in plant biotechnology is associated with genetic improvement through PTC. This technique gains importance in the face of the imminent effects and challenges of CC, as it will require the selection of crops adapted to new local conditions for the conservation and development of agricultural resources (Andrade *et al.*, 2017). One economically significant crop in North Patagonia is hops (*Humulus lupulus*), primarily used in brewing beer. This region has optimal natural conditions for its cultivation (photoperiod, temperatures, and precipitation). However, this crop has not experienced significant advances since its inception concerning the adoption of cutting-edge technologies and the lack of plant breeding programs. Additionally, this crop faces a series of CC challenges related to the ideal conditions of light, temperature, and water for successful growth, and is also susceptible to certain fungal and viral diseases that affect yield (Agehara *et al.*, 2020; Nievas *et al.*, 2021).

Hops have been successfully introduced to PTC (Fig. 9.2F) (Lagos *et al.*, 2022; Ha *et al.*, 2023; Iacuzzi *et al.*, 2023; Di Sario *et al.*, 2024), which allows for the future application and combination of biotechnological techniques to sustainably produce high-quality plants, ensuring a consistent supply of material. Furthermore, the incorporation of genetic improvement techniques, particularly regarding the tolerance of hops to different types of stress, is highly relevant for North Patagonia, as they could have a direct and indirect impact, both in the medium and long term, on the current commercial and productive chains of the bioeconomic region.

These developments facilitate research and the creation of new varieties adapted to the needs of the productive sector and the food market, as well as to new scenarios. As with hops, these techniques can be used to improve the cultivation conditions of other economically important species. In this sense, biotechnology aims to increase investments in rural infrastructure and agricultural research and extension services associated with techniques that make the cultivation of certain strategic and innovative crops more efficient.

Plant biotechnology can be expanded to species found in North Patagonia, such as those belonging to *Neltuma*, *Geoffroea*, and *Berberis* genera, among others. Concerning this, the most common use of PTC is mass multiplication of plant material (micropropagation) to acquire a high number of plants that can be utilized for ecosystem restoration, conservation of genotypes of interest, or domestication of native species for use without exerting pressure on natural populations. In recent years, PTC propagation and conservation methodologies have been applied to native species found in North Patagonia, like *N. alpataco*, *N. caldenia* (Burkart C.E. Hughes and G.P. Lewis), *Larrea divaricata*, and *Berberis microphylla* (Verdes, 2007; Palacio *et al*., 2012; McLeod *et al*., 2015; Boeri and Sharry, 2018). Through these biotechnological innovations, strategies focused on enhancing agricultural productivity can be formulated, as well as diversifying the market for small producers by cultivating high-quality native species with food potential, as mentioned earlier, and with the ability to adapt to challenges linked with climate change.

Another relevant aspect for the region is the utilization of agroindustrial waste, particularly in peripheral territories, due to its potential for producing high-quality products with territorial identity. Grape pomace (Fig. 9.2G), the primary solid waste of the wine industry, constitutes up to 30% of the total weight of grapes produced in vineyards (Moro *et al*., 2021). It is mainly composed of skin waste, pulp, seeds, and stems (Beres *et al*., 2017; Bordiga *et al*., 2019; Maicas and Mateo, 2020). Its valorization is associated with its potential as a source of nutrients and bioactive compounds and the reduction of environmental pollution (Beres *et al*., 2017; Averilla *et al*., 2019; Antonić *et al*., 2020; Sridhar and Charles, 2021; Ferrer-Gallego and Silva, 2022). It is noteworthy for its high content of polyphenols, retaining approximately 70% of the phenolic compounds from the grape (Moro *et al*., 2021). Additionally, it is rich in dietary fiber, vitamins (C, E, A), and minerals such as iron, phosphorus, potassium, zinc, manganese, calcium, magnesium, and copper. Grape pomace seeds are especially rich in proteins, peptides, fatty acids, phytosterols, and carotenoids, linked with beneficial health effects, such as antiinflammatory, cardioprotective, antimicrobial, and anticancer properties (Sousa *et al*., 2014; Nayak *et al*., 2018; Mohamed Ahmed *et al*., 2020; Spinei and Oroian, 2021; Varedesara *et al*., 2021; Yang *et al*., 2021; Chakka and Babu, 2022; Alvarez-Ossorio *et al*., 2022).

Another fruit commonly used in the Patagonian region for artisanal product manufacturing is the calafate (*B. microphylla*). The seeds of this species are one of the primary industrial discards but it has been identified that they contain a substantial protein content, which possesses antioxidant activity that increases following a hydrolysis process. In this sense, this by-product could be considered as a potential functional ingredient (Boeri *et al*., 2020; Dalzotto *et al*., 2021). By employing agroindustrial waste, the efficiency and sustainability of agricultural production are elevated, waste and environmental pollution are diminished, and new products with high nutritional and functional value are created. Consequently, these practices aim to ensure the sustainability of food production systems.

The integration of the strategies described above represents both potential and a challenge for the global bioeconomy to ensure food security. The scientific knowledge generated in the North Patagonia region concerning this economic model not only constitutes a transformative approach to redefine the relationships between agriculture, biomass, and regional industry but also contributes to achieving most of the SDG 2 targets. However, for all these contributions to materialize into commercial reality, it is necessary to reinforce the regulatory framework, such as the Código Alimentario Argentino, and to implement comprehensive policies, such as the Argentina National Plan Against Hunger (Plan NacionalArgentina Contra el Hambre, DGISE, 2021). The success of these measures hinges on effective coordination among the different levels of government, active participation from society, and constant investment in research and development.

9.5 Conclusion

Due to the natural conditions of North Patagonia and the intensification of CC effects in this region, agriculture continues to play a crucial role in the economy and society, particularly in achieving SDG 2. The agricultural production of this region not only significantly contributes to the national and international food supply but also helps stabilize markets and reduce price volatility, which is essential for ensuring affordable and healthy products. The region presents multiple opportunities and challenges derived from its geography and climate, generating possibilities for technological innovation and sustainable agricultural practices, which contribute to the four pillars of food security (availability, access, utilization, and stability) and the three dimensions of sustainability (environmental, social, and economic).

The bioeconomy approach and the integration of biotechnologies have proven to be viable strategies globally for increasing productivity, diversifying the food supply, and reducing environmental impact. Initiatives such as the bioprospecting of native species and the utilization of agroindustrial waste not only improve the sustainability of production systems but also promote health. The bioeconomy provides safe and nutritious food to a growing population, offering development opportunities and new jobs while reducing environmental damage and waste. Moreover, its potential extends well beyond the reduction of effects of climate change and biodiversity conservation, as it opens new possibilities for rural livelihoods, the lives of indigenous peoples and local communities, and innovation at all levels of society. The policies and actions adopted by Río Negro and Neuquén, as well as the knowledge generated in the region, reflect a commitment to the 2030 Agenda. The collaboration between businesses, public organizations, and academia constitutes an integrated model that not only strengthens the implementation of sustainable solutions in North Patagonia but also offers a valuable example for other regions facing similar challenges, highlighting the importance of collaboration in achieving the SDGs.

References

Agehara, S., Acosta-Rangel, A., Gallardo, M. and Vallad, G. (2020) Selection and preparation of planting material for successful hop production in Florida. *EDIS* 2020(5).

Alvarez-Ossorio, C., Orive, M., Sanmartín, E., Alvarez-Sabatel, S., Labidi, J. *et al.* (2022) Composition and techno-functional properties of grape seed flour protein extracts. *ACS Food Science & Technology* 2(1), 125–135.

Andrade, F., Taboada, M., Lema, D., Maceira, N., Echeverría, H. *et al.* (2017) *Los desafíos de la agricultura argentina*. Ediciones INTA, Buenos Aires.

Antonić, B., Jančíková, S., Dordević, D. and Tremlová, B. (2020) Grape pomace valorization: A systematic review and meta-analysis. *Foods* 9(11), 1627.

Arana, M.D., Martínez, G.A., Oggero, A.J., Natale, E.S. and Morrone, J.J. (2017) Map and shapefile of the biogeographic provinces of Argentina. *Zootaxa* 4341, 420–422.

Averilla, J.N., Oh, J., Kim, H.J., Kim, J.S. and Kim, J.S. (2019) Potential health benefits of phenolic compounds in grape processing by-products. *Food Science and Biotechnology* 28, 1607–1615.

Bendini, M. (2002) La configuración de una región agrícola dinámica en Argentina: Actores sociales en la negociación local. In: Perez, E. and Sumpsi, J.M. (eds) *Políticas, Instrumentos y Experiencias de Desarrollo Rural En América Latina y Europa. Ministerio de Agricultura, Alimentación y Medio Ambiente.* Ministerio de Agricultura, Alimentación y Medio Ambiente, Madrid, pp. 357–370.

Beres, C., Costa, G.N.S., Cabezudo, I., Silva-James, N.K., Teles, A.S.C. *et al.* (2017) Towards integral utilization of grape pomace from winemaking process: A review. *Waste Management* 68, 581–594.

Bidinost, F., Gaetano, A., Alvarez, R., Reuque, R., Ojeda, J. *et al.* (2020) Escenario prospectivo del entramado cárnico de la Región Sur de Río Negro: ¿Oportunidad para fortalecer la soberanía alimentaria y generar trabajo genuino? Technical report. INTA, CR Patagonia Norte, EEA Bariloche.

Blanco, G. and Tozzini, M.A. (2017) La historia agraria norpatagónica en clave regional. In: Bandieri, S. and Fernández, S. (eds) *La Historia Argentina En Perspectiva Local y Regional*. Teseo, Buenos Aires, pp. 437–466.

Boeri, P., Piñuel, L., Sharry, S. and Barrio, D. (2017) Caracterización nutricional de la harina integral de algarroba (*Prosopis alpataco*) de la patagonia Norte de Argentina. *Revista de La Facultad de Agronomía* 116, 129–140.

Boeri, P. and Sharry, S. (2018) Somatic embryogenesis of alpataco (*Prosopis alpataco* L.). In: Jain, S. and Gupta, P. (eds) *Step Wise Protocols for Somatic Embryogenesis of Important Woody Plants*, Vol. II. Springer, Dordrecht, pp. 189–198.

Boeri, P., Piñuel, L., Dalzotto, D., Monasterio, R., Fontana, A. *et al.* (2020) Argentine Patagonia barberry chemical composition and evaluation of its antioxidant capacity. *Journal of Food Biochemistry* 44(7), e13254.

Bordiga, M., Travaglia, F. and Locatelli, M. (2019) Valorisation of grape pomace: An approach that is increasingly reaching its maturity – A review. *International Journal of Food Science & Technology* 54(4), 933–942.

Bran, D., Ayesa, J. and Lopez, C. (2000) Regiones ecológicas de Río Negro. *Technical Communication* 59.

Brendel, A.S., Del Barrio, R.A., Mora, F., León, E.A.O., Flores, J.R. *et al.* (2020) Current agro-climatic potential of Patagonia shaped by thermal and hydric patterns. *Theoretical and Applied Climatology* 142(3–4), 855–868.

Cabrera, A.A. (2023) Valorización de la "jarilla rastrera" (*Larrea ameghinoi* speg.): Introducción al cultivo de tejidos *in vitro* y bioprospección química. Undergraduate dissertation, Universidad Nacional de Río Negro, Río Negro, Argentina.

Chakka, A.K. and Babu, A.S. (2022) Bioactive compounds of winery by-products: Extraction techniques and their potential health benefits. *Applied Food Research* 2(1), 100058.

Constanzo Caso, C. (2020) Apoyo técnico para la implementación de los objetivos de desarrollo sostenible de la agenda 2030 en la provincia de Río Negro. Available at: https://rionegro.gov.ar/contenido/Hidrogeno%20Verde/agenda2030.pdf (accessed 14 June 2024).

COPADE (2020) La Agenda 2030 y los ODS en la Provincia del Neuquén. Available at: www.copade.gob.ar/wp-content/uploads/2020/12/La-agenda-2030-y-los-ODS-en-la-Provincia-del-Neuquen-v9.pdf (accessed 14 June 2024).

Dalzotto, D., Piñuel, L., Boeri, P. and Sharry, S. (2021) El potencial de la biodiversidad nativa para el desarrollo de la bioeconomía regional – el caso del calafate. *Bionatura* 1659–1663.

Del Valle, H.F.D., Elissalde, N.O., Gagliardini, D.A. and Milovich, J. (1998) Status of desertification in the Patagonian region: Assessment and mapping from satellite imagery. *Arid Land Research and Management* 12(2), 95–121.

DGISE (2021) Plan Nacional Argentina Contra el Hambre. Available at: www.argentina.gob.ar/sites/default/files/informe_plan_argentina_contra_el_hambre_primer_semestre_de_2021.pdf (accessed 17 June 2024).

Di Sario, L., Zubillaga, M.F., Zepeda Moreno, C.F., Pizzio, G.A. and Boeri, P.A. (2024) Micropropagation and synthetic seeds production of *Humulus lupulus* L. var. Mapuche: Optimization, advances, and agricultural applications. Available at: www.preprints.org/manuscript/202405.1906/v1

FAO (n.d.) Reduce rural poverty: Overview. Available at: www.fao.org/reduce-rural-poverty/overview/es/ (accessed 17 June 2024).

Ferrer-Gallego, R. and Silva, P. (2022) The wine industry by-products: Applications for food industry and health benefits. *Antioxidants* 11, 10. DOI: 10.3390/antiox11102025.

Franzese, J. and Ghermandi, L. (2014) Early competition between the exotic herb *Rumex acetosella* and two native tussock grasses with different palatability and water stress tolerance. *Journal of Arid Environments* 106, 58–62.

Gaitán, J.J., Bran, D.E. and Oliva, G.E. (2020) Patagonian desert. In: Goldstein, M.I. and DellaSala, D.A. (eds) *Encyclopedia of the World's Biomes*. Elsevier, St Louis, pp. 163–180.

Garcia, R.A., Cabeza, M., Rahbek, C. and Araújo, M.B. (2014) Multiple dimensions of climate change and their implications for biodiversity. *Science* 344(6183), 1247579.

Garcia Martinez, G.C., Ciari, G., Gaitan, J.J., Opazo, W.J., Caruso, C.A. *et al.* (2017) Diagnóstico de la evolución del clima, los pastizales naturales y las cargas ganaderas para el noroeste de la provincia de Chubut en el período 2000-2014. Technical report. INTA, CR Patagonia Sur EAA, Esquél.

Gaur, M.K. and Squires, V.R. (2018) Geographic extent and characteristics of the world's arid zones and their peoples. In: Gaur, M.K. and Squires, V.R. (eds) *Climate Variability Impacts on Land Use and Livelihoods in Drylands*. Springer, Cham, pp. 3–20.

Ha, T.H., Lyu, J.I., Lee, J.H., Ryu, J., Park, S.H. *et al.* (2023) Studies on growth characteristics and propagation method of introduced hop (*Humulus lupulus* L.) cultivars. *Korean Journal of Plant Research* 36, 181–190.

Halder, M., Sarkar, S. and Jha, S. (2019) Elicitation: A biotechnological tool for enhanced production of secondary metabolites in hairy root cultures. *Engineering in Life Sciences* 19(12), 880–895.

Hoffmann, E., Boeri, P., Monasterio, R., Fontana, A., Puppo, M.C. *et al.* (2024) Bioactivity of *Prosopis alpataco* and *Prosopis flexuosa* flours: Healthy alternatives as ingredients for nutritional foods. *Food Bioscience* 58, 103778.

Huang, J., Ji, M., Xie, Y., Wang, S., He, Y. *et al.* (2016) Global semi-arid climate change over last 60 years. *Climate Dynamics* 46(3–4), 1131–1150.

Iacuzzi, N., Salamone, F., Farruggia, D., Tortorici, N., Vultaggio, L. *et al.* (2023) Development of a new micropropagation protocol and transfer of *in vitro* plants to *in vivo* conditions for cascade hop. *Plants* 12(15), 2877.

INV (2023) Región Patagónica, provincias de Río Negro y Neuquén. In: *Laboratorio Estadístico Del Instituto Nacional de Viticultura*. Mendoza, Argentina.

IPCC (2021) Climate change 2021: The physical science basis. In: *Contribution of Working Group I to the Sixth Assessment Report of the Intergovernmental Panel on Climate Change*. Cambridge University Press, Cambridge, p. 2391.

IPCC (2022) *Climate change and land*. Available at: www.ipcc.ch/srccl/ (accessed 6 April 2025).

Jaramillo, E., Trigo, E., Henry, G., Aramendis Ramírez, R.H. and Castaño, A. (2019) *La bioeconomía. Nuevo marco para el crecimiento sostenible en América Latina*. Available at: http://hdl.handle.net/10554/43705 (accessed 9 April 2025).

Jocou, A. and Rodríguez, A. (2024) Síntesis de aspectos del sistema productivo hortícola en la Norpatagonia.Technical report. INTA, Alto Valle.

Lagos, F.S., Zuffellato-Ribas, K.C. and Deschamps, C. (2022) Vegetative propagation of hops (*Humulus lupulus* L.): Historical approach and perspectives. *Semina Ciências Agrárias* 43(3), 1373–1394.

Laterra, P., Nahuelhual, L., Sirimarco, X., Monjeau, A., Gluch, M. *et al.* (2021) Natural capital and local employment in Argentine Patagonia. In: Peri, P.L., Martínez Pastur, G. and Nahuelhual, L. (eds) *Ecosystem Services in Patagonia. Natural and Social Sciences of Patagonia*. Springer, Cham, pp. 451–467.

Magrin, G.O., Marengo, J.A., Boulanger, J.-P., Buckeridge, M.S., Castellanos, E. *et al.* (2014) Central and South America. In: Barros, V.R., Field, C.B., Dokken, D.J., Mastrandrea, M.D. and Mach, K.J. (eds) *Climate Change 2014: Impacts, Adaptation and Vulnerability. Part B: Regional Aspects*. Cambridge University Press, Cambridge, pp. 1499–1566.

Maicas, S. and Mateo, J.J. (2020) Sustainability of wine production. *Sustainability* 12(2), 559.

Maschio, J.I., Boeri, P., Dalzotto, D., Piñuel, L. and Sharry, S. (2021) Caracterización química y nutricional de harina de chañar de diferente distribución geográfica. *Bionatura* 6(1), 1563–1568.

MAyDS (2020) Informe del estado del ambiente 2020. Available at: www.argentina.gob.ar/sites/default/files/iea_2020_digital.pdf (accessed 25 June 2024).

McLeod, C., Ojeda, A., Moreno, S., Molina, A. and Pino, M.T. (2015) Seed germination and *in vitro* propagation of different *Berberis microphylla G* Forst accessions from Magellan's Region in Chile. In: *XIV EUCARPIA Symposium on Fruit Breeding and Genetics*. Bologna, Italy, pp. 225–230.

Michael, C.L. and Núñez, P.G. (2020) Globalization in Northern Andean Patagonia: A reading from family farming. *Revista Austral de Ciencias Sociales* 38, 247–268.

MINCyT (2023) Mapa del Potencial Bioeconómico: Primera estimación georeferenciada del potencial bioeconómico del país. Available at: www.argentina.gob.ar/sites/default/files/presentacion_mpba.pdf (accessed 25 June 2024).

Ministerio de Economía (2023) Provincial production reports. Available at: www.argentina.gob.ar/economia/politicaeconomica/informes-productivos-0/informes-provinciales/caracterizacion-productiva (accessed 5 June 2024).

Mohamed Ahmed, I.A., Özcan, M.M., Al Juhaimi, F., Babiker, E.F.E., Ghafoor, K. *et al.* (2020) Chemical composition, bioactive compounds, mineral contents, and fatty acid composition of pomace powder of different grape varieties. *Journal of Food Processing and Preservation* 44(7), e14539. DOI: 10.1111/jfpp.14539.

Morello, J., Matteucci, S.D., Rodriguez, A.F. and Silva, M.E. (2012) *Ecorregiones y complejos ecosistémicos argentinos*. Orientación Gráfica Editora, Buenos Aires.

Moro, K.I.B., Bender, A.B.B., Silva, L.P. and Penna, N.G. (2021) Green extraction methods and microencapsulation technologies of phenolic compounds from grape pomace: a review. *Food and Bioprocess Technology* 14(8), 1407–1431.

Nayak, A., Bhushan, B., Rosales, A., Turienzo, L.R. and Cortina, J.L. (2018) Valorisation potential of cabernet grape pomace for the recovery of polyphenols: Process intensification, optimisation and study of kinetics. *Food and Bioproducts Processing* 109, 74–85.

Nievas, W.E., Villarreal, P., Rosati, A., Rodriguez, A.B. and Lago, J. (2021) *El cultivo del lúpulo. Aspectos agroambientales y económicos para el alto valle del Río Negro*. Ediciones INTA, Buenos Aires.

Oliva, G., Gaitan, J. and Ferrante, D. (2016) *Humans Cause Deserts: Evidence of Irreversible Changes in Argentinian Patagonia Rangelands*. Springer Earth System Sciences, Cham, pp. 363–386.

Palacio, L., Cantero, J.J., Cusidó, R.M. and Goleniowski, M.E. (2012) Phenolic compound production in relation to differentiation in cell and tissue cultures of *Larrea divaricata* (Cav.). *Plant Science* 193–194, 1–7.

PNUD (2015) Informe Sobre Desarrollo Humano 2015: Trabajo al servicio del desarrollo humano. Available at: https://frh.cvg.utn.edu.ar/pluginfile.php/20196/mod_resource/content/2/Informe%20Naciones%20Unidas%20Indice%20de%20Desarrollo%20Humano%202015.pdf (accessed 25 June 2024).

Quintin, D., Garcia-Gomez, P., Ayuso, M. and Sanmartin, A.M. (2019) Active biocompounds to improve food nutritional value. *Trends in Food Science & Technology* 84, 19–21.

Ramarao, M.V.S., Sanjay, J., Krishnan, R., Mujumdar, M., Bazaz, A. *et al.* (2019) On observed aridity changes over the semiarid regions of India in a warming climate. *Theoretical and Applied Climatology* 136(1–2), 693–702.

Rodríguez, A.G., Mondaini, A.O. and Hitschfeld, M.A. (2017) *Bioeconomía en américa latina y el caribe: contexto regional y perspectivas*, Serie Desarrollo Productivo y Empresarial, Vol. 215. CEPAL, Santiago de Chile.

Romitelli, M., Raffo, D., Curetti, M., Magdalena, C. and Fernández, D. (2017) Mecanización de la fruticultura en los valles irrigados de la Norpatagonia. Dissemination report. INTA, Alto Valle.

Sheridan, M. and Vásquez, P. (2023) Contribuyendo a consolidar la expansión agroecológica en el Alto Valle. Dissemination report. INTA, Bahía Blanca.

Sousa, E., Uchoa-Tomaz, A.M., Carioca, J.O., Morais, S.M., Lima, A. *et al.* (2014) Chemical composition and bioactive compounds of grape pomace (*Vitis vinifera* L.). *Food Science and Technology* 34, 135–142.

Spinei, M. and Oroian, M. (2021) The potential of grape pomace varieties as a dietary source of pectic substances. *Foods* 10(4), 867.

Sridhar, K. and Charles, A.L. (2021) Fortification using grape extract polyphenols – a review on functional food regulations. *International Journal of Food Science & Technology* 56(8), 3742–3751.

UN (2023) Sustainable Development Goals Report 2023: Special Edition. Available at: https://unstats.un.org/sdgs/report/2023/ (accessed 13 June 2024).

Valenta, M.D., Golluscio, R.A., Frey, A.L., Garibaldi, L.A. and Cipriotti, P.A. (2020) Short-term responses to sheep grazing in a Patagonian steppe. *Rangeland Journal* 42(1), 1.

Varedesara, M.S., Ariaii, P. and Hesari, J. (2021) The effect of grape seed protein hydrolysate on the properties of stirred yogurt and viability of *Lactobacillus casei* in it. *Food Science & Nutrition* 9(4), 2180–2190.

Verdes, P. (2007) Micropropagación de *Prosopis caldenia* burk: Estado actual y perspectivas. *Revista Científica Agropecuaria* 11, 45–51.

Villagra, S., Mañueco, L., Caballe, G., Martínez, R.S., Buda, V. *et al.* (2023) Aportes para repensar los sistemas productivos Norpatagónicos en un contexto de déficit hídrico y cambio climático.Technical report. INTA, Alto Valle.

Yang, C., Shang, K., Lin, C., Wang, C., Shi, X. *et al.* (2021) Processing technologies, phytochemical constituents, and biological activities of grape seed oil (GSO): A review. *Trends in Food Science & Technology* 116, 1074–1083.

10 Biofortification: Improving Nutritional Content for Human Health and Food Security

Rakesh Kumar Bairwa[1], Om Prakash Gupta[1*], Arun Gupta[1], Antim Kundu[2], Anamika Pandey[3], Dharmendra Kumar[4] and Mohd. Kamran Khan[3]
[1]ICAR-Indian Institute of Wheat and Barley Research, Karnal, India; [2]ICAR-Central Soil Salinity Research Institute, Karnal, India; [3]Department of Soil Science and Plant Nutrition, Faculty of Agriculture, Selcuk University, Turkiye; [4]ICAR-Central Potato Research Institute, Shimla, India

Abstract
Micronutrients such as Zn, Fe, and Se are essential for optimal human health. Wheat is a staple food for a large section of the global population, and its adaptability to different climatic conditions makes it crucial in addressing food security challenges. Malnutrition is a global issue, particularly affecting large sections of people in the resource-poor and developing nations. The level of these micronutrients can be enhanced by fortification and biofortification strategies, which have the advantage of cost-effectiveness, sustainability, and the potential to address malnutrition on a larger scale. Biofortification involving conventional breeding, agronomic practices, microbe-mediated methods, and modern biotechnological techniques can enhance the nutrient content of food crops. Agronomic biofortification and microbe-mediated biofortification are promising, but they should be used in a specific situation or in conjunction with other strategies. Genetic strategies involve both conventional and molecular breeding techniques to enhance the nutritional quality of crops. These strategies include characterizing and evaluating germplasm to identify superior lines, discovering genes or quantitative trait loci (QTLs), and incorporating beneficial genes into elite genetic backgrounds or popular varieties through both conventional and molecular breeding methods. Genetic biofortification is more promising in the long term. The integration of advanced biotechnological strategies, particularly CRISPR-Cas genome editing, represents a significant advancement in wheat biofortification. Despite challenges, the benefits of these technologies in combating malnutrition and improving global food security are substantial, paving the way for nutritionally enriched wheat varieties that can help address global nutritional needs. In this chapter, all these techniques of biofortification are discussed with their significant impact.

Keywords: micronutrient deficiency, biofortification approaches, phytic acid, wheat, Zn and Fe biofortification

10.1 Introduction

Micronutrients are vital for optimal health, as the body cannot synthesize them. Therefore, it relies completely on external food sources to obtain these essential elements. The human body needs micronutrients like zinc, iron, and selenium to function properly. However, most of

*Corresponding author: opguptaiari@gmail.com

DOI: 10.1079/9781800627659.0010

the cereal varieties currently grown are low in these essential nutrients. As a result, consuming these micronutrient-poor cereals can lead to 'hidden hunger,' a form of malnutrition that predominantly impacts women and children (Jawaldeh *et al.*, 2019). Micronutrient deficiency is a pressing health issue, particularly in developing nations (Wakeel *et al.*, 2018).

Iron (Fe) and zinc (Zn) deficiencies are the most prevalent micronutrient deficiencies globally, affecting 38% of pregnant women and 43% of children under 5 years of age (Gupta *et al.*, 2024).

Zinc deficiency leads to a range of physiological problems, including growth retardation, increased susceptibility to infections, abnormal immune responses, and impaired learning abilities (Wessels and Rink, 2020). In Africa, zinc deficiency contributes to 14.4% of deaths from diarrhea, 10.4% of deaths from malaria, and 6.7% of deaths from pneumonia among children aged 6 months to 5 years (Berhe *et al.*, 2019). It also affects pregnancy outcomes and impairs linear growth in children (Terrin *et al.*, 2015). Furthermore, zinc influences immune defense, growth, bone health, and cognitive function (Huang *et al.*, 2015). Zinc is crucial for metabolic functions, with approximately 10% of human proteins being zinc dependent (Kambe *et al.*, 2015; Krężel and Maret, 2016). Additionally, it is involved in the metabolic processes of over 300 enzymes in the body (Lu *et al.*, 2021).

According to the World Health Organization (WHO), the recommended daily zinc intake is 5 mg for infants (7–12 months), 10 mg for children (1–10 years), 15 mg for adults (11–51 years), and 20–25 mg for pregnant women. The average zinc concentration in wheat is only 28.48 mg/kg (Wang *et al.*, 2020). However, the recommended daily zinc intake is relatively low, at 11 mg for men and 8 mg for women (De Groote *et al.*, 2021). To meet these daily requirements through wheat alone, the zinc concentration would need to be increased to 45 mg/kg.

Iron deficiency is the most common of all micronutrient deficiencies, affecting approximately one-third of the world's population (Mantadakis *et al.*, 2020). Iron is essential for human health, playing a crucial role in the body's growth and development. It is integral to several physiological mechanisms, notably including the facilitation of oxygen transport from the pulmonary system to peripheral tissues, the development of the immune system, and the synthesis of oxygen-carrying proteins like hemoglobin and myoglobin (Jomova *et al.*, 2022). Anemia prevalence is more common in Africa, affecting 106 million women and 103 million children, and in South-East Asia, where it impacts 244 million women and 83 million children (WHO, 2023). Iron is essential for respiration, photosynthesis, antioxidant defenses, various biochemical pathways, and symbiotic nitrogen fixation (Rubio and Ludden, 2008; Kobayashi *et al.*, 2019). Iron deficiency can cause a range of health issues, including increased mortality risk, restricted physical and mental development, weakened immunity, anemia, and fatigue (Black *et al.*, 2008). In pregnant women, iron deficiency is often associated with premature births, low birthweight infants, impaired growth and development in newborns, and reduced cognitive abilities (Bailey *et al.*, 2015).

Human iron (Fe) nutrition is closely linked to the Fe concentration in plants, which in turn depends on the Fe content in the soil (Li *et al.*, 2023). In saline, alkaline, sodic, and calcareous soils, Fe transforms into insoluble Fe–hydroxyl complexes, leading to natural Fe deficiencies in these soils. In calcareous soils, which cover approximately 30% of the Earth's land surface, Fe deficiency is commonly observed within a pH range of 7.4–8.5 (Fageria *et al.*, 2008; Prasad and Djanaguiraman, 2017). Additionally, the removal of aleurone and embryonic tissues during postharvest processing can significantly reduce Fe content in food, contributing to lower Fe levels (Connorton and Balk, 2019; Narwal *et al.*, 2020). Therefore, the deficiency of micronutrients depends on a variety of factors.

Globally, over 40% of the population relies on wheat as a staple food (Giraldo *et al.*, 2019). Its adaptability to different climates and environmental stresses makes wheat a crucial component in addressing food security challenges (Qadir *et al.*, 2018). It provides about 20% of the total dietary calories and proteins consumed worldwide (Shiferaw *et al.*, 2013). In South Asian countries like India, Pakistan, Bangladesh, and Nepal, wheat provides 70% of the daily calorie intake for the rural population (Zhang *et al.*, 2010).

10.2 Fortification

Fortification is the process of improving the nutritional status of food by adding essential micronutrients that are deficient in the food during processing or at the point of use (WHO, 2017). Fortification involves adding nutrients directly to food during processing. This differs from biofortification, which enhances nutrient levels at the production level through breeding or agricultural practices. In biofortification, nutrient levels in crops are increased during their growth period, whereas in conventional fortification, nutrients are added manually during postharvest processing of the crops (Mannar and Hurrell, 2018). It is an economically feasible and socially acceptable method for improving nutritional status, particularly in developed countries. It can be employed on a large scale or used to target specific groups such as children and pregnant women (WHO, 2017). Fortification strategies include diversifying the diet, providing supplements, and food fortification by adding micronutrients and vitamins to foods (Yadav *et al.*, 2020).

Diversifying diets can effectively address malnutrition, but making diverse, affordable diets available to everyone presents significant challenges. Achieving widespread access to varied diets is a long-term goal that requires a fundamental overhaul of the global agri-food system to reduce inequalities. Consequently, dietary diversification is often impractical for large populations in developing countries, which limits its effectiveness (Lowe, 2024). Supplementation involves distributing dietary supplements through various welfare programs, which has proven effective in the short term. However, when these programs face funding issues and are discontinued, people may stop using the supplements due to their cost and availability. Supplementation can effectively address specific instances of malnutrition, but it is often impractical on a large scale due to the high recurring costs and logistical distribution challenges (Lowe, 2024). In contrast, food fortification adds essential minerals and vitamins directly to food products before consumption. In industrialized countries, fortification is the most cost-effective and sustainable method for delivering micronutrients to large populations. This approach has been successfully implemented in Asia, Latin America, and Africa.

During fortification, key considerations include the stability of the product, its safety, and the bioavailability of the nutrients. Enhancing the bioavailability of micronutrients can sometimes reduce the stability of the product. Fortification can also impact the color, taste, odor, and appearance of the product, which may affect consumer acceptability. Increasing nutrient solubility can improve bioavailability, but this can also lead to reactions with other food components, potentially altering product quality and stability. However, recent technological advancements have made it possible to boost bioavailability without affecting reactivity or significantly altering the food's quality (Hurrel and Cook, 1992). The stability of nutrients is a crucial aspect, influenced by factors such as temperature, pH, light, and oxygen during processing and storage. These parameters must be carefully managed, otherwise nutrients may degrade during these stages, leading to their loss and reduced availability at the time of consumption.

Food fortification offers several advantages. It does not require significant changes in eating habits or dietary patterns, nor does it depend on individual compliance. It can simultaneously provide a substantial portion of the recommended dietary allowances (RDAs) for various micronutrients in a continuous manner. Additionally, food fortification can be seamlessly integrated into existing food production and distribution systems, making it a sustainable option over the long term. However, there are notable disadvantages. First, fortified foods cannot substitute for a well-balanced diet. Fortification may also be unsuitable for all individuals within a target population. Moreover, due to limited purchasing power, fortified foods often fail to reach the poorest segments of society, who are most vulnerable to micronutrient deficiencies. In very low-income groups, multiple micronutrient deficiencies often coexist, and relying solely on fortified foods to meet all micronutrient needs is unrealistic, even with advanced fortification technologies. Technological challenges further complicate the effectiveness of fortification.

Despite being more cost-effective compared to other interventions, the overall expense of implementing food fortification programs can be substantial, potentially limiting their effectiveness (Yadav *et al.*, 2020). Therefore, the main challenges to the broader adoption of fortification are the stability of nutrients in the fortified product and financial constraints.

10.3 Biofortification Approaches

Biofortification employs various techniques, including conventional breeding, agronomic practices, microbe-mediated methods, and modern biotechnology, to enhance the nutrient content of food crops while maintaining the traits valued by consumers and farmers (Nestel *et al.*, 2006). It aims to enhance the micronutrient content in the edible parts of food crops (White and Broadley, 2009). Beyond just increasing micronutrient levels, improving the bioavailability of these nutrients is also crucial. Unlike fortification, this approach is cost-effective, sustainable, and has the potential to address malnutrition on a larger scale, especially in resource-poor and developing countries. Additionally, it offers an advantage over fortification by reducing recurring expenses.

10.3.1 Agronomic biofortification approach to enhance Fe and Zn content

For centuries, mineral fertilizers have been used to enhance the nutrient content of soil and crops. Similarly, agronomic biofortification has been employed to boost the mineral levels in cereal grains (Aciksoz *et al.*, 2011). This technique includes applying organic manures and synthetic fertilizers to the soil and foliage, as well as priming the seeds (Shivay *et al.*, 2016). Agronomic biofortification has proven effective in significantly increasing Zn and Fe content in wheat grain through the application of these micronutrients. Various experimental studies have demonstrated the benefits of this approach, highlighting its effectiveness in boosting Zn and Fe levels (Ram *et al.*, 2024).

Micronutrients can be applied either directly to the soil or via foliar application. Foliar application tends to be more efficient than soil-based methods because it allows for better uptake by plant foliage and reduces the risk of soil immobilization. Numerous studies have demonstrated that Zn fertilization effectively increases the Zn content in grains (Table 10.1). Recently, ferti-fortification has gained attention for its rapid and effective enhancement of Zn and Fe content in wheat grains. Iron can be applied from various sources, and a positive correlation with grain iron content has been observed with the application of different sources. Different wheat varieties respond differently to micronutrient applications. Genetic diversity plays a crucial role in determining the success of foliar enrichment of Fe and Zn in wheat grains. The use of multiple micronutrients, including selenium (Se) and iodine (I), along with Fe and Zn, is gaining popularity due to its potential benefits (Zou *et al.*, 2019). In addition to increasing Fe and Zn content in grain, foliar application of Fe and Zn has also shown notable improvements in grain yield and protein content (Dapkekar *et al.*, 2018).

While the precise mechanisms behind these advantages are not fully deciphered, they may be attributed to increased enzymatic activity, which could enhance photosynthesis and improve the translocation of assimilates to the seed. Applying these micronutrients to the soil also boosts the overall uptake of nitrogen, phosphorus, and potassium (Meena and Fathima, 2017).

Recent studies have demonstrated that agronomic biofortification can further enhance micronutrient levels in newly developed Fe and Zn biofortified wheat varieties. For instance, the Zn content in the Zn-biofortified wheat cultivar Zincol-2016 increased by approximately 43%, reaching 53 mg/kg from an initial 37 mg/kg, following the application of Zn to the soil at a rate of 6 mg/kg (Yaseen and Hussain, 2021). In the study conducted by Zou *et al.* (2019), the effects of applying Zn, I alone, and a micronutrient cocktail (containing I, Zn, Se, and Fe) were evaluated. The results indicated a significant increase in grain Zn content, rising from 28.6 mg/kg to 46.0 mg/kg with the micronutrient application alone. Similarly, grain iodine content increased from 24 μg/kg to 361 μg/kg. When using the micronutrient cocktail, grain Zn levels rose from 28.6 mg/kg to 47.1 mg/kg, while grain iodine increased from 24 μg/kg to 249 μg/kg. Additionally, selenium levels in the grain increased from 90 μg/kg to 338 μg/kg,

Table 10.1. Effect of soil and foliar application of Zn on grain Zn content.

Zn source	Dose/conc. of Zn fertilizer	Stage of foliar application	Increase in grain zinc content (%)	Reference
$ZnSO_4.7H_2O$	15 kg/ha + 0.3%	Booting stage	48	Ranjbar and Bahmaniar (2007)
Zn	0.5%	Stem elongation and flowering stage	99	Pahlavan-Rad and Pessarakli (2009)
$ZnSO_4.7H_2O$	0.5%	Heading and milking stage	83.5	Zou *et al.* (2012)
	50 kg/ha + 0.5%		87.48	
Zn	0.3%	-	84	Wang *et al.* (2015)
Zn + N	0.3%	-	83	
Zn + P	0.3%	-	55	
$ZnSO_4.7H_2O$	30 kg/ha + 0.4%		56.5	Barut *et al.* (2017)
$ZnSO_4.7H_2O$	50 kg/ha + 0.5%	Booting and milking	88	Chattha *et al.* (2017)
	0.5%		81	
$ZnSO_4$	0.3%	Three foliar application at 7-day intervals during early milk stage	95.7	Wang *et al.* (2017)
$ZnSO_4$ + N	0.3%		101	
$ZnSO_4$ + P	0.3%		67.9	
Zn SO_4 + K	0.3%		121	
Zn-glycine	0.23%	Early grain-filling stage	108	Li *et al.* (2018)
$ZnSO_4$	0.3%		102	
Zn-glycine plus urea	0.23%		90	
$ZnSO_4$ combined with N (urea)	0.3%		65.6	
Zn-complexed chitosan nanoparticles plus urea	0.4%	5 foliar applications at 7 days interval during grain-development stage	35	Dapkekar *et al.* (2018)
U+$ZnSO_4$	400 mg/l		50	
$ZnSO_4$.7 H_2O	0.5%	Maximum tillering stage, anthesis, and initiation of grain filling stage	34	Nayak *et al.* (2023)
$ZnSO_4$.7 H_2O	2.5 kg/ ha +0.5%	Maximum tillering and anthesis stage	47	
$ZnSO_4.7H_2O$	50 kg/ha + 0.5%	Flowering and milking stage	39.6	Ram *et al.* (2024)
$FeSO_4.7H_2O + ZnSO_4.7H_2O$	0.25% + 0.50%		43.8	

and there was approximately a 12% increase in grain Fe content.

Micronutrients can also be applied in the form of nanoparticles (Khan *et al.*, 2021). One study demonstrated that using Zn complexed with chitosan nanoparticles (Zn-CNP) for ferti-fortification of durum wheat resulted in about a 35% increase in Zn content (Dapkekar *et al.*, 2018). Seed priming involves soaking seeds in nutrient-rich solutions before sowing, which enhances germination, seedling establishment, and yield (Raj and Raj, 2019). Notably, priming wheat seeds with a zinc sulfate heptahydrate solution led to a significant increase in the Zn concentration of wheat grains (Praharaj *et al.* 2019). Several key factors, such as soil characteristics, genetic variability among wheat varieties, the composition of applied fertilizers, and their interactions, significantly affect the variability in Fe and Zn content in wheat grains. These findings underscore the considerable potential of agronomic biofortification as a rapid method to enhance Fe and Zn levels in wheat.

This approach offers a promising solution to swiftly address the global risk of micronutrient deficiencies, particularly in resource-constrained populations, making it a viable strategy for tackling this critical public health issue. Agronomic biofortification, like supplements and fortification, is probably best used in a specific situation or in conjunction with other strategies. It is used as a foliar application when minerals cannot be easily translocated to edible tissues.

10.3.2 Microbe-mediated biofortification for micronutrients

Microbe-mediated biofortification is a promising strategy that uses soil and plant-associated microorganisms to enhance the uptake and accumulation of essential micronutrients in plants (Singh and Prasanna, 2020). This section explores the potential of microbe-mediated biofortification in addressing micronutrient deficiencies, discusses various types of microorganisms involved, and highlights the current state of research and future prospects.

Microbe-mediated biofortification offers a unique opportunity to enhance the micronutrient content of crops by leveraging the natural interactions between plants and microorganisms. Microorganisms, including bacteria, fungi, and mycorrhizae, play a crucial role in the soil ecosystem, promoting nutrient cycling and plant growth. These microorganisms can enhance the bioavailability of micronutrients in the soil, facilitating their uptake by plants. The microbe-mediated biofortification process involves the use of beneficial microorganisms to improve the micronutrient content of crops, thereby enhancing their nutritional value (Verma *et al.*, 2021). Different types of microorganisms play a key role in increasing the availability of micronutrients to food crops.

- *Rhizobacteria*: plant growth-promoting rhizobacteria (PGPR) are known to enhance nutrient uptake and plant growth by various mechanisms, including solubilization of nutrients, production of phytohormones, and suppression of plant pathogens. Rhizobacteria have been used to increase the bioavailability of micronutrients such as zinc, iron, and selenium in crops. For instance, *Pseudomonas fluorescens* and *Bacillus subtilis* have been shown to enhance the uptake of zinc and iron in wheat and rice, respectively (Backer *et al.*, 2018).
- *Mycorrhizal fungi*: mycorrhizal fungi form symbiotic associations with plant roots, enhancing nutrient and water uptake. Arbuscular mycorrhizal (AM) fungi, in particular, are known to improve the uptake of phosphorus, zinc, and iron by plants. Studies have shown that the application of AM fungi can significantly increase the micronutrient content of crops such as maize, wheat, and soybean (Kumar *et al.*, 2021a).
- *Endophytic bacteria and fungi*: endophytes, which reside within plant tissues, have been recognized for their potential to enhance plant growth and nutrient uptake. Certain endophytic bacteria and fungi can produce siderophores, organic compounds that bind to iron and make it more available for plant uptake. The use of endophytes has been shown to increase the iron content in crops such as maize and chickpea (Singh *et al.*, 2022).
- *Biofertilizers*: biofertilizers are formulations containing live microorganisms that can enhance plant growth and nutrient uptake. Biofertilizers containing a combination of rhizobacteria, mycorrhizal fungi, and other beneficial microorganisms have been developed to improve the micronutrient content of various crops. For example, biofertilizers containing *Azospirillum brasilense* and *Glomus*

intraradices have been shown to increase the iron and zinc content in maize and rice, respectively (Kumar *et al.*, 2021b).

Microbe-mediated biofortification can occur through several mechanisms.

- *Solubilization of micronutrients:* certain microorganisms can solubilize micronutrients, making them more available for plant uptake. For example, phosphate-solubilizing bacteria and fungi can release phosphorus from insoluble compounds in the soil, enhancing its availability to plants. Similarly, zinc-solubilizing bacteria can convert insoluble zinc compounds into soluble forms, increasing their bioavailability (Kumar *et al.*, 2024).
- *Production of siderophores:* siderophores are organic compounds produced by microorganisms that bind to iron and other micronutrients, forming soluble complexes that can be easily taken up by plants. Siderophore-producing bacteria and fungi have been used to enhance the iron content of crops, improving their nutritional value (Aloo *et al.*, 2020).
- *Phytohormone production:* certain microorganisms can produce phytohormones such as auxins, gibberellins, and cytokinins, which promote plant growth and nutrient uptake. These phytohormones can enhance root growth and nutrient absorption, leading to increased micronutrient content in crops (Santos *et al.*, 2021).
- *Suppression of plant pathogens:* microorganisms can suppress the growth of plant pathogens, reducing plant stress and improving nutrient uptake. By promoting healthier plant growth, these microorganisms can enhance the accumulation of micronutrients in crops (Vitorino and Bessa, 2017).

Recent research has focused on exploring the potential of microbe-mediated biofortification in various crops and environments. The use of siderophore-producing bacteria has been explored to increase the iron content of wheat grains. The application of *Bacillus subtilis* enhanced the iron content of wheat grains by solubilizing iron in the soil and facilitating its uptake by plants (Dhiman *et al.*, 2023).

Despite the promising potential of microbe-mediated biofortification, several challenges need to be addressed to ensure its widespread adoption and success.

- *Variability in microbial efficacy:* the efficacy of microbe-mediated biofortification can vary depending on factors such as soil type, environmental conditions, and crop species. Further research is needed to understand the interactions between microorganisms, plants, and the environment to optimize biofortification strategies.
- *Regulatory and safety concerns:* the use of genetically modified microorganisms and biofertilizers raises concerns about their safety and impact on the environment. Regulatory frameworks need to be developed to ensure the safe and responsible use of microbe-mediated biofortification techniques.
- *Farmer adoption and awareness:* farmer awareness and acceptance of microbe-mediated biofortification techniques are crucial for their successful implementation. Extension services and educational programs are needed to promote the benefits of biofortification and encourage farmer adoption.
- *Research and development:* continued research and development are needed to explore the potential of new microorganisms and biofertilizers for microbe-mediated biofortification. Collaborative efforts between researchers, policymakers, and industry stakeholders are essential to advance the field and promote sustainable agricultural practices.

Microbe-mediated biofortification offers a sustainable and cost-effective approach to enhancing the micronutrient content of food crops, addressing the global challenge of micronutrient deficiencies. By leveraging the natural interactions between plants and microorganisms, microbe-mediated biofortification can improve the nutritional value of crops and contribute to food security and human health. Continued research and development, along with supportive policies and farmer education, are crucial to realizing the full potential of microbe-mediated biofortification.

10.3.3 Genetic approaches of biofortification

Genetic strategies involve both conventional and molecular breeding techniques to enhance the nutritional quality of crops. These strategies include characterizing and evaluating germplasm to identify superior lines, discovering genes or quantitative trait loci (QTLs), and incorporating beneficial genes into elite genetic backgrounds or popular varieties through both conventional and molecular breeding methods (Saini *et al.*, 2020). Conventional breeding utilizes the existing genetic diversity found within the primary, secondary, and tertiary gene pools to increase the Fe and Zn content in wheat grains. Therefore, a key requirement for enhancing Fe and Zn levels in wheat is having genetic diversity in the traits of interest (Khan *et al.*, 2020). Mutagenesis can also be leveraged to accelerate conventional breeding by generating novel alleles. Mutation breeding can increase trait variability through the use of physical and chemical mutagens or by employing methods like TILLING and EcoTILLING (Maqbool *et al.*, 2018). Molecular breeding techniques involve identifying QTLs and marker–trait associations (MTAs) through genome-wide association studies (GWAS) and QTL mapping. These loci are then incorporated into agronomically superior lines using marker-assisted selection (MAS) and marker-assisted recurrent selection (MARS).

Modern wheat varieties are deficient in micronutrient content and not able to provide the level of micronutrient considered appropriate for human diets (Khokhar *et al.*, 2020). In contrast, wheat's wild relatives, synthetics, and landraces possess ample genetic variation for enhancing the micronutrient content in wheat grain (Table 10.2). Diploid progenitors of wheat, including *Aegilops tauschii*, einkorn (*Triticum monococcum*), wild emmer (*T. dicoccoides*), *T. polonicum*, *T. spelta*, and various landraces of *T. aestivum*, have shown potential as rich sources of iron and zinc (Çakmak *et al.*, 2000; Goel *et al.*, 2018; Arora *et al.*, 2019). These genetic resources can be exploited by developing amphidiploids and synthetic hexaploids, introducing chromosomal segments from related species, and screening landraces for desirable quality traits.

Genes associated with Zn efficiency are essential for improving quality. Research has shown that disomic wheat-rye addition lines and an octoploid triticale line exhibit greater Zn efficiency compared to wheat parents, including increased shoot Zn concentration and higher grain Zn content. The rye chromosomes, particularly 1R (which has the highest Zn efficiency at approximately 80%), 2R, and 7R, contain these Zn efficiency genes (Çakmak *et al.*, 1997). Among cultivated wheat varieties, hexaploid wheat shows the highest Zn efficiency at 64%, followed by diploid wheat at 60% and tetraploid wheat at 36%. The superior Zn efficiency of hexaploid wheat is attributed to the DD genome of *Aegilops tauschii* and the AA genome of *Triticum monococcum*. This is further supported by the development of synthetic hexaploid wheat, which demonstrates improved Zn levels under both Zn-deficient and Zn-sufficient soil conditions. These synthetic hexaploid lines offer a way to boost Zn content in elite wheat varieties while avoiding the linkage drag often associated with gene transfer from wild relatives (Çakmak *et al.*, 1997; Calderini and Ortiz-Monasterio, 2003; Velu *et al.*, 2017; Kumar *et al.*, 2019). Positive correlation among grain Zn, Fe (Peleg *et al.*, 2008; Velu *et al.*, 2017; Amiri *et al.*, 2020) also enables breeders to simultaneously enhance both elements.

Genetic strategies are widely acknowledged as effective methods for enhancing iron and zinc content in wheat grain. The success of these strategies is evident from the biofortified wheat varieties that have been developed and released for commercial cultivation in India, Pakistan, Bangladesh, Nepal, Mexico, and Bolivia (Table 10.3). Over the past decade, the positive impact of these biofortified wheat varieties on enhancing Fe and Zn content has been evident. These varieties can play a key role in reducing malnutrition in developing countries (Velu *et al.*, 2020; Khan *et al.*, 2021; Ibba *et al.*, 2022). They have the potential to address specific nutrient deficiencies while allowing people to maintain their current dietary habits and cropping practices (Kamble *et al.*, 2022).

However, the long-term performance of biofortified varieties across various environments and locations is a subject of investigation. Key factors to be studied include the stability of Fe and Zn content, yield, end-product quality, *in vitro* and *in vivo* bioavailability, and the overall health impact on the target population. The performance of

Table 10.2. Micronutrient content of wild relatives and landraces of modern wheat.

Species/ landrace (no. of genotypes)	Fe range (mg/kg)	Mean (mg/kg)	Zn range (mg/kg)	Mean (mg/kg)	Instrument used for Fe and Zn estimation	Reference
A. longissima (5)	59.12–81.59	73.24	24.99–50.52	41.66	Atomic absorption spectrophotometer	Rawat *et al.* (2009)
A. kotschyi (14)	22.89–90.96	67.46	22.29–58.61	49.27		
A. peregrina (10)	34.37–82.32	52.85	33.13–49.49	39.54		
A. cylindrica (3)	52.21–93.27	66.76	32.38–52.18	38.51		
A. ventricosa (3)	55.41–93.52	65.75	24.01–39.08	33.81		
A. geniculata (3)	52.25–81.97	69.95	31.93–40.81	37.7		
A. tauschii	30.33–69.44	47.26	17.54–49.78	30.73	Inductively coupled plasma-optical emission spectrometer	Arora *et al.* (2019)
A. umbellulata (45)	69.74–348.09	137.45	49.83–101.60	70.42	Atomic absorption spectrophotometer	Song *et al.* (2023)
A. tauschii (7)	45.88–93.41	72.70	23.71 -–47.11	34.41		
A. comosa (5)	77.81–108.38	95.43	44.97–60.50	53.57		
A. markgrafii (2)	99.79–129.17	114.48	95.89–100	97.95		
T. boeoticum (12)	41–92	79	45–177	89	–	Çakmak *et al.* (2000)
T. monococcum (13)	34–85	48	29–89	56		
T. dicoccoides (46)	20–159	75	24–78	40		
Triticum turgidum subsp. *dicoccoides* (825)	15–109	46	14–190	61	Inductively coupled plasma-atomic emission spectrometry	Çakmak *et al.* (2004)
Triticum turgidum subsp. *dicoccoides* (22)	48–88	67	71–134	102	Inductively coupled plasma-optical emission spectroscopy	Peleg *et al.* (2008)
T. boeoticum (13)	23.88–40.50	30.91	22.12–39.06	29.27	Atomic absorption spectrophotometer	Rawat *et al.* (2009)
T. araraticum (6)	23.10–59.06	29.85	19.27–30.54	23.52		

Continued

Table 10.2. Continued

Species/ landrace (no. of genotypes)	Fe range (mg/kg)	Mean (mg/kg)	Zn range (mg/kg)	Mean (mg/kg)	Instrument used for Fe and Zn estimation	Reference
Triticum monococcum var. *monococcum* (5)	38.2–55.5 SE 178.5–440	45.9 278.9	20.1–27.8	22.4	Inductively coupled plasma-atomic emission spectrometry	Zhao *et al.* (2009)
Triticum turgidum var. *dicoccum* (5)	31–39.6 SE 150.6–325.8	34.1 229.2	15.8–30.3	22.8		
T. aestivum var. *spelta* (5)	37.9–44.1 SE 125.1–244	41.8 209	16.8–28	22.9		
Triticum turgidum subsp. *dicoccoides (19)*	27–86	44.4	39–115	66	Inductively coupled plasma-optical emission spectrometer	Gomez-Becerra *et al.* (2010)
Iranian landraces (50)	23.74–67.6	49.32	12.4–39.45	27.98	Atomic absorption spectrophotometer	Heidari *et al.* (2016)
Turkish landraces (86)	35.53–53.08		22.66–38.57		Atomic absorption spectrophotometer	Akcura and Kokten (2017)
Indian landraces (108)	32.7–54.5	40.6	15.8–66.3	45	Energy-dispersive X-ray fluorescence spectrophotometer	Goel *et al.* (2018)
Watkins landraces and the UK elite cultivar Paragon (245)	19.4–71.2	34.3	15.6–60.1	33.6 mg/kg	Inductively coupled plasma-mass spectrometry	Khokhar *et al.* (2020)
Iranian wheat landraces (158)	30.2–52.1	38.6	27.9–65.0	43.6	Energy dispersive X-ray fluorescence spectrometry	Maryami *et al.* (2020)

biofortified varieties is also influenced by environmental factors, including soil type, fertility levels, and environmental interactions (Badakhshan *et al.*, 2013; Khokhar *et al.*, 2018). Therefore, evaluating these genotypes over multiple years and across various locations is essential before they are released for commercial cultivation.

Addressing these aspects will be crucial for the ongoing success and expansion of biofortified wheat programs. In the past, the HarvestPlus project and CIMMYT developed synthetic lines by crossing *Triticum aestivum* subsp. *spelta* with *T. turgidum* subsp. *durum* and subsequently transferred quality traits to elite lines. The wild wheat species and landraces exhibit variation in grain zinc concentration, which could be utilized to meet the breeding goal of 40 mg/kg zinc in wheat grain. Genetic biofortification of wheat tends to be more advantageous in the long term compared to agronomic biofortification, provided that there is sufficient genetic variation within the wheat germplasm.

10.3.4 Biotechnological approaches for wheat biofortification

Biofortification aims to enhance the concentration of essential minerals such as iron and zinc, as well as vitamins in crops through various conventional and nonconventional breeding

Table 10.3. List of bio-fortified wheat varieties developed by plant breeding and released for commercial cultivation around the globe (Gupta *et al*., 2024; Mishra *et al*., 2024; HarvestPlus, 2003).

Variety	Nutritional quality	Other features	Year of release	Developer/sources
India				
WB 02	Zn: 42 ppm; Fe: 40 ppm; Protein: 12.4%	Yield: 51.6 q/ha	2017	ICAR-IIWBR, Karnal
DBW173	Fe: 40.7 ppm; Protein: 12.5%	Yield: 47.2 q/ha	2018	
DBW187 (Karan Vandana)	Fe: 43.1 ppm	Yield 61.3 q/ha	2018	
DDW47 (d)	Fe: 40.1 ppm; Protein: 12.7%	Yield: 37.3 q/ha	2020	
DBW 303 (Karan Vaishnavi)	Fe: 35.8 ppm; Zn: 36.9 ppm; Protein: 12.1%	Yield: 81.2 q/ha	2020	
DDW48 (d)	Fe: 38.8 ppm; Zn: 39.7 ppm; Protein: 12.1%	Yield: 47.4 q/ha	2020	
DBD332 (Karan Aditya)	Protein: 12.2%	Yield: 78.3 q/ha	2021	
DBD327 (Karan Shivani)	Zn: 40.6 ppm	Yield: 79.4 q/ha	2021	
DBW316 (Karan Prema)	Fe: 38.2 ppm; Zn: 38.2 ppm; Protein: 13.2%	Yield: 41 q/ha	2023	
DDW55 (d)	Zn: 43.3 ppm	Yield: 35.6 q/ha	2023	
DBW371 (Karan Vrinda)	Fe: 44.9 ppm; Protein: 12.2%	Yield: 75.2 q/ha	2023	
DBW372 (Karan Varuna)	Fe: 40.8 ppm Protein: 12.2%	Yield: 75.2 q/ha	2023	
HPBW01 (Ankur Shiva)	Zn: 40.6 ppm; Fe: 40 ppm; Protein: 12.3%	Yield: 51.7 q/ha	2017	PAU, Ludhiana
PBW 752	Fe: 37.1 ppm; Zn: 38.7 ppm; Protein: 12.4%	Yield: 49.7 q/ha	2018	
PBW 757	Zn: 42.3 ppm	Yield: 36.7 q/ha	2018	
PBW 771	Zn: 41.4 ppm	Yield: 50.3 q/ha	2020	
PBW833	Protein: 12.9%	Yield: 42.7 q/ha	2023	
PBW872	Zn: 40.7 ppm; Fe: 42.3 ppm	Yield: 75.2 q/ha	2023	
PBW Zinc 2	High Zn	Yield: 58 q/ha	2023	

Continued

Table 10.3. Continued

Variety	Nutritional quality	Other features	Year of release	Developer/sources
HI8759 (Pusa Tejas) (d)	Zn: 42.8 ppm; Fe: 42.1 ppm; Protein: 12.0%	Yield: 56.9 q/ha	2017	IARI Indore
HI1605 (Pusa Ujala)	Fe: 43 ppm; Protein: 13.1%	Yield: 30.0 q/ha	2017	
HI8777 (d)	Fe: 48.7 ppm; Zn: 43.6 ppm; Protein: 14.3%	Yield: 18.5 q/ha	2018	
HI8802 (d)	Fe: 39.5 ppm; Zn: 35.9 ppm; Protein: 13.0 %	Yield: 29.1 q/ha	2020	
HI8805 (d)	Fe: 40.4 ppm; Protein: 12.8%	Yield: 30.4 q/ha	2020	
HI1633 (Pusa Vani)	Fe: 41.6 ppm; Zn: 41.1 ppm; Protein: 12.4%	Yield: 41.7 q/ha	2020	
HI1636 (Pusa Vakula)	Zn: 44.4 ppm	Yield: 56.6 q/ha	2021	
HI8823 (Pusa Prabhat) (d)	Zn: 40.1 ppm; Protein: 12.1%	Yield: 38.5 q/ha	2021	
HI1650 (Pusa Ojaswi)	Zn: 42.7 ppm	Yield: 57.2 q/ha	2023	
HD3171	Zn: 47.1 ppm	Yield: 28.01 q/ha	2017	IARI, Delhi
HD 3249	Fe: 42.5 ppm	Yield: 48.8 q/ha	2020	
HD 3298	Fe: 43.1 ppm; Protein: 12.1%	Yield: 39 q/ha	2020	
HD3369 (Pusa Wheat)	Fe: 40.6 ppm	Yield: 50.6 q/ha	2023	
HD3406	Protein: 12.2%	Yield: 54.7 q/ha	2023	
Unnat HD3407	Protein: 12%	Yield: 46.7 q/ha	2023	
HD3411 (NICRA PUSA wheat 3411)	Protein: 12.4%	Yield: 46.7 q/ha	2023	
HD3386	Zn: 41.8 ppm		2024	
MACS4028 (d)	Zn: 40.3 ppm; Fe: 46.1 ppm; Protein: 14.7%	Yield: 19.3 q/ha	2018	Agharkar Research Institute, Pune, Maharashtra
MACS4058 (d)	Fe: 39.5 ppm; Zn: 37.8 ppm; Protein: 12.82%	Yield: 29.6 q/ha	2020	
MACS6768	Fe: 41.2 ppm; Zn: 45.1 ppm; Protein: 12%	Yield: 56.6 q/ha	2023	
UAS375	Protein: 13.8%	Yield: 21.4 q/ha	2018	University of Agricultural Sciences, Dharwad, India

Continued

Table 10.3. Continued

Variety	Nutritional quality	Other features	Year of release	Developer/sources
HUW838	Zn: 41.8 ppm	Yield: 51.3 q/ha	2021	Banaras Hindu University, Varanasi
BHU-1				Banaras Hindu University, Varanasi
MP1358 (JW1358)	Fe: 40.6 ppm; Protein: 12.1%	Yield: 30.9 q/ha	2021	Jawaharlal Nehru Krishi Vishwavidyalaya, jabalpur
Shalimar wheat 3 (SKW356)	Fe: 40.5 ppm; Zn: 43.8 ppm	Yield: 32.3 q/ha		SKUAS&T, Srinagar
Abhay (Zinc Shakthi)	High Zn	Registered by private seed companies and grower	2015	Nirmal Seeds, Harvest Plus and participatory variety selection
Zinc Shakthi (Chitra)	High Zn	Registered by private seed companies and grower	2016	Participatory variety selection
Pakistan				
NR- 421 (Zincol-16)	High Zn	Contains up to 30% more zinc than traditional wheat varieties	2015	Pakistan Agriculture Research Council/CIMMYT
Akbar-19	High Zn	Contains up to 26% more zinc than traditional wheat varieties	2019	Faisalabad Agricultural Research Institute/CIMMYT
Nawab 21	High Zn		2021	Regional Agricultural Research Institute (RARI), Bahawalpur
Tarnab Renbar	High Zn			ARI Tarnab, Peshawar
Tarnab Gandum 1	High Zn			ARI Tarnab, Peshawar
Bangladesh				
BARI Gom 33	High Zn	7–8 ppm Zn, and also resistance to wheat blast	2017	CIMMYT, Mexico
Mexico				
Nohely-F2018 2018	High Zn	Nohely-F2018 released in Mexico for the Mexicali valley of northern Sonora region		CIMMYT, Mexico
Bolivia				
Iniaf-Okinawa	High Zn		2018	INIAF, Bolivia and CIMMYT, Mexico

Continued

Table 10.3. Continued

Variety	Nutritional quality	Other features	Year of release	Developer/sources
Nepal				
Zinc Gahun 1	High Zn		2020	NARC, Nepal and CIMMYT, Mexico
Zinc Gahun 2	High Zn		2020	
Bheri-Ganga	High Zn		2020	
Himganga	High Zn		2020	
Khumal-Shakti	High Zn		2020	

approaches (MAS, transgenic methods, genome editing, e.g. CRISPR), and agronomic practices. This section will delve into the key biotechnological strategies employed in wheat biofortification, which primarily target the improvement of nutritional content by increasing levels of critical micronutrients. The approaches discussed include MAS, QTL mapping, GWAS, genomic-assisted selection, transgenic techniques, RNA interference (RNAi), and genome editing technologies like ZFN, TALEN, and CRISPR-Cas9.

Marker-assisted selection and QTL mapping

Marker-assisted selection (MAS), quantitative trait loci (QTL) mapping and genome-wide association studies (GWAS) are integral to modern plant breeding, especially for traits controlled by multiple genes, such as those targeted in biofortification. QTL mapping and GWAS identify genomic regions associated with specific quantitative traits by correlating genetic markers with phenotypic characteristics (Tanksley and McCouch, 1997; Huang and Han, 2014). QTL mapping involves crossing individuals with differing traits, analyzing their offspring, and using statistical methods to pinpoint marker–trait associations while GWAS analyzes naturally occurring genetic variation in a large, diverse population without controlled crosses. It relies on the existing genetic diversity within a population, which could be human, animal, or plant populations. Finally, the QTL locations identified (through QTL mapping or GWAS) on chromosomes help elucidate the genetic basis of the traits.

Once QTLs are identified and associated markers are established, MAS can be employed to enhance breeding efficiency. MAS uses these markers to select individuals that carry the beneficial QTL alleles, facilitating earlier and more accurate selection of plants with desired traits. By incorporating genetic markers into the selection process, MAS speeds up breeding compared to traditional phenotypic selection, enabling breeders to identify and select plants with advantageous traits, quality traits like high Fe, Zn or vitamin concentration in wheat, at an earlier stage in development (Collard and Mackill, 2008; Korte and Farlow, 2013; Singh and Kumar, 2016). Subsequent research efforts built upon these findings by identifying key QTLs on chromosomes such as 2A and 3A, which were instrumental in breeding programs aimed at increasing iron levels in wheat (Tiwari *et al.*, 2009). QTLs associated with both iron and zinc content were often found in similar genomic regions, allowing for the simultaneous enhancement of these micronutrients within the same breeding lines (Velu *et al.*, 2018). Various studies which had identified QTLs for grain iron and zinc content are given in Table 10.4. Additionally, markers linked to lower phytic acid content were identified, which facilitated the development of wheat varieties with improved mineral bioavailability by reducing antinutrient levels (Raboy, 2007).

Enhancing wheat's provitamin A content is a critical biofortification goal. Distelfeld *et al.* (2007) identified key molecular markers linked to carotenoid biosynthesis, facilitating targeted breeding for increased provitamin A levels. These advancements in molecular breeding techniques have also enabled the introgression of beneficial traits from nutrient-rich wild relatives and landraces into commercial wheat cultivars, further boosting their nutritional value (Monasterio and Graham, 2000).

Table 10.4. List of QTLs identified for grain zinc and iron content.

Species/cross	No. of total QTLs	Chromosome/chromosome arm	Reference
Bread wheat	GFe-1	5A	Kato *et al.* (2000)
Bread wheat	Zn conc.-4; Zn content-7	1A, 2D, 3A, 4A, 4D, 5A, 7A	Shi *et al.* (2008)
Bread wheat	GZn-4; GFe-1	3D,4B,6B,7A; 3D	Genc *et al.* (2009)
Einkorn wheat	GZn-2; GFe-3	7A; 2A,7A	Tiwari *et al.* (2009)
Durum and wild emmer	GZn-6; GFe-11	2A, 5A, 6B, 7A, 7B; 2A, 2B, 3A,3B, 4B, 5A, 6A, 6B, 7A, 7B	Peleg *et al.* (2009)
Bread wheat	GZn-2; GFe-2	4B, 5A; 5A	Xu *et al.* (2012)
Bread wheat	GFe-4	4D, 5A, 7A,7B	Shi *et al.* (2013)
Bread wheat	GZn-2; GFe-6	1A, 4A; 2A, 3D, 4D, 7B, 7D	Roshanzamir *et al.* (2013)
Bread wheat	GZn-3	1B, 2B, 3A	Hao *et al.*, 2014
Bread wheat	GZn-3	1B, 2B, 3A	Hao *et al.* (2014)
Synthetic hexaploid wheat (SHW)	GZn-4; GFe-4	2D, 3D, 4D, 5D; 2B, 5B, 5D, 7D	Pu *et al.* (2014)
Bread wheat	GZn-3; GFe-4	3D, 4D, 5B; 4A, 4D, 5A, 5B	Pu *et al.* (2014)
Spelta and bread wheat cross	GZn-5; GFe-5	2A, 2B, 3D, 6A, 6B; 1A, 2A, 3B	Srinivasa *et al.* (2014)
SHW	GZn-3; GFe-3	1D, 2D, 5D; 2D, 4D, 5D	Gorafi *et al.* (2016)
Bread wheat	GZn-2; GFe-1	1B, 2B; 2B	Tiwari *et al.* (2016)
SHW and *spelta* cross	GZn-4; GFe-3	1B, 6A,7B; 3A, 4B, 5B,	Crespo-Herrera *et al.* (2016)
Sphaerococcum and bread wheat cross	GZn-10; GFe-7	1B, 1D, 2B, 3A, 3D, 6A, 6B, 7A, 7B; 1B, 2A, 2B, 3A, 6B, 7B	Velu *et al.* (2017)
SHW and *spelta* cross	GZn-12; GFe-7	1A, 1B, 3B, 3D, 4A, 5B, 6A, 7B, 7D; 2A, 2B, 3B, 4A, 4D, 5B	Crespo-Herrera *et al.* (2017)
SHW and bread wheat cross	GZn-3; GFe-5	4BS, 6AL, 6BL; 2BL, 2DS, BS, 6AL, 7DS	Crespo-Herrera *et al.* (2017)
SHW and bread wheat cross	GZn-5; GFe-4	2A, 4A, 5A, 7A, 7B; 2A, 5A, 7A, 7B	Krishnappa *et al.* (2017)
Ae. tauschii	GZn-4; GFe-5	4D, 2D, 6D, 7D; 4D, 2D, 1D, 7D, 3D	Arora *et al.* (2019)
Panel of SHW, *spelta* and *polonicum* wheat	GZn-3; GFe-4	1A, 2A, 5B; 1A, 2A, 3B, 5B	Cu *et al.* (2020)
Bread wheat	GZn-7; GFe-4	1DS, 2AS, 3BS, 4DS, 6AS, 6DL, 7BL; 3BL, 4DS, 6AS,7BL	Wang *et al.* (2021)
Bread wheat	GFe-5	1D	Wang *et al.* (2022)
Bread wheat	GZN-2; GFe-3	7B, 1A; 2B, 3B, 6A	Rathan *et al.* (2022)
Bread wheat	GZn-11; GFe-9	3B, 4B, 7B, 3A, 7A, 2D, 7D, 5A, 5D; 1D, 2D, 4D, 4B, 5B, 6B, 6D	Manjunath *et al.* (2023)

Genomics-assisted selection

Genomic selection (GS) utilizes genome-wide marker data to estimate genomic estimated breeding values (GEBVs), accelerating the genetic improvement of nutrient-dense wheat varieties (Pérez-Rodríguez *et al.*, 2012). Following the foundational work of Meuwissen *et al.* (2001), extensive research has evaluated GS's predictive accuracy across various crops using diverse marker platforms, densities, and both parametric and nonparametric models (De Los Campos *et al.*, 2010; Pérez-Rodríguez *et al.*, 2012; Crossa *et al.*, 2014). High-throughput genomics has been applied to select and breed wheat varieties with enhanced nutritional profiles, focusing on increasing zinc and iron concentrations – critical biofortification targets. Velu *et al.* (2016) demonstrated the effectiveness of genomic prediction in selecting wheat lines with higher grain zinc (GZnC) and iron content (GFeC). Genotypic values were estimated as Best Linear Unbiased Estimators (BLUEs), and the Genomic Best Linear Unbiased Predictor (GBLUP) model was used to account for interactions among lines, environments, markers, and pedigree. The genotype × environment models showed moderate to high correlations between observed and predicted GZnC and GFeC values, highlighting GS's potential to enhance these traits and accelerate breeding cycles.

Transgenic approaches

Genetic engineering is a pivotal approach in developing nutrient-enriched staple crops. While conventional and molecular breeding have been extensively used to create biofortified varieties and understand the genetic basis of key nutrient traits, transgenic technology enables the direct transfer of desired genes into elite varieties to elevate micronutrient levels, ensuring accumulation in the edible part without negatively affecting plant physiology or yield. Although limited work has been done to enhance iron (Fe) content in wheat through genetic engineering, notable efforts include Borg *et al.* (2012), who overexpressed the wheat ferritin gene (TaFer1-A), resulting in a 50–85% increase in grain Fe content. Similarly, overexpression of the VACUOLAR IRON TRANSPORTER (VIT) gene (TaVIT2) under an endosperm-specific promoter doubled Fe levels in white flour fractions without increasing antinutrient phytate content (Connorton *et al.*, 2017). While transgenic efforts to increase zinc (Zn) content in wheat are lacking, the QTL responsible for grain protein content (Gpc-B1) also enhances Fe and Zn remobilization, offering potential for biofortification.

The introduction of nutrient-enhancing genes from other organisms, such as ferritin for iron or carotenoid biosynthesis genes for provitamin A, has proven effective in improving the nutritional quality of wheat. Studies by Bai *et al.* (2013) highlight the potential of transgenic approaches in boosting nutrient uptake and content in wheat. Wirth *et al.* (2009) further illustrate the application of these methods in enhancing iron and folate levels in cereals, providing a robust foundation for future wheat biofortification efforts.

Although transgenic methods offer a precise way to enhance crop varieties, their adoption in India faces significant challenges. India's strict regulatory framework for genetically modified (GM) crops leads to prolonged approval processes, often influenced by public concerns and political factors. This environment, coupled with strong public resistance due to fears over food safety, environmental impact, and corporate control, hinders the commercialization of transgenic wheat. Concerns also exist regarding the potential ecological impact, including the unintended spread of transgenes to wild relatives, which could disrupt ecosystems. Moreover, the introduction of transgenic wheat raises intellectual property issues, potentially increasing farmers' dependence on seed companies. Additionally, India risks losing key export markets, especially in Europe, due to global restrictions on GM crops. To address these challenges, scientists and researchers are increasingly focusing on nontransgenic approaches, such as RNAi and genome editing techniques like CRISPR, which do not involve introducing foreign genes.

Genome editing using CRISPR

Genome editing with CRISPR-Cas (clustered regularly interspaced short palindromic repeats) is a novel technique used to make specific changes to an organism's DNA. It consists of two key elements: guide RNA, which directs the Cas enzymes to a specific DNA sequence, and the Cas enzymes (Cas9/Cpf1 or Cas12a/C2c2 or Cpf1 or Cas13/Cas14/Caspase/Cas3) which create a double-strand break at the target site. The break is then repaired by adding, removing, or altering bases/genetic material. This method allows for accurate and efficient genetic modifications, offering significant potential for applications in crop improvement.

CRISPR-Cas technology significantly advances wheat biofortification by enabling precise gene editing to enhance essential micronutrients such as iron, zinc, and provitamin A, while also addressing antinutritional factors like phytic acid, boosting protein content, and improving disease resistance (Gupta *et al.*, 2022). Techniques are being developed, such as those by Liang *et al.* (2018), to edit the wheat genome without inserting foreign genes, reducing off-target effects and ensuring precise modifications. Ibrahim *et al.* (2022) highlighted the potential of CRISPR-Cas9 for rapidly creating biofortified wheat with lower phytic acid and higher iron and zinc content. Camerlengo *et al.* (2020) demonstrated CRISPR's ability to efficiently knock out immunogenic proteins, confirming the edits at both molecular and biochemical levels.

In conclusion, the integration of advanced biotechnological strategies, particularly CRISPR-Cas genome editing, represents a significant advancement in wheat biofortification. Techniques such as MAS, QTL mapping, GWAS, genomic-assisted selection, and RNAi offer a robust method for enhancing wheat's nutritional content, specifically increasing levels of essential micronutrients like iron, zinc, and provitamin A. Among these, CRISPR-Cas stands out for its precision, efficiency, and ability to make heritable and stable genetic changes, crucial for sustainable improvements. Its potential to develop wheat varieties with enhanced micronutrient content and reduced antinutritional factors like phytic acid, while minimizing off-target effects, makes it a powerful tool in breeding programs. Despite challenges, the benefits of these technologies in combating malnutrition and improving global food security are substantial, paving the way for nutritionally enriched wheat varieties that can help address global nutritional needs.

10.4 Phytic Acid and Phytase Ratio and Their Effect on Bioavalability of Fe and Zn

Wheat is a staple food for a large portion of the world's population, providing essential proteins, carbohydrates, fats, minerals, and vitamins. The nutrient distribution within a wheat seed mirrors that of many other cereals. The germ contains vitamins B and E, unsaturated fats, high-quality proteins, carbohydrates, and minerals. The bran is rich in insoluble carbohydrates, proteins, minerals, and antinutritional factors like phytic acid. The endosperm mainly consists of starch and protein (Ram *et al.*, 2020).

The bioavailability of Fe and Zn in wheat is influenced by several factors, including the Fe and Zn content of different wheat varieties, the ratio of phytic acid to phytase, and host factors affecting mineral absorption, utilization, and excretion in the human body (Gupta *et al.*, 2024). Generally, the bioavailability of Fe and Zn in cereals is low due to the low levels of these minerals in cultivated varieties and the presence of antinutritional factors (Das *et al.*, 2012). Phytic acid, which is the primary storage form of phosphorus in cereals, has a negative charge that allows it to form complexes with Fe and Zn, thereby reducing their bioavailability (Kishor *et al.*, 2019). This issue is compounded by the lack of phytase enzymes in monogastric animals, which further impairs mineral absorption.

The bioavailability of Fe and Zn is often assessed using the molar ratio of phytic acid to Fe and phytic acid to Zn (Magallanes-López *et al.*, 2017). Phytic acid has a higher affinity for chelating Fe^{3+} compared to Fe^{2+}, further decreasing Fe bioavailability. To address this problem, breeding programs are working on developing wheat varieties with low phytate levels and high phytase content. Mutational breeding also offers a solution to this problem by production of nonlethal recessive low phytate mutants using physical and chemical mutagens (Kenzhebayeva *et al.*, 2019). These low phytate mutants can be used in breeding programmes for introgression.

Advanced biotech techniques can be used to lower the phytic acid content in wheat seed. RNAi was used to downregulate the expression of genes responsible for antinutritional factors like phytic acid, thereby improving the bioavailability of minerals like iron and zinc. These references provide insight into the use of RNAi technology for reducing antinutritional factors in wheat, particularly focusing on lowering phytic acid content to enhance the nutritional quality of the grain. Ibrahim *et al.* (2022) regulated the auxin signaling pathway (gene *TaIPK*1) using RNAi technology in wheat. Another study by Bhati *et al.* (2016) reported regulation of the *TaABCC13* gene (myoinositol pathway) using RNAi technology. This study focuses on using RNAi technology to reduce phytate levels in wheat, thereby enhancing the bioavailability of important minerals like iron and zinc. Aggarwal *et al.* (2018) utilized RNAi-mediated gene silencing of the *IPK1* (pentakisphosphate kinase) gene in hexaploid wheat to elucidate its role in phytic acid (PA) biosynthesis. This study marks a significant advancement in enhancing micronutrient content in wheat grains. The reduction of PA, a key antinutrient that chelates essential micronutrients and restricts their bioavailability, was accompanied by a substantial increase in Fe and Zn levels, thereby improving their respective molar ratios. These results highlight the *IPK1* gene as a promising target for genome editing aimed at boosting mineral accumulation in wheat grains.

Despite its applications, RNAi has several limitations, including incomplete gene silencing, which often results in only partial knockdown rather than complete inhibition. Additionally, RNAi can produce off-target effects, unintentionally silencing unintended genes, and its transient nature limits its effectiveness for long-term trait improvement. In contrast, new gene editing techniques like CRISPR-Cas9 offer precise gene editing by introducing targeted mutations, deletions, or insertions directly at the DNA level. This method ensures permanent and heritable changes, providing stable trait enhancement across generations. In contrast to RNAi, CRISPR-Cas9 can completely knock out gene function, which is beneficial for both removing undesirable traits and enhancing nutrient accumulation, overcoming many of the limitations.

10.5 Conclusions and Prospects

Malnutrition is a global issue affecting large groups of people particularly in resource-poor and developing nations. This is a very complex problem and to overcome it, active participation of all strata in the society is essential, including farmers, consumers, researchers, nutritionists, industrialists, and policymakers. To address this issue, the immediate focus should be on children, pregnant women, and the elderly, since micronutrients are essential for growth and reproductive health.

Wheat, a primary staple for over 40% of the global population, exhibits diverse adaptability across various climates and environments. It plays a crucial role in addressing food security challenges and has the potential to mitigate malnutrition. However, most of the currently grown wheat varieties are deficient in iron and zinc. The micronutrient content of wheat can be enhanced through fortification and biofortification approaches. Biofortification, in particular, offers several advantages over traditional fortification, being a more sustainable and cost-effective solution for alleviating malnutrition. It includes approaches such as conventional breeding, biotechnological methods, mineral fertilization, and microbe-mediated techniques.

In the past decade, significant progress has been made in this area, thanks to the availability of molecular markers for mapping major QTLs, along with increased research funding and policy support for biofortification. Wild relatives and landraces have significantly contributed to enhancing the iron and zinc content in wheat. The success of breeding methods in enhancing micronutrient content is evident from the release of biofortified varieties in various countries. Breeding strategies should be complemented by additional biofortification methods, such as agronomic and microbe-mediated approaches, which have proven effective in numerous studies. To expand the coverage of biofortified varieties, they should be integrated into the seed chain to enhance both formal and informal farming systems. To increase the adoption of biofortified varieties, initially these varieties should be made available to farmers at subsidized cost. Governments also need to make intensive efforts to develop policies and guidelines that will promote consumer acceptance of biofortified varieties.

Future research should focus on uncovering the mechanisms that influence mineral homeostasis in plant cells, with the goal of enhancing micronutrient concentrations in edible crops.

References

Aciksoz, S.B., Yazici, A., Ozturk, L. and Çakmak, I. (2011) Biofortification of wheat with iron through soil and foliar application of nitrogen and iron fertilizers. *Plant and Soil* 349(1–2), 215–225.

Aggarwal, S., Kumar, A., Bhati, K.K., Kaur, G., Shukla, V. *et al.* (2018) RNAi-mediated downregulation of inositol pentakisphosphate kinase (*IPK1*) in wheat grains decreases phytic acid levels and increases Fe and Zn accumulation. *Frontiers in Plant Science* 9, 259.

Akcura, M.E.V.L. and Kokten, K. (2017) Variations in grain mineral concentrations of Turkish wheat landraces germplasm. *Quality Assurance and Safety of Crops and Foods* 9(2), 153–159.

Aloo, B.N., Mbega, E.R. and Makumba, B.A. (2020) Rhizobacteria-based technology for sustainable cropping of potato (*Solanum tuberosum* L.). *Potato Research* 63(2), 157–177.

Amiri, R., Bahraminejad, S., Cheghamirza, K. and Arzani, A. (2020) Genetic analysis of iron and zinc concentrations in bread wheat grains. *Journal of Cereal Science* 95, 103077.

Arora, S., Cheema, J., Poland, J., Uauy, C. and Chhuneja, P. (2019) Genome-wide association mapping of grain micronutrients concentration in *Aegilops tauschii*. *Frontiers in Plant Science* 10, 54.

Backer, R., Rokem, J.S., Ilangumaran, G., Lamont, J., Praslickova, D. *et al.* (2018) Plant growth-promoting rhizobacteria: Context, mechanisms of action, and roadmap to commercialization of biostimulants for sustainable agriculture. *Frontiers in Plant Science* 9, 1473.

Badakhshan, H., Moradi, N., Mohammadzadeh, H. and Zakeri, M.R. (2013) Genetic variability analysis of grains Fe, Zn and beta-carotene concentration of prevalent wheat varieties in Iran. *International Journal of Agriculture and Crop Sciences* 6(2), 57.

Bai, C., Liang, Y. and Hawkesford, M.J. (2013) Identification of QTLs associated with seedling root traits and their correlation with plant height in wheat. *Journal of Experimental Botany* 64(6), 1745–1753.

Bailey, R.L., West, Jr., K.P. and Black, R.E. (2015) The epidemiology of global micronutrient deficiencies. *Annals of Nutrition and Metabolism* 66(Suppl. 2), 22–33.

Barut, H., Şimşek, T., Irmak, S., Sevilmiş, U. and Aykanat, S. (2017) The effect of different zinc application methods on yield and grain zinc concentration of bread wheat varieties. *Turkish Journal of Agriculture - Food Science and Technology* 5(8), 898–907.

Berhe, K., Gebrearegay, F. and Gebremariam, H. (2019) Prevalence and associated factors of zinc deficiency among pregnant women and children in Ethiopia: A systematic review and meta-analysis. *BMC Public Health* 19(1), 1663.

Bhati, K.K., Alok, A., Kumar, A., Kaur, J., Tiwari, S. *et al.* (2016) Silencing of ABCC13 transporter in wheat reveals its involvement in grain development, phytic acid accumulation and lateral root formation. *Journal of Experimental Botany* 67(14), 4379–4389.

Black, R.E., Allen, L.H., Bhutta, Z.A., Caulfield, L.E., Onis, M. *et al.* (2008) Maternal and child undernutrition: Global and regional exposures and health consequences. *Lancet* 371(9608), 243–260.

Borg, S., Brinch-Pedersen, H., Tauris, B., Madsen, L.H., Darbani, B. *et al.* (2012) Wheat ferritins: Improving the iron content of the wheat grain. *Journal of Cereal Science* 56(2), 204–213.

Çakmak, I., Derici, R., Torun, B., Tolay, I., Braun, H.J. *et al.* (1997) Role of rye chromosomes in improvement of zinc efficiency in wheat and triticale. In: *Plant Nutrition for Sustainable Food Production and Environment: Proceedings of the XIII International Plant Nutrition Colloquium*, Tokyo, Japan, September 13–19, pp. 237–241.

Çakmak, I., Ozkan, H., Braun, H.J., Welch, R.M. and Romheld, V. (2000) Zinc and iron concentrations in seeds of wild, primitive, and modern wheats. *Food and Nutrition Bulletin* 21(4), 401–403.

Çakmak, İ., Torun, A., Millet, E., Feldman, M., Fahima, T. *et al.* (2004) *Triticum dicoccoides*: An important genetic resource for increasing zinc and iron concentration in modern cultivated wheat. *Soil Science and Plant Nutrition* 50(7), 1047–1054.

Calderini, D.F. and Ortiz-Monasterio, I. (2003) Are synthetic hexaploids a means of increasing grain element concentrations in wheat? *Euphytica* 134(2), 169–178.

Camerlengo, F., Frittelli, A., Sparks, C., Doherty, A., Martignago, D. *et al.* (2020) CRISPR-Cas9 multiplex editing of the α-amylase/trypsin inhibitor genes to reduce allergen proteins in durum wheat. *Frontiers in Sustainable Food Systems* 4, 104.

Chattha, M.U., Hassan, M.U., Khan, I., Chattha, M.B., Mahmood, A. *et al.* (2017) Biofortification of wheat cultivars to combat zinc deficiency. *Frontiers in Plant Science* 8, 281.

Collard, B.C.Y. and Mackill, D.J. (2008) Marker-assisted selection: An approach for precision plant breeding in the twenty-first century. *Philosophical Transactions of the Royal Society B* 363(1491), 557–572.

Connorton, J.M. and Balk, J. (2019) Iron biofortification of staple crops: Lessons and challenges in plant genetics. *Plant & Cell Physiology* 60(7), 1447–1456.

Connorton, J.M., Jones, E.R., Rodríguez-Ramiro, I., Fairweather-Tait, S., Uauy, C. *et al.* (2017) Wheat vacuolar iron transporter TaVIT2 transports Fe and Mn and is effective for biofortification. *Plant Physiology* 174(4), 2434–2444.

Crespo-Herrera, L.A., Velu, G. and Singh, R.P. (2016) Quantitative trait loci mapping reveals pleiotropic effect for grain iron and zinc concentrations in wheat. *Annals of Applied Biology* 169(1), 27–35.

Crespo-Herrera, L.A., Govindan, V., Stangoulis, J., Hao, Y. and Singh, R.P. (2017) QTL mapping of grain Zn and Fe concentrations in two hexaploid wheat RIL populations with ample transgressive segregation. *Frontiers in Plant Science* 8, 1800.

Crossa, J., Pérez, P., Hickey, J., Burgueño, J., Ornella, L. *et al.* (2014) Genomic prediction in CIMMYT maize and wheat breeding programs. *Heredity* 112(1), 48–60.

Cu, S.T., Guild, G., Nicolson, A., Velu, G., Singh, R. *et al.* (2020) Genetic dissection of zinc, iron, copper, manganese and phosphorus in wheat (*Triticum aestivum* L.) grain and rachis at two developmental stages. *Plant Science* 291, 110338.

Dapkekar, A., Deshpande, P., Oak, M.D., Paknikar, K.M. and Rajwade, J.M. (2018) Zinc use efficiency is enhanced in wheat through nanofertilization. *Scientific Reports* 8(1), 6832.

Das, A., Raychaudhuri, U. and Chakraborty, R. (2012) Cereal based functional food of Indian subcontinent: A review. *Journal of Food Science and Technology* 49, 665–672.

De Groote, H., Tessema, M., Gameda, S. and Gunaratna, N.S. (2021) Soil zinc, serum zinc, and the potential for agronomic biofortification to reduce human zinc deficiency in Ethiopia. *Scientific Reports* 11(1), 8770.

De Los Campos, G., Gianola, D., Rosa, G.J.M., Weigel, K.A. and Crossa, J. (2010) Semi-parametric genomic-enabled prediction of genetic values using reproducing kernel Hilbert spaces methods. *Genetics Research* 92(4), 295–308.

Dhiman, K., Sharma, D., Kumari, R. and Tomar, P. (2023) Biofortification of crops using microbes – A promising sustainable agriculture strategy. *Journal of Plant Nutrition* 46(12), 2912–2935.

Distelfeld, A., Çakmak, I. and Peleg, Z. (2007) An integrative view of biofortification: The combined use of agronomic and genetic approaches to biofortify crops. *Plant and Soil* 302(1–2), 1–18.

Fageria, N.K., Baligar, V.C. and Li, Y.C. (2008) The role of nutrient efficient plants in improving crop yields in the twenty first century. *Journal of Plant Nutrition* 31(6), 1121–1157.

Genc, Y., Verbyla, A.P., Torun, A.A., Çakmak, I., Willsmore, K. *et al.* (2009) Quantitative trait loci analysis of zinc efficiency and grain zinc concentration in wheat using whole genome average interval mapping. *Plant and Soil* 314(1–2), 49–66.

Giraldo, P., Benavente, E., Manzano-Agugliaro, F. and Gimenez, E. (2019) Worldwide research trends on wheat and barley: A bibliometric comparative analysis. *Agronomy* 9(7), 352.

Goel, S., Singh, B., Grewal, S., Jaat, R.S. and Singh, N.K. (2018) Variability in Fe and Zn content among Indian wheat landraces for improved nutritional quality. *Indian Journal of Genetics and Plant Breeding* 78(4), 426–432.

Gomez-Becerra, H.F., Yazici, A., Ozturk, L., Budak, H., Peleg, Z. *et al.* (2010) Genetic variation and environmental stability of grain mineral nutrient concentrations in *Triticum dicoccoides* under five environments. *Euphytica* 171(1), 39–52.

Gorafi, Y.S.A., Ishii, T., Kim, J.-S., Elbashir, A.A.E. and Tsujimoto, H. (2016) Genetic variation and association mapping of grain iron and zinc contents in synthetic hexaploid wheat germplasm. *Plant Genetic Resources* 16(1), 9–17.

Gupta, O.P., Singh, A.K., Singh, A., Singh, G.P., Bansal, K.C. *et al.* (2022) Wheat biofortification: Utilizing natural genetic diversity, genome-wide association mapping, genomic selection, and genome editing technologies. *Frontiers in Nutrition* 9, 826131.

Gupta, O.P., Singh, A., Pandey, V., Sendhil, R., Khan, M.K. *et al.* (2024) Critical assessment of wheat biofortification for iron and zinc: A comprehensive review of conceptualization, trends, approaches, bioavailability, health impact, and policy framework. *Frontiers in Nutrition* 10, 1310020.

Hao, Y., Velu, G., Peña, R.J., Singh, S. and Singh, R.P. (2014) Genetic loci associated with high grain zinc concentration and pleiotropic effect on kernel weight in wheat (*Triticum aestivum* L.). *Molecular Breeding* 34(4), 1893–1902.

HarvestPlus (2003). Available at: www.harvestplus.org/

Heidari, B., Padash, S. and Dadkhodaie, A. (2016) Variations in micronutrients, bread quality and agronomic traits of wheat landrace varieties and commercial cultivars. *Australian Journal of Crop Science* 10(3).

Huang, X. and Han, B. (2014) Natural variations and genome-wide association studies in crop plants. *Cell* 157(3), 522–530.

Huang, L., Drake, V.J. and Ho, E. (2015) Zinc. *Advances in Nutrition* 6(2), 224–226.

Hurrel, R.F. and Cook, J.D. (1992) Micronutrient fortification of food: Technology and quality control. Strategies for iron fortification of foods. *FAO/WHO* 1, 56–61.

Ibba, M.I., Gupta, O.P., Govindan, V., Johnson, A.A.T., Brinch-Pedersen, H. *et al.* (2022) Wheat biofortification to alleviate global malnutrition. *Frontiers in Nutrition* 9, 1001443.

Ibrahim, S., Saleem, B., Rehman, N., Zafar, S.A., Naeem, M.K. *et al.* (2022) CRISPR/cas9 mediated disruption of inositol pentakisphosphate 2-kinase 1 (*TaIPK1*) reduces phytic acid and improves iron and zinc accumulation in wheat grains. *Journal of Advanced Research* 37, 33–41.

Jawaldeh, A.A., Pena-Rosas, J.P., McColl, K., Johnson, Q., Elmadfa, I. *et al.* (2019) *Wheat Flour Fortification in the Eastern Mediterranean Region*. WHO Regional Office for the Eastern Mediterranean, Cairo.

Jomova, K., Makova, M., Alomar, S.Y., Alwasel, S.H., Nepovimova, E. *et al.* (2022) Essential metals in health and disease. *Chemico-Biological Interactions* 367, 110173.

Kambe, T., Tsuji, T., Hashimoto, A. and Itsumura, N. (2015) The physiological, biochemical, and molecular roles of zinc transporters in zinc homeostasis and metabolism. *Physiological Reviews* 95(3), 749–784.

Kamble, U., Mishra, C.N., Govindan, V., Sharma, A.K., Pawar, S. *et al.* (2022) Ensuring nutritional security in India through wheat biofortification: A review. *Genes* 13(12), 2298.

Kato, K., Miura, H. and Sawada, S. (2000) Mapping QTLs controlling grain yield and its components on chromosome 5A of wheat. *Theoretical and Applied Genetics* 101(7), 1114–1121.

Kenzhebayeva, S., Abekova, A., Atabayeva, S., Yernazarova, G., Omirbekova, N. *et al.* (2019) Mutant lines of spring wheat with increased iron, zinc, and micronutrients in grains and enhanced bioavailability for human health. *BioMed Research International* 1, 9692053.

Khan, M.K., Pandey, A., Hamurcu, M., Hakki, E.E., Gezgin, S. *et al.* (2020) Role of molecular approaches in improving genetic variability of micronutrients and their utilization in breeding programs. In: Gupta, O.P., Pandey, V., Narwal, S., Sharma, P. and Ram, S. (eds) *Wheat and Barley Grain Biofortification*. Woodhead Publishing, Cambridge, pp. 27–52.

Khan, M.K., Pandey, A., Hamurcu, M., Gezgin, S., Athar, T. *et al.* (2021) Insight into the prospects for nanotechnology in wheat biofortification. *Biology* 10(11), 1123.

Khokhar, J.S., Sareen, S., Tyagi, B.S., Singh, G., Wilson, L. *et al.* (2018) Variation in grain Zn concentration, and the grain ionome, in field-grown Indian wheat. *PLoS One* 13(1), e0192026.

Khokhar, J.S., King, J., King, I.P., Young, S.D., Foulkes, M.J. *et al.* (2020) Novel sources of variation in grain zinc (Zn) concentration in bread wheat germplasm derived from watkins landraces. *PLoS One* 15(2), e0229107.

Kishor, D.S., Lee, C., Lee, D., Venkatesh, J., Seo, J. *et al.* (2019) Novel allelic variant of Lpa1 gene associated with a significant reduction in seed phytic acid content in rice (*Oryza sativa* L.). *PLoS One* 14(3), e0209636.

Kobayashi, T., Nozoye, T. and Nishizawa, N.K. (2019) Iron transport and its regulation in plants. *Free Radical Biology & Medicine* 133, 11–20.

Korte, A. and Farlow, A. (2013) The relative abundance of QTLs for complex traits depends on the genetic architecture. *Heredity* 110(6), 518–523.

Krężel, A. and Maret, W. (2016) The biological inorganic chemistry of zinc ions. *Archives of Biochemistry and Biophysics* 611, 3–19.

Krishnappa, G., Singh, A.M., Chaudhary, S., Ahlawat, A.K., Singh, S.K. *et al.* (2017) Molecular mapping of the grain iron and zinc concentration, protein content and thousand kernel weight in wheat (*Triticum aestivum* L.). *PLoS One* 12(4), e0174972.

Kumar, A., Kapoor, P., Chunduri, V., Sharma, S. and Garg, M. (2019) Potential of *Aegilops* sp. for improvement of grain processing and nutritional quality in wheat (*Triticum aestivum*). *Frontiers in Plant Science* 10, 308.

Kumar, D., Dutt, S., Raigond, P., Changan, S.S., Lal, M.K. *et al.* (2021a) Beneficial microorganisms in crop growth, soil health, and sustainable environmental management: Current status and future perspectives. In: Tyagi, S., Kumar, R., Saharan, B. and Nadda, A. (eds) *Plant-Microbial Interactions and Smart Agricultural Biotechnology*. Taylor and Francis, Abingdon, pp. 11–32.

Kumar, D., Dutt, S., Raigond, P., Changan, S.S., Lal, M.K. *et al.* (2021b) Different biofertilizers and their application for sustainable development. In: Bhatt, P., Gangola, S., Udayanga, D. and Kumar, G. (eds) *Microbial Technology for Sustainable Environment*. Springer, Cham, pp. 31–48.

Kumar, D., Dutt, S., Changan, S.S., Lal, M.K., Tiwari, R.K. *et al.* (2024) Role of soil microbes in modulating the physiological attributes of plants under extreme environmental conditions. In: Ahanger, M., Bhatt, J., Ahmad, P. and John, R. (eds) *Improving Stress Resilience in Plants*. Academic Press, New York, pp. 15–34.

Li, M., Wang, S., Tian, X. and Huang, Y. (2018) Improving nutritional quality of wheat grain through foliar zinc combined with macronutrients. *Agronomy Journal* 110(1), 38–46.

Li, M., Watanabe, S., Gao, F. and Dubos, C. (2023) Iron nutrition in plants: Towards a new paradigm? *Plants* 12(2), 384.

Liang, Z., Chen, K., Zhang, Y., Liu, J., Yin, K. *et al.* (2018) Genome editing of bread wheat using biolistic delivery of CRISPR/Cas9 in vitro transcripts or ribonucleoproteins. *Nature Protocols* 13(3), 413–430.

Lowe, N.M. (2024) Fortification or biofortification: Complimentary strategies or duplication of effort? *Proceedings of the Nutrition Society* 1–10.

Lu, J., Hu, Y., Li, M., Liu, X., Wang, R. *et al.* (2021) Zinc nutritional status and risk factors of elderly in the China adult chronic disease and nutrition surveillance 2015. *Nutrients* 13(9), 3086.

Magallanes-López, A.M., Hernandez-Espinosa, N., Velu, G., Posadas-Romano, G., Ordoñez-Villegas, V.M.G. *et al.* (2017) Variability in iron, zinc and phytic acid content in a worldwide collection of commercial durum wheat cultivars and the effect of reduced irrigation on these traits. *Food Chemistry* 237, 499–505.

Manjunath, K.K., Krishna, H., Devate, N.B., Sunilkumar, V.P., Chauhan, D. *et al.* (2023) Mapping of the QTLs governing grain micronutrients and thousand kernel weight in wheat (*Triticum aestivum* L.) using high density SNP markers. *Frontiers in Nutrition* 10, 1105207.

Mannar, M.V. and Hurrell, R.F. (eds) (2018) Food fortification: Past experience, current status, and potential for globalization. In: *Food Fortification in a Globalized World*. Academic Press, New York, pp. 3–11.

Mantadakis, E., Chatzimichael, E. and Zikidou, P. (2020) Iron deficiency anemia in children residing in high and low-income countries: Risk factors, prevention, diagnosis and therapy. *Mediterranean Journal of Hematology and Infectious Diseases* 12(1), e2020041.

Maqbool, M.A., Aslam, M., Beshir, A. and Khan, M.S. (2018) Breeding for provitamin a biofortification of maize (*Zea mays* L.). *Plant Breeding* 137(4), 451–469.

Maryami, Z., Huertas-García, A.B., Azimi, M.R., Hernández-Espinosa, N., Payne, T. *et al.* (2020) Variability for glutenins, gluten quality, iron, zinc and phytic acid in a set of one hundred and fifty-eight common wheat landraces from Iran. *Agronomy* 10(11), 1797.

Meena, N. and Fathima, P.S. (2017) Nutrient uptake of rice as influenced by agronomic biofortification of Zn and Fe under methods of rice cultivation. *International Journal of Pure & Applied Bioscience* 5(5), 456–459.

Meuwissen, T.H., Hayes, B.J. and Goddard, M.E. (2001) Prediction of total genetic value using genome-wide dense marker maps. *Genetics* 157(4), 1819–1829.

Mishra, C.N., Kumar, S., Gupta, V., Tyagi, B.S., Ram, S. *et al.* (2024) Glimpses of biofortified wheat varieties of India. *Technical Bulletin* 1–26.

Monasterio, I. and Graham, R.D. (2000) Breeding for trace minerals in wheat. *Food and Nutrition Bulletin* 21(4), 392–396.

Narwal, S., Gupta, O.P., Pandey, V., Kumar, D., Ram, S. *et al.* (2020) Effect of storage and processing conditions on nutrient composition of wheat and barley. In: Gupta, O.P., Pandey, V., Narwal, S., Sharma, P. and Ram, S. (eds) *Wheat and Barley Grain Biofortification*. Woodhead Publishing, Cambridge, pp. 229–256.

Nayak, S., Shivay, Y.S., Prasanna, R., Mandi, S., Parveen, S. *et al.* (2023) Soil and foliar application of Zn enhances its biofortification, bioavailability and productivity in both biofortified and non-biofortified wheat varieties. *Journal of Food Composition and Analysis* 124, 105691.

Nestel, P., Bouis, H.E., Meenakshi, J.V. and Pfeiffer, W. (2006) Biofortification of staple food crops. *Journal of Nutrition* 136(4), 1064–1067.

Pahlavan-Rad, M.R. and Pessarakli, M. (2009) Response of wheat plants to zinc, iron, and manganese applications and uptake and concentration of zinc, iron, and manganese in wheat grains. *Communications in Soil Science and Plant Analysis* 40(7–8), 1322–1332.

Peleg, Z., Saranga, Y., Yazici, A., Fahima, T., Ozturk, L. *et al.* (2008) Grain zinc, iron and protein concentrations and zinc-efficiency in wild emmer wheat under contrasting irrigation regimes. *Plant and Soil* 306(1–2), 57–67.

Peleg, Z., Çakmak, I., Ozturk, L., Yazici, A., Jun, Y. *et al.* (2009) Quantitative trait loci conferring grain mineral nutrient concentrations in durum wheat x wild emmer wheat RIL population. *Theoretical and Applied Genetics* 119(2), 353–369.

Pérez-Rodríguez, P., Gianola, D., González-Camacho, J.M., Crossa, J., Manès, Y. *et al.* (2012) Comparison between linear and non-parametric regression models for genome-enabled prediction in wheat. *G3: Genes Genomes Genetics* 2(12), 1595–1605.

Praharaj, S., Singh, R., Singh, V.K., Chandra, R., Guru, S.K. *et al.* (2019) Yield and grain zinc concentration of wheat as affected by nutri priming and foliar application of zinc. *Journal of Pharmacognosy and Phytochemistry* 8(5), 503–505.

Prasad, P.V.V. and Djanaguiraman, M. (2017) Iron chlorosis. In: Thomas, B., Murray, B.G. and Murphy, D.J. (eds) *Encyclopedia of Applied Plant Sciences*. Academic Press, New York, pp. 246–255.

Pu, Z., Yu, M., He, Q., Chen, G., Wang, J. *et al.* (2014) Quantitative trait loci associated with micronutrient concentrations in two recombinant inbred wheat lines. *Journal of Integrative Agriculture* 13(11), 2322–2329.

Qadir, M., Wang, X., Baloch, A.H., Baloch, I.A., Azeem, M. *et al.* (2018) The impact of drought on phenotypic characters of five advance bread wheat genotypes. *Pure and Applied Biology* 7(2), 635–642.

Raboy, V. (2007) Seed phosphorus and the development of low-phytate crops. In: Turner, B.L., Richardson, A.E. and Mullaney, E.J. (eds) *Inositol Phosphates: Linking Agriculture and the Environment*. CABI, Wallingford, pp. 111–132.

Raj, A.B. and Raj, S.K. (2019) Seed priming: An approach towards agricultural sustainability. *Journal of Applied and Natural Science* 11(1), 227–234.

Ram, S., Narwal, S., Gupta, O.P., Pandey, V., Singh, G.P. *et al.* (2020) Anti-nutritional factors and bioavailability: Approaches, challenges, and opportunities. In: Gupta, O.P., Pandey, V., Narwal, S., Sharma, P. and Ram, S. (eds) *Wheat and Barley Grain Biofortification*. Woodhead Publishing, Cambridge, pp. 101–128.

Ram, S., Malik, V.K., Gupta, V., Narwal, S., Sirohi, M. *et al.* (2024) Impact of foliar application of iron and zinc fertilizers on grain iron, zinc, and protein contents in bread wheat (*Triticum aestivum* L.). *Frontiers in Nutrition* 11, 1378937.

Ranjbar, G.A. and Bahmaniar, M.A. (2007) Effects of soil and foliar application of Zn fertilizer on yield and growth characteristics of bread wheat (*Triticum aestivum* L.) cultivars. *Asian Journal of Plant Sciences* 6(6), 1000–1005.

Rathan, N.D., Krishna, H., Ellur, R.K., Sehgal, D., Govindan, V. *et al.* (2022) Genome-wide association study identifies loci and candidate genes for grain micronutrients and quality traits in wheat (*Triticum aestivum* L.). *Scientific Reports* 12(1), 7037.

Rawat, N., Tiwari, V.K., Singh, N., Randhawa, G.S., Singh, K. *et al.* (2009) Evaluation and utilization of *Aegilops* and wild *Triticum* species for enhancing iron and zinc content in wheat. *Genetic Resources and Crop Evolution* 56(1), 53–64.

Roshanzamir, H., Kordenaeej, A. and Bostani, A. (2013) Mapping QTL related to Zn and Fe concentrations in bread wheat (*Triticum aestivum*) grain using microsatellite markers. *Iranian Journal of Genetics and Plant Breeding* 2, 10–17.

Rubio, L.M. and Ludden, P.W. (2008) Biosynthesis of the iron-molybdenum cofactor of nitrogenase. *Annual Review of Microbiology* 62, 93–111.

Saini, D.K., Devi, P. and Kaushik, P. (2020) Advances in genomic interventions for wheat biofortification: A review. *Agronomy* 10(1), 62.

Santos, M.S., Rodrigues, T.F., Nogueira, M.A. and Hungria, M. (2021) The challenge of combining high yields with environmentally friendly bioproducts: A review on the compatibility of pesticides with microbial inoculants. *Agronomy* 11(5), 870.

Shi, R., Li, H., Tong, Y., Jing, R., Zhang, F. *et al.* (2008) Identification of quantitative trait locus of zinc and phosphorus density in wheat (*Triticum aestivum* L.) grain. *Plant and Soil* 306(1–2), 95–104.

Shi, R., Tong, Y., Jing, R., Zhang, F. and Zou, C. (2013) Characterization of quantitative trait loci for grain minerals in hexaploid wheat (*Triticum aestivum* L.). *Journal of Integrative Agriculture* 12(9), 1512–1521.

Shiferaw, B., Smale, M., Braun, H.J., Duveiller, E., Reynolds, M. *et al.* (2013) Crops that feed the world 10. Past successes and future challenges to the role played by wheat in global food security. *Food Security* 5(3), 291–317.

Shivay, Y.S., Singh, U., Prasad, R. and Kaur, R. (2016) Agronomic interventions for micronutrient biofortification of pulses. *Indian Journal of Agronomy* 61, 161–172.

Singh, D. and Prasanna, R. (2020) Potential of microbes in the biofortification of Zn and Fe in dietary food grains. *A Review. Agronomy for Sustainable Development* 40(2), 15.

Singh, R.K. and Kumar, V. (2016) Marker-assisted selection for improving crop yield and quality: From QTL discovery to application. *Frontiers in Plant Science* 7, 1284.

Singh, P., Kumar, S., Srivastava, S., Nagargade, M., Kumar, D. *et al.* (2022) Endophytes for sustainable sugar beet production. In: Misra, V., Srivastava, S. and Mall, A. (eds) *Sugar Beet Cultivation, Management and Processing*. Springer Nature, Singapore, pp. 777–792.

Song, Z., Zuo, Y., Xiang, Q., Li, W., Li, J. *et al.* (2023) Investigation of *Aegilops* umbellulata for stripe rust resistance, heading date, and the contents of iron, zinc, and gluten protein. *Journal of Integrative Agriculture* 22(4), 1258–1265.

Srinivasa, J., Arun, B., Mishra, V.K., Singh, G.P., Velu, G. *et al.* (2014) Zinc and iron concentration QTL mapped in a *Triticum spelta* × *T. aestivum* cross. *Theoretical and Applied Genetics* 127(7), 1643–1651.

Tanksley, S.D. and McCouch, S.R. (1997) Seedling and plant QTL analysis in maize and rice. *Theoretical and Applied Genetics* 95(6), 1195–1205.

Terrin, G., Berni Canani, R., Chiara, M., Pietravalle, A., Aleandri, V. *et al.* (2015) Zinc in early life: A key element in the fetus and preterm neonate. *Nutrients* 7(12), 10427–10446.

Tiwari, C., Wallwork, H., Arun, B., Mishra, V.K., Velu, G. *et al.* (2016) Molecular mapping of quantitative trait loci for zinc, iron and protein content in the grains of hexaploid wheat. *Euphytica* 207(3), 563–570.

Tiwari, V.K., Rawat, N., Chhuneja, P., Neelam, K., Aggarwal, R. *et al.* (2009) Mapping of quantitative trait loci for grain iron and zinc concentration in diploid a genome wheat. *Journal of Heredity* 100(6), 771–776.

Velu, G., Crossa, J., Singh, R.P., Hao, Y., Dreisigacker, S. *et al.* (2016) Genomic prediction for grain zinc and iron concentrations in spring wheat. *Theoretical and Applied Genetics* 129(8), 1595–1605.

Velu, G., Tutus, Y., Gomez-Becerra, H.F., Hao, Y., Demir, L. *et al.* (2017) QTL mapping for grain zinc and iron concentrations and zinc efficiency in a tetraploid and hexaploid wheat mapping populations. *Plant and Soil* 411(1–2), 81–99.

Velu, G., Singh, R.P. and Joshi, A.K. (2020) A decade of progress on genetic enhancement of grain zinc and iron in CIMMYT wheat germplasm. In: Gupta, O.P., Pandey, V., Narwal, S., Sharma, P. and Ram, S. (eds) *Wheat and Barley Grain Biofortification*. Woodhead Publishing, Cambridge, pp. 129–138.

Velu, G., Crespo Herrera, L., Guzman, C., Huerta, J., Payne, T. *et al.* (2018) Assessing genetic diversity to breed competitive biofortified wheat with enhanced grain Zn and Fe concentrations. *Frontiers in Plant Science* 9.

Verma, S., Chakdar, H., Kumar, M., Varma, A. and Saxena, A.K. (2021) Microorganisms as a sustainable alternative to traditional biofortification of iron and zinc: Status and prospect to combat hidden hunger. *Journal of Soil Science and Plant Nutrition* 21(2), 1700–1717.

Vitorino, L.C. and Bessa, L.A. (2017) Technological microbiology: Development and applications. *Frontiers in Microbiology* 8, 827.

Wakeel, A., Farooq, M., Bashir, K. and Ozturk, L. (2018) Micronutrient malnutrition and biofortification: Recent advances and future perspectives. In: Hossain, M., Kamiya, T., Burritt, D., Tran, L. and Fujiwara, T. (eds) *Plant Micronutrient Use Efficiency*. Academic Press, New York, pp. 225–243.

Wang, J., Shi, X., Zhou, Z., Qin, M., Wang, Y. *et al.* (2022) Genetic dissection of grain iron concentration in hexaploid wheat (*Triticum aestivum* L.) using a genome-wide association analysis method. *PeerJ* 10, e13625.

Wang, M., Kong, F., Liu, R., Fan, Q. and Zhang, X. (2020) Zinc in wheat grain, processing, and food. *Frontiers in Nutrition* 7, 124.

Wang, S., Li, M., Tian, X., Li, J., Li, H. *et al.* (2015) Foliar zinc, nitrogen, and phosphorus application effects on micronutrient concentrations in winter wheat. *Agronomy Journal* 107(1), 61–70.

Wang, S., Li, M., Liu, K., Tian, X., Li, S. *et al.* (2017) Effects of Zn, macronutrients, and their interactions through foliar applications on winter wheat grain nutritional quality. *PLoS One* 12(7), e0181276.

Wang, Y., Xu, X., Hao, Y., Zhang, Y., Liu, Y. *et al.* (2021) QTL mapping for grain zinc and iron concentrations in bread wheat. *Frontiers in Nutrition* 8, 680391.

Wessels, I. and Rink, L. (2020) Micronutrients in autoimmune diseases: Possible therapeutic benefits of zinc and vitamin D. *Journal of Nutritional Biochemistry* 77, 108240.

White, P.J. and Broadley, M.R. (2009) Biofortification of crops with seven mineral elements often lacking in human diets – iron, zinc, copper, calcium, magnesium, selenium and iodine. *New Phytologist* 182(1), 49–84.

Wirth, J., Poletti, S., Aeschlimann, B., Yakandawala, N., Drosse, B. *et al.* (2009) Rice endosperm iron biofortification by targeted and synergistic action of nicotianamine synthase and ferritin. *Plant Biotechnology Journal* 7(7), 631–644.

World Health Organization (2017) Nutritional anaemias: Tools for effective prevention. Available at: www.who.int/publications/i/item/9789241513067 (accessed August 2024).

World Health Organization (2023) Anaemia. Available at: www.who.int/news-room/fact-sheets/detail/anaemia (accessed 9 April 2025).

Xu, Y., An, D., Liu, D., Zhang, A., Xu, H. *et al.* (2012) Molecular mapping of QTLs for grain zinc, iron and protein concentration of wheat across two environments. *Field Crops Research* 138, 57–62.

Yadav, D.N., Bansal, S., Tushir, S., Kaur, J., Sharma, K. *et al.* (2020) Advantage of biofortification over fortification technologies. In: Gupta, O.P., Pandey, V., Narwal, S., Sharma, P. and Ram, S. (eds) *Wheat and Barley Grain Biofortification*. Woodhead Publishing, Cambridge, pp. 257–273.

Yaseen, M.K. and Hussain, S. (2021) Zinc-biofortified wheat required only a medium rate of soil zinc application to attain the targets of zinc biofortification. *Archives of Agronomy and Soil Science* 67(4), 551–562.

Zhang, Y., Shi, R., Rezaul, K.M., Zhang, F. and Zou, C. (2010) Iron and zinc concentrations in grain and flour of winter wheat as affected by foliar application. *Journal of Agricultural and Food Chemistry* 58(23), 12268–12274.

Zhao, F.J., Su, Y.H., Dunham, S.J., Rakszegi, M., Bedo, Z. *et al.* (2009) Variation in mineral micronutrient concentrations in grain of wheat lines of diverse origin. *Journal of Cereal Science* 49(2), 290–295.

Zou, C., Du, Y., Rashid, A., Ram, H., Savasli, E. *et al.* (2019) Simultaneous biofortification of wheat with zinc, iodine, selenium, and iron through foliar treatment of a micronutrient cocktail in six countries. *Journal of Agricultural and Food Chemistry* 67(29), 8096–8106.

Zou, C.Q., Zhang, Y.Q., Rashid, A., Ram, H., Savasli, E. *et al.* (2012) Biofortification of wheat with zinc through zinc fertilization in seven countries. *Plant and Soil* 361, 119–130.

11 Sustainable Crop Management: Agroecology of the Future

Braulio Edgar Herrera-Cabrera[1], Rafael Salgado Garciglia[2], Alejandra García Hernández[2], Adriana Delgado Alvarado[1], Luis Germán López Valdez[3], Francisco Javier Verduzco Miramón[4] and Hebert Jair Barrales-Cureño[4]*

[1]Colegio de Postgraduados, Campus Puebla, Puebla, México; [2]Instituto de Investigaciones Químico Biológicas, Universidad Michoacana de San Nicolás de Hidalgo, Morelia, México; [3]Laboratorio de Productos Naturales, Texcoco, México; [4]Tecnológico Nacional de México, Zamora, México

Abstract

Agroecology is emerging as a new field of scientific knowledge, with different epistemological, methodological, theoretical and practical implications, which delineate a transdisciplinary approach by addressing ecological, ethical, social, and political issues in order to solve rural problems. Agroecology maximizes and optimizes the production of agroecosystems economically, socially, and ecologically, incorporating an approach to agriculture linked to the natural environment and socially sensitive, focused on sustainable production, integrating the ecological phenomena that occur in a crop field. Agroecology is defined as the sustainable ecological management of agroecosystems, through collective social action, as an alternative to the agro-industrial management model, with participatory development proposals in the areas of production and circulation of its products, establishing forms of production and consumption that contribute to combat the ecological and social crisis generated by neoliberalism and economic globalization. Therefore, the objectives of this chapter are to discuss agroecology as a tool for the sustainable development of agriculture in general, indicate the ten elements of agroecology that can help countries implement it, as well as the advantages of sustainable agroecology and the necessary instruments it uses to compete and maintain itself in the future.

Keywords: food self-sufficiency, precision agroecology, sustainable agriculture

11.1 Introduction

Sustainable agriculture focuses on maintaining long-term productivity without depleting natural resources. Examples include crop rotation, efficient use of water, and reducing dependence on agrochemicals (Liu *et al.*, 2024). Sustainable agriculture focuses on meeting current needs without compromising future generations. It manifests in the protection of biodiversity, the preservation of native seeds, and the promotion of organic agriculture. Agroecology has emerged as a new approach to agricultural development, more sensitive to the complexities of local agriculture, embracing properties of sustainability, such as security of sustainability, biological food security, biological stability, resource conservation, and equity

*Corresponding author: hebert.bc@zamora.tecnm

 Plant Biotechnology and Sustainable Agriculture (ed. Tariq Aftab)
DOI: 10.1079/9781800627659.0011

(Velten *et al.*, 2015). Thus, the agroecological paradigm views the agricultural process as an integrated system with environmental, economic, social, and cultural aspects and its purpose is not only to increase the productivity of one component but to optimize the system as a whole and maintain sustainability in time and space (Valenzuela, 2016). Scientists believe that farmers already produce enough food to feed 10 billion people, far more than the current population of 7.3 billion. However, 800 million people go hungry every day and many more are malnourished.

Research consistently shows that global hunger is not a problem of supply but of poverty, lack of democracy and unequal access to land, water, and other resources, especially for women. Therefore, the objective of this chapter is to discuss agroecology as a tool for the sustainable development of agriculture in general, indicate the ten elements of agroecology that can help countries implement agroecology, and enumerate the advantages of sustainable agroecology and the necessary instruments it uses to compete and maintain itself in the future.

11.2 Agroecology

Agroecology makes maximum use of nature's resources and minimum use of external inputs (fertilizers and pesticides, i.e. insecticides, fungicides and herbicides). It favors biodiversity and relies on the association of various plants, complementary in their agronomic functioning, the valorization of organic matter (composting) and the association between agriculture and livestock farming. Agroecology includes organic farming; it sometimes differs from it in that it may, under certain circumstances, use fertilizers or pesticides in moderate quantities (Diacono *et al.*, 2021).

The term 'agroecology' in fact encompasses three main areas.

- A technical approach to sustainable agricultural production activities. The aim is to make optimal use of the local ecosystem (nature is the farmer's 'helper'), seeking to reduce pressure on the environment and preserve natural resources and their capacity for renewal.
- A domain of observation and research for agronomists (literally, agroecology is the science of agricultural ecosystems) and other scientists of natural mechanisms and interactions within agricultural ecosystems. It is especially about understanding, developing, and optimizing agroecological techniques that use these interactions.
- A social movement that promotes an agriculture based on the principles of agroecology: agricultural biodiversity; efficiency in the use of resources and the circularity of their uses; natural balances, regulations, and synergies (Dönmez *et al.*, 2024).

Conventional agriculture has won the battle so far, demonstrating its production capacity and profitability, but at an extremely dangerous cost for the continuity of life on Earth. The absolute management and control of food production, the commoditization of the life process (biotechnology), and the homogenization of agriculture on a planetary scale are issues on which there are many criticisms. The sustainability of this type of agricultural production could provoke an incalculable ecological disaster, making it impossible for life to continue on the planet and/or to produce enough food for the billions of hungry people in the world (Diacono *et al.*, 2021).

Faced with this situation, an alternative, sustainable agriculture is being promoted, with diametrically opposed parameters, which places emphasis on the relationship with the elements in nature. It is not possible to think exclusively of a 'technical' response to a complex problem, where the interactions and synergies between the various biological components of agroecosystems are ignored. In the alternative proposal for sustainable agriculture, we must look at integrality, incorporating cultural, social, economic, political, and environmental dimensions (Viccaro *et al.*, 2024).

Agroecology is a scientific discipline oriented to agricultural, livestock, and forestry practices, whose principles are based on the maintenance of natural resources or causing the least possible damage to the environment. It is based on the conservation of biodiversity in agriculture and on the restoration of the ecological balance of agroecosystems, with the intention of achieving sustainable food production. It respects forests, environmental health,

Table 11.1. Advantages of agroecology.

Industrial agriculture	Agroecology	Added value of agroecology	Reference
Very expensive inputs	Less expensive and more profitable inputs for farmers	Agriculture as a more remunerative activity	Herman (2024)
Soil depletion and biodiversity loss	Improving soil and diversity	Best yield per hectare	Telo da Gama (2023)
Soil improvement and diversity	Reduced and limited quantities of water	Optimal utilization and conservation of water	Del Duca *et al.* (2024)
Continuous environmental pollution	Protects the environment from pollution	A healthy and safe environment	Wang *et al.* (2021)
Depends on a single product in a surface unit	It is based on agricultural diversity in the same area unit	It implies certain security in terms of income and harvesting	LaFevor (2022)
Dependence on foreign companies for inputs	Independence from foreign companies	Financial independence for farmers and the homeland	Mokgomo *et al.* (2022)

diversification, and energy efficiency, uses indigenous knowledge (crop diversity, organic poisons, knowledge of the moon phases) and natural cycles; it dispenses with chemical inputs (Zenda and Rudolph, 2024). Table 11.1 shows the major advantages of agroecology.

Agroecology requires much more work than chemical agriculture. Organic waste must be collected, composted, and applied to the land, plants must be found that act as ecological pesticides, weeding must be done by hand, and so on. For example, 40 wheelbarrows of decomposed organic fertilizer are needed for 1 hectare of land. In chemical agriculture, it is sufficient to buy and transport one sack of fertilizer.

Solving the problem of world hunger requires policies and programs that democratize access to food, arable land, water, credit, and fair markets, especially for women. To combat hunger and poverty, we must increase public investment in agroecological agriculture, especially among small-scale food producers who constitute 90% of all farmers worldwide. We must also reduce global food waste and move towards a plant-based food model (especially in rich countries that consume large quantities of meat), away from the growing demand for livestock feed and biofuels (Diacono and Montemurro, 2023). A growing body of research shows that organic farming and agroecological agriculture provide more than enough food to feed the world's growing population, while generating beneficial effects on the economy, health, and the environment. Organic farming methods create more ecological resilience to climate change than industrial agriculture (Cleves *et al.*, 2022).

Calorie production and economic efficiency may make industrial agriculture seem more efficient, but this approach does not take into account the environmental, social, and health degradation associated with industrial food production, processing, distribution, consumption, and waste. Considering the damage it causes, the dominant system of food production is very expensive and inefficient. Rather than feeding the world sustainably in the future, the industrial food production system is closing the branch on which it is established, as the ecosystem functions to allow food produced to be degraded (Oteros-Rozas *et al.*, 2019).

The methods employed by agroecological agriculture are the best path to long-term sustainable food production. They provide large harvests while protecting human and ecosystem health. Policymakers must expand regulation of industrial agriculture, eliminate support that encourages destructive industrial farming practices, and invest in diversified organic farming systems. The Common Agricultural Policy (CAP) plays a key role in this transition. It should be reformed to promote a food production system with new central axes: farmers practicing sustainable agriculture, small businesses associated with them, and community-supported food production projects that connect producers and consumers (Cárceles Rodríguez *et al.*, 2022).

11.3 Sustainable Agriculture

Methods of sustainable agriculture include crop rotation and efficient use of water. Farmers are adopting crop rotation, a practice that helps maintain soil fertility and reduces pressure on a single crop (Gorooei *et al.*, 2023). In countries where water scarcity is a major problem, sustainable agriculture promotes the efficient use of water through techniques such as drip irrigation. The National Soil Strategy for Sustainable Agriculture (ENASAS) promotes sustainable soil management and restoration practices, which allow better use of this resource and boost productivity in the agricultural field. In addition, small-scale farmers should be made aware of the risks of incidental forest fires and how to avoid damage to agricultural land, starting with community workshops, if necessary (García-Tejero and Durán-Zuazo, 2022). Table 11.2 shows the environmental benefits of organic farming practices used in agroecology.

11.3.1 Objectives and needs of sustainable agriculture

The main issue in sustainable agriculture is not to achieve maximum yields but rather to achieve long-term stabilization. The development of

Table 11.2. Environmental benefits of organic farming practice.

Organic farming practice	Environmental benefits	Reference
Crop rotation	Enhances soil quality, disrupts weed, insect, and disease life cycles. Retains carbon and nitrogen and diversifies production	Darguza and Gaile (2023)
Use of manure, compost, and green manure	Enhances soil quality, retains carbon and nitrogen and contributes to productivity	Zhao *et al.* (2024)
Cover crop	Enhances soil quality, reduces erosion, retains carbon and nitrogen, prevents dust (protects air quality), improves soil nutrients and contributes to productivity	Quintarelli *et al.* (2022)
Avoid use of synthetic fertilizers	Prevents contamination of surface and groundwater, enhances soil quality, sequesters carbon, reduces salinization (in many cases)	Santillano-Cázares *et al.* (2022)
Avoid use of synthetic pesticides	Promotes biodiversity, improves air quality and soil quality, aids in pest control, prevents damage to pollinators, reduces the cost of chemicals, and reduces exposure of farm workers and rural communities to poisonous pesticides	Leskovac and Petrović (2023)
Planting of habitat corridors, boundaries, and/or plants to encourage insects	Promotes biodiversity, biological control of pests, provides habitat for wildlife	Marcantonio *et al.* (2023)
Buffer zones (neutrality)	Improves water quality, enhances biodiversity, prevents erosion caused by wind	Alharbi *et al.* (2024)

small-scale, economically viable, diversified, and self-sufficient agroecosystems comes from new designs of cropping and/or livestock systems, which are managed with technologies adapted to the local environments found within the farmers' resources (Krebs and Bach, 2018).

Energy and resource conservation, environmental quality, public health, and equitable socioeconomic development should be taken into consideration in order to make decisions about crop species, rotations, row spacing, fertilization, pest control, and harvesting. From a management point of view, the basic components of a sustainable agroecosystem include the following.

- Vegetative cover as an effective soil and water conservation measure, achieved through the use of no-tillage practices, mulch-based agriculture, cover crops, etc.
- Regular supply of organic matter through the continuous addition of organic manure and/or compost and the promotion of soil biotic activity.
- Nutrient recycling mechanisms through the use of crop rotations, combined crop/livestock systems, agroforestry, and intercropping systems based on leguminous plants, etc.
- Pest regulation, ensured by increasing the activity of biological control agents, obtained through biodiverse manipulations, and the introduction and/or conservation of natural enemies.
- Increased multiple-use capacity of the landscape.
- Sustained crop production without the use of chemical inputs that degrade the environment (Weih *et al.*, 2022).

The above components are organized into a strategy that stresses conservation and management of local agricultural resources following a development methodology that emphasizes participation, traditional knowledge, and adaptation to local conditions (Dhillon and Moncur, 2023). In the participatory agroecological approach, economic, social, and environmental objectives are defined through the local community, and low-input technologies are implemented to harmonize economic growth, social equity, and environmental preservation (Diacono *et al.*, 2017). Finally, in addition to the development and dissemination of agroecological technologies, the motivation for sustainable agriculture requires changes in research agendas, agricultural policies, and economic systems, including markets and fair prices, as well as government incentives (García-Llorente *et al.*, 2019).

11.3.2 The transition to sustainable agriculture

The structure of corporate agriculture and the organization of agricultural research (which focuses on short-term problems and the many modifications of existing technology) prevents the recommendations of ecological research from being incorporated into agricultural management systems. It is obvious that agribusinesses will not invest in sustainable technology where profits cannot be realized immediately (Khan *et al.*, 2021). Table 11.3 shows the elements for generating agroecological strategies.

11.4 Agroecology and Sustainability: Advantages and Obstacles

Agroecology proposes an alternative ecological agrarian model that generates sustainable development models, using local knowledge as the main element: integrality, harmony, balance, autonomy of management and control, minimization of negative externalities in agro-productive activity, maintenance and strengthening of short circuits, management of agroecosystems, pluriactivity, selectivity, complementarity, and knowledge linked to traditional systems.

Development with high-input and capital-intensive technologies (mechanization, agrochemicals, imported seeds) is ecologically inadequate and socially unequal, benefiting only a small part of the population. The Green Revolution and agrobiotechnology make farmers lose their autonomy and become dependent (seeds, sterile seeds, credits, markets). Agroecological development improves production, income distribution, and employment and does not deteriorate ecosystems, because it demands more time for action than for research and publication. Thus, agroecology offers other advantages.

Table 11.3. Basic technical elements of an agroecological strategy.

Basic technical element	Examples	Reference
Conservation and regeneration of natural resources	Soil (erosion, fertility and plant health) Water (harvesting, *in situ* conservation, management, irrigation) Germplasm (native animal and plant species, soil types, adapted germplasm) Beneficial fauna and flora (natural enemies, pollinating agents, multiple use of vegetation)	O'Donoghue *et al.* (2022)
Management of productive resources	Diversification – temporal (rotations, sequences, etc.), spatial (polycultures, agroforestry, combined crop/livestock systems), genetic (multilines, etc.), regional (zoning, mosaics, etc.) Recycling of nutrients and organic matter – plant biomass (green manure, crop residues, N fixation), animal biomass (manure, urine, etc.), reuse of nutrients and resources internal and external to the farm Biotic regulation (crop protection and animal health) – natural biological control (improvement of natural control agents, importation and augmentation of natural enemies, botanical insecticides, alternative veterinary products, etc.)	Kountios *et al.* (2024)
Implementation of technical elements	Definition of resource regeneration, conservation and management techniques adapted to local needs and agroecological and socioeconomic circumstances Implementation can be at the level of micro-region divisions, at the farm level and at the cropping system level Implementation is guided by a holistic (integrated) concept and therefore does not emphasize isolated elements The strategy should be in accordance with farmer rationality and should incorporate elements of traditional management	Kalfas *et al.* (2024)

- Provides an alternative path, based on agricultural knowledge and techniques relevant to local conditions, in the management of diverse resources and inputs, and incorporates the principles and biological resources that can be used in agricultural systems.
- Offers the only concrete way of real recovery for arable soils that have been degraded by conventional practices.
- It is a safe way forward for the environment, and for small producers in marginal areas who could intensify their production in a sustainable way.
- Values the assets possessed by small producers, such as local knowledge and low-cost labor. Thus, this system reduces inequalities and improves ecological sustainability.

Sustainable ecosystem management is the application of an ecological conception to the different fields of natural appropriation, through four forms: use, conservation, management, and restoration of the ecosystem (Altieri and Masera, 1993).

From the indigenous point of view, this does not lead to a conservationist imposition but to the establishment of an intercultural dialogue that takes into account the different cultural-epistemological interpretation patterns. Agroecology evaluates the efficiency of rural production from an ecological and social perspective, where comparative analysis is an obligatory goal. Thus, agroecological research studies a wide range of rural producers, from the most isolated indigenous peoples or rural communities integrated into the market to modern farmers practicing industrialized agriculture with a very mercantile orientation. An important agroecological aspect is that the spiritual dimension manifests itself only implicitly, through agricultural practices which are derived from the spiritual.

The practical significance of agroecological research resides in the realization of productive models useful in the implementation of sustainable (rural) development. Sustainability refers to the capacity to maintain production over time. This means replacing the maximization of production (profit) in the short term, as the primary goal, by a new perspective that considers the ability to maintain production over time and space. Such production implies the creation of productive systems that do not destroy ecosystems, the material basis of production. Under the paradigm of sustainability, agroecologists validate the forms of appropriation of the ecosystem among each human culture studied, through comparative analysis.

11.5 Ten Elements of Agroecology

In order to provide guidance to countries to transform their agricultural and food systems, integrate large-scale sustainable agriculture and achieve the Zero Hunger Challenge and many other Sustainable Development Goals (SDGs), the FAO regional agroecology workshops identified ten important elements.

The ten elements of agroecology are interrelated and interdependent. Why are the ten elements useful and how will they be used? As an analytical tool, the ten elements can help countries implement agroecology. By identifying important properties of agroecological systems and approaches, as well as key considerations for the development of an enabling environment for agroecology, the ten elements provide a guide for policymakers, practitioners, and stakeholders in planning, managing, and evaluating agroecological transitions.

Here are the ten elements of agroecology that will enable agricultural innovation.

11.5.1 Diversification

Diversification is essential in agroecological transitions to ensure food security and nutrition while conserving, protecting, and enhancing natural resources (Di Bene *et al.*, 2022).

Agroecological systems are completely diverse. From a biological point of view, agroecological systems enhance species biodiversity and genetic resources in different ways. For example, agroforestry systems organize crops, shrubs, livestock, and trees of different heights and shapes in different strata, which increases vertical diversity. Intercropping combines complementary species with the objective of increasing spatial diversity. Crop rotation, often including legumes, increases temporal diversity.

Integrated agricultural production systems rely on the diversity of local breeds adapted to specific environments.

In the aquatic world, traditional fish polyculture, integrated multitrophic aquaculture or rotational aquaculture systems follow the same principles to maximize biodiversity.

Increasing biodiversity contributes to a range of production, socioeconomic, nutritional, and environmental benefits.Through diversity planning and management, agroecological approaches enhance the provision of ecosystem services, particularly pollination and soil health, on which agricultural production primarily depends. Diversification can increase productivity and resource use efficiency by optimizing biomass harvesting and water harvesting.

Agroecological diversification also strengthens ecological and socioeconomic resilience by creating new market opportunities. For example, crop and animal diversity reduces the risk of failure in the face of climate change. Mixed grazing of different ruminant species reduces health risks from parasitism, while the coexistence of diverse species or local breeds increases their ability to survive, produce, and maintain reproductive levels in hostile environments. In turn, having a variety of income sources from new and differentiated markets, such as diverse products, local food processing and agro-tourism, helps stabilize household incomes. Varied consumption of cereals, legumes, fruits, vegetables, and animal products contributes to improved nutritional outcomes.

In addition, the genetic diversity of different varieties, breeds, and species is important in contributing macronutrients, micronutrients, and other bioactive compounds to human diets. For example, in Micronesia, the reintroduction of an underutilized traditional variety of orange-fleshed banana with 50 times more β-carotene than the commercial white-fleshed banana proved to be instrumental in improving health and nutrition.

Globally, three cereal crops provide nearly 50% of all calories consumed, while genetic diversity of crops, livestock, aquatic animals, and trees continues to be rapidly lost. Agroecology can reverse these trends by managing and conserving agrobiodiversity, as well as responding to the growing demand for diverse products that are environmentally friendly. One example is 'fish-stock friendly' rice production that takes place in irrigated, rainfed and deepwater rice ecosystems, where the diversity of aquatic species and their importance to rural livelihoods are valued.

11.5.2 Cocreation and knowledge sharing

Agricultural innovations are more responsive to local challenges when they are cocreated through participatory processes (Rossi, 2020).

Agroecological practices are adapted to the environmental, social, economic, cultural, and political context. Cocreation and knowledge sharing play a key role in the process of developing and implementing agroecological innovations to address food system challenges, including adaptation to climate change. Through the cocreation process, agroecology combines traditional and indigenous knowledge, the know-how of producers and traders, and global scientific knowledge. The agricultural biodiversity knowledge and management experience of producers in specific contexts, as well as their knowledge of markets and institutions, are absolutely indispensable in this process.

Education (formal or informal) is important for the exchange of agroecological innovations gained from the cocreation process. For example, for more than 30 years, the horizontal 'farmer-to-farmer' movement has played a critical role in agroecological knowledge sharing and established relationships among hundreds of thousands of producers in Latin America. Promoting participatory processes and institutional innovations that nurture mutual trust enables cocreation and knowledge sharing, which contributes to relevant and inclusive agroecological transition processes.

11.5.3 Synergies

Creating synergies enhances the main functions of food systems, which enhances production and multiple ecosystem services (Bartolini and Vergamini, 2023).

Agroecology pays careful attention to the design of diversified systems that selectively combine annual and perennial crops, livestock, aquatic animals, trees, soils, water, and other components in farms and agricultural territories with a view to increasing synergies in the context of an increasingly changing climate. Creating synergies in food systems brings multiple benefits. By optimizing biological synergies, agroecological practices enhance ecological functions, resulting in increased resource use efficiency and resilience. For example, biological nitrogen fixation by legumes in intercropping systems or rotations saves about US$10 million in nitrogen fertilizer globally per annum, while contributing to soil health, climate change mitigation, and adaptation.

In addition, about 15% of the nitrogen applied to crops comes from animal manure, highlighting the synergies arising from the integration of agriculture and livestock. In Asia, integrated rice paddies combine rice cultivation with fish, duck, and tree production. By maximizing synergies, integrated rice fields significantly improve yields, dietary diversity, weed suppression, soil structure, and fertility, as well as providing habitat for biodiversity and pest control. At the territorial level, it is necessary to synchronize productive activities in time and space to enhance synergies. In integrated agroecological systems in the East African highlands, it is common to use hedgerows of the genus *Calliandra* to combat soil erosion. In this example, the management practice of periodic pruning reduces tree competition with crops growing between the hedgerows and, at the same time, provides fodder for animals, thus creating synergies between the different components.

Livestock husbandry and extensive grazing systems manage complex interactions between people, multispecies herds and variable environmental conditions, which build resilience and contribute to the provision of ecosystem services such as dispersal, habitat conservation, and soil fertility. As agroecological approaches seek to maximize synergies, there are also trade-offs between natural and human systems. For example, the allocation of resource use or access rights often involves trade-offs. To promote synergies in the broader food system and better manage trade-offs, agroecology emphasizes the importance of partnerships, cooperation, and responsible governance, involving different actors at multiple scales.

11.5.4 Efficiency

Innovative agroecological practices produce more using fewer external resources (Wang *et al.*, 2022).

Increased resource use efficiency is an emerging property of agroecological systems that carefully plan and manage diversity with the goal of creating synergies between different components of the system. For example, one of the main efficiency challenges is that less than 50% of the nitrogen fertilizer added globally to cropland is converted into harvested products and the rest is released into the environment, causing significant environmental problems.

Agroecological systems improve the use of natural resources, especially those that are abundant and free, such as carbon, nitrogen from the atmosphere, and solar radiation. By improving biological processes and recycling biomass, nutrients, and water, producers can use fewer external resources, reducing the costs and negative environmental effects of their use. Ultimately, reducing dependence on external resources empowers producers by increasing their autonomy and resilience to natural or economic shocks.

One way to measure the efficiency of integrated systems is by using the land equivalent ratio (LER). The LER compares crop yields of two or more components (e.g. crops, trees, and animals) with those obtained by growing the same components in monocultures. Integrated agroecological systems often demonstrate higher LERs. Consequently, agroecology promotes farming systems with the biological, socioeconomic, and institutional diversity and the adjustment to time and space needed to support greater efficiency.

11.5.5 Recycling

Recycling more means agricultural production with lower economic and environmental costs (Diacono *et al.*, 2019).

Waste is a human concept – it does not exist in natural ecosystems. By mimicking natural ecosystems, agroecological practices favor biological processes that drive the recycling of biomass, water, and nutrients from production systems, thereby increasing resource use efficiency and minimizing waste and pollution.

Recycling can take place both on farms and in territories through diversification and the creation of synergies between different components and activities. For example, agroforestry systems that include deep-rooted trees can capture nutrients that are not reached by the roots of annual crops. Agroforestry systems promote the recycling of organic matter by using manure for composting or directly as fertilizer and crop residues and by-products for animal feed. The cycling of nutrients accounts for 51% of the economic value of all nonsupply ecosystem services and the integration of livestock plays an important role in this regard.

Similarly, in rice-fish farming systems, aquatic animals help to fertilize the rice crop and reduce pests, thus reducing the need for external fertilizers or pesticides. Recycling brings multiple benefits by closing cycles and reducing waste, resulting in less dependence on external resources, which in turn increases producers' autonomy and reduces their vulnerability to market and climate shocks. Recycling organic materials and by-products holds enormous potential for agroecological innovations.

11.5.6 Resilience

Improving the resilience of people, communities, and ecosystems is critical to ecosystems and fundamental to achieving sustainable food and agricultural systems (Roosevelt *et al.*, 2023).

11.5.7 Human and social values

Diversified agroecological systems are more resilient, i.e. they have a greater capacity to recover from shocks, particularly extreme weather events such as drought, floods or hurricanes, and to resist attack by pests and diseases. Following the 1998 Hurricane Mitch in Central America, biodiversity-based farms, particularly agroforestry, contour tillage and cover crops, retained 20–40% more topsoil, suffered less erosion and experienced fewer economic losses than neighboring farms with conventional monocultures.

By maintaining a functional balance, agroecological systems are better able to resist attack by pests and diseases. Agroecological practices restore the biological complexity of agricultural systems and promote the necessary community of interacting organisms to self-regulate pest outbreaks. At a territorial scale, diversified agricultural systems have greater potential to contribute to pest and disease control functions.

Agroecological approaches can also improve socioeconomic resilience. Through diversification and integration, farmers reduce their vulnerability in the event of failure of one crop, livestock species or other product. By reducing dependence on external inputs, agroecology can reduce producers' vulnerability to economic risk. Improving ecological resilience goes hand in hand with improving socioeconomic resilience; after all, humans are an integral part of ecosystems (Campos *et al.*, 2021).

Agroecology places particular emphasis on human and social values, such as dignity, equity, inclusion, and justice, which all contribute to the livelihoods dimension of the SDGs. It places the aspirations and needs of food producers, distributors, and consumers at the heart of food systems. Empowering adaptive capacities to manage their agroecosystems, agroecological approaches equip people and communities with the means to overcome poverty, hunger, and malnutrition, while promoting human rights, such as the right to adequate food, as well as environmental stewardship, so that future generations can also lead prosperous lives.

Agroecology aims to address gender inequalities by creating opportunities for women. Globally, women make up almost half of the agricultural workforce. They also play an important role in household food security, dietary diversity, and health, as well as in the conservation and sustainable use of biodiversity. Despite this, women remain economically marginalized, vulnerable to violations of their rights, and their contributions often go unrecognized. Agroecology can help rural women in family farming to achieve a greater degree of autonomy through the promotion of knowledge, collective action, and the creation of opportunities for commercialization. It can also open up spaces for women to become more autonomous and empower them at the household, community, and larger scale, for example through participation in producer groups. Women's participation is fundamental to agroecology and it is often women who are in charge of agroecology projects.

In several parts of the world, rural youth are facing an employment crisis. Agroecology offers a promising solution as a source of decent work. It is based on a different kind of agricultural production that is environmentally friendly, socially responsible, innovative, knowledge-intensive, and dependent on skilled labor. Meanwhile, rural youth have energy, creativity, and a desire to bring about positive change in their world. What they need is support and opportunity. As a bottom-up paradigm for sustainable development, agroecology empowers people to become their own agents of change.

11.5.8 Culture and food traditions

By supporting healthy, diversified and culturally appropriate diets, agroecology increases food security and nutrition while maintaining healthy ecosystems (Zhu *et al.*, 2024).

Agriculture and food are essential components of humanity's heritage. Therefore, food culture and traditions play a fundamental social role, as well as in shaping human behavior. However, in many cases, our current food systems have created a disconnect between food habits and culture. This disconnect has contributed to a situation where hunger and obesity coexist in a world that produces enough food to feed its entire population. Globally, almost 800 million people suffer from chronic hunger and 2 billion from micronutrient deficiencies.

Meanwhile, there has been an uncontrolled increase in obesity and diet-related diseases; 1.9 billion people are overweight or obese and noncommunicable diseases (e.g. cancer, cardiovascular disease, and diabetes) are the leading cause of global mortality.

To correct the imbalances in our food systems and move towards the eradication of hunger in the world, it is not enough to simply increase production. Agroecology therefore plays an important role in rebalancing tradition and modern food habits, bringing them together in a harmonious way that promotes the production and consumption of healthy food and supports the right to adequate food. In this sense, agroecology seeks to cultivate a healthy relationship between people and food. Cultural identity and a sense of belonging are closely linked to territories and food systems. Given that people and ecosystems have evolved together, cultural practices and indigenous and traditional knowledge offer extensive experience that can inspire agroecological solutions. For example, it has been estimated that in India there are 50,000 indigenous rice varieties that have been cultivated for centuries for their organoleptic, nutritional, and pest resistance properties, as well as their adaptability to different conditions.

Taking this accumulation of traditional knowledge as a guide, agroecology can help to realize the potential of territories to maintain their populations based on the culinary traditions of their different varieties while taking advantage of their different properties.

11.5.9 Responsible governance

To achieve sustainable food and agriculture, it is necessary to have responsible and effective governance mechanisms at different scales, from local to national to global (Rodorff *et al.*, 2019).

Agroecology requires responsible and effective governance to support the transition to sustainable food and farming systems. Transparent, inclusive, and accountability-based governance mechanisms are needed to create an enabling environment that helps producers transform their systems following agroecological concepts and practices. Examples of successful cases include school feeding and public procurement programs, market regulations that allow for branding to differentiate agroecological products, and subsidies and incentives for ecosystem services. Land and natural resource governance is an excellent example.

Most of the world's poor and vulnerable rural populations rely heavily on terrestrial and aquatic biodiversity and ecosystem services for their livelihoods but lack secure access to these resources. Agroecology depends on equitable access to land and natural resources; it is key to social justice but also to incentivizing the investments that must be made over the long term to protect soils, biodiversity, and ecosystem services. Agroecology can best be promoted by adopting responsible governance mechanisms at different scales. Many countries have already developed laws, policies, and programs at the national scale that reward models of agricultural management that enhance biodiversity and the provision of ecosystem services.

Community and territorial-scale governance, such as traditional and customary governance models, is extremely important in fostering cooperation between different actors, maximizing synergies and minimizing trade-offs.

11.5.10 Circular and solidarity economy

Circular and solidarity economies that reconnect producers and consumers offer innovative solutions for living within the limits of our planet and at the same time strengthen the social foundations for inclusive and sustainable development (Landaburú-Mendoza *et al.*, 2024).

Agroecology seeks to reconnect producers and consumers through a circular and solidarity economy that prioritizes local markets and supports local economic development by creating virtuous circles.

Agroecological approaches promote fair solutions based on local needs, resources, and capacities and create more equitable and sustainable markets. Strengthening short food circuits can increase the income of food producers while maintaining a fair price for consumers. Some of these circuits are the new innovative markets, in addition to the more traditional territorial markets where most small producers market their products. Social and institutional innovations play a key role in promoting agroecological production and consumption.

Examples of innovations that help establish relationships between producers and consumers include e-commerce systems, local farmers' markets, appellation of origin labeling, participatory guarantee systems and community-supported agriculture.

These innovative markets respond to growing consumer demand for healthier diets. Reformulating food systems based on circular economy principles can help address the challenge of global food waste by shortening food value chains and increasing resource efficiency. Currently, one-third of all food produced is lost or wasted, thereby failing to contribute to food security and nutrition and exacerbating pressures on natural resources. The energy used to produce food that is lost or wasted accounts for approximately 10% of total global energy consumption and the footprint of food waste is equivalent to 3.5 Gt CO_2 of greenhouse gas emissions per year.

11.6 Conclusion

Sustainable agriculture and agroecology are two complementary perspectives that work in harmony to achieve an agricultural future with ecological impact. To make progress in the fight against climate change, we must move towards sustainable agroecological agriculture. To reap great benefits and protect natural resources, more public money must be invested in agroecological farming research, technical assistance, access to credit, and other incentives to grow agroecological farming systems for the future.

References

Alharbi, S., Felemban, A., Abdelrahim, A. and Al-Dakhil, M. (2024) Agricultural and technology-based strategies to improve water-use efficiency in arid and semiarid areas. *Water* 16(13), 1842.

Altieri, M.A. and Masera, O. (1993) Sustainable rural development in Latin America: Building from the bottom-up. *Ecological Economics* 7(2), 93–121.

Bartolini, F. and Vergamini, D. (2023) Trade-offs and synergies between ecosystem services provided by different rural landscape. *Agronomy* 13(4), 977.

Campos, V., Sanchis, J.R. and Talavera, C. (2021) The importance of social value in agroecological farms: Adjusting the common good balance sheet to improve their sustainable management. *Sustainability* 13(3), 1184.

Cárceles Rodríguez, B., Durán-Zuazo, V.H., Soriano Rodríguez, M., García-Tejero, I.F., Gálvez Ruiz, B. *et al*. (2022) Conservation agriculture as a sustainable system for soil health: A review. *Soil Systems* 6(4), 87.

Cleves, A., Youkhana, E. and Toro, J. (2022) A method to assess agroecosystem resilience to climate variability. *Sustainability* 14(14), 8588.

Darguza, M. and Gaile, Z. (2023) The productivity of crop rotation depending on the included plants and soil tillage. *Agriculture* 13(9), 1751.

Del Duca, S., Mocali, S., Vitali, F., Fabiani, A., Cucu, M.A. *et al*. (2024) Impacts of soil management and sustainable plant protection strategies on soil biodiversity in a Sangiovese vineyard. *Land* 13(5), 599.

Dhillon, R. and Moncur, Q. (2023) Small-scale farming: A review of challenges and potential opportunities offered by technological advancements. *Sustainability* 15(21), 15478.

Di Bene, C., Francaviglia, R., Farina, R., Álvaro-Fuentes, J. and Zornoza, R. (2022) Agricultural diversification. *Agriculture* 12(3), 369.

Diacono, M. and Montemurro, F. (2023) Achieving agroecosystem resilience through an agroecological approach. *Agronomy* 13(8).

Diacono, M., Persiani, A., Fiore, A., Montemurro, F. and Canali, S. (2017) Agro-ecology for potential adaptation of horticultural systems to climate change: Agronomic and energetic performance evaluation. *Agronomy* 7(2), 35.

Diacono, M., Persiani, A., Testani, E., Montemurro, F. and Ciaccia, C. (2019) Recycling agricultural wastes and by-products in organic farming: Biofertilizer production, yield performance and carbon footprint analysis. *Sustainability* 11(14), 3824.

Diacono, M., Trinchera, A. and Montemurro, F. (2021) An overview on agroecology and organic agriculture strategies for sustainable crop production. *Agronomy* 11(2), 223.

Dönmez, D., Isak, M.A., İzgü, T. and Şimşek, Ö. (2024) Green horizons: Navigating the future of agriculture through sustainable practices. *Sustainability* 16(8), 3505.

García-Llorente, M., Pérez-Ramírez, I., Portilla, C., Haro, C. and Benito, A. (2019) Agroecological strategies for reactivating the agrarian sector: The case of Agrolab in Madrid. *Sustainability* 11(4), 1181.

García-Tejero, I.F. and Durán-Zuazo, V.H. (2022) Plant water use efficiency for a sustainable agricultural development. *Agronomy* 12(8), 1806.

Gorooei, A., Gaiser, T., Aynehband, A., Rahnama, A. and Kamali, B. (2023) The effect of farming management and crop rotation systems on chlorophyll content, dry matter translocation, and grain quantity and quality of wheat (*Triticum aestivum* L.) grown in a semi-arid region of Iran. *Agronomy* 13(4), 1007.

Herman, E. (2024) Sustainable agriculture and its impact on the rural development in EU countries: A multivariate analysis. *Land* 13(7), 947.

Kalfas, D., Kalogiannidis, S., Papaevangelou, O., Melfou, K. and Chatzitheodoridis, F. (2024) Integration of technology in agricultural practices towards agricultural sustainability: A case study of Greece. *Sustainability* 16(7), 2664.

Khan, N., Ray, R.L., Sargani, G.R., Ihtisham, M., Khayyam, M. *et al.* (2021) Current progress and future prospects of agriculture technology: Gateway to sustainable agriculture. *Sustainability* 13(9), 4883.

Kountios, G., Chatzis, I. and Papadavid, G. (2024) Agricultural knowledge and innovation systems and sustainable management of natural resources. *Proceedings* 94, 10.

Krebs, J. and Bach, S. (2018) Permaculture—scientific evidence of principles for the agroecological design of farming systems. *Sustainability* 10(9), 3218.

LaFevor, M.C. (2022) Characterizing agricultural diversity with policy-relevant farm typologies in Mexico. *Agriculture* 12(9), 1315.

Landaburú-Mendoza, J., Arguello, L., Montalván-Burbano, N., Chunga-Montalván, L. and Pico-Saltos, R. (2024) Development of community and agricultural associations through social and solidarity economy with collaboration of university. *Social Sciences* 13(6), 306.

Leskovac, A. and Petrović, S. (2023) Pesticide use and degradation strategies: Food safety, challenges and perspectives. *Foods* 12(14), 2709.

Liu, M., Chen, X. and Jiao, Y. (2024) Sustainable agriculture: Theories, methods, practices and policies. *Agriculture* 14(3), 473.

Marcantonio, M., Voda, R., Re, D., Igot, Q., Dennis, R.L.H. *et al.* (2023) The effect of habitat on insect movements: Experimental evidence from wild-caught butterflies. *Insects* 14(9), 737.

Mokgomo, M.N., Chagwiza, C. and Tshilowa, P.F. (2022) The impact of government agricultural development support on agricultural income, production and food security of beneficiary small-scale farmers in South Africa. *Agriculture* 12(11), 1760.

O'Donoghue, T., Minasny, B. and McBratney, A. (2022) Regenerative agriculture and its potential to improve farmscape function. *Sustainability* 14(10), 5815.

Oteros-Rozas, E., Ravera, F. and García-Llorente, M. (2019) How does agroecology contribute to the transitions towards social-ecological sustainability? *Sustainability* 11(16), 4372.

Quintarelli, V., Radicetti, E., Allevato, E., Stazi, S.R., Haider, G. *et al.* (2022) Cover crops for sustainable cropping systems: A review. *Agriculture* 12(12), 2076.

Rodorff, V., Siegmund-Schultze, M., Guschal, M., Hölzl, S. and Köppel, J. (2019) Good governance: A framework for implementing sustainable land management, applied to an agricultural case in Northeast-Brazil. *Sustainability* 11(16), 4303.

Roosevelt, M., Raile, E.D. and Anderson, J.R. (2023) Resilience in food systems: Concepts and measurement options in an expanding research agenda. *Agronomy* 13(2), 444.

Rossi, A. (2020) From co-learning to shared commitment to agroecology. Some insights from initiatives aimed at reintroducing agrobiodiversity. *Sustainability* 12(18), 7766.

Santillano-Cázares, J., Turmel, M.-S., Cárdenas-Castañeda, M.E., Mendoza-Pérez, S., Limón-Ortega, A. *et al.* (2022) Can biofertilizers reduce synthetic fertilizer application rates in cereal production in mexico? *Agronomy* 12(1), 80.

Telo da Gama, J. (2023) The role of soils in sustainability, climate change, and ecosystem services: Challenges and opportunities. *Ecologies* 4(3), 552–567.

Valenzuela, H. (2016) Agroecology: A global paradigm to challenge mainstream industrial agriculture. *Horticulturae* 2(1), 2.

Velten, S., Leventon, J., Jager, N. and Newig, J. (2015) What is sustainable agriculture? A systematic review. *Sustainability* 7(6), 7833–7865.

Viccaro, M., Romano, S., Rosalia, I. and Cozzi, M. (2024) Searching for the profitability of energy crops: An agroecological–economic land use suitability (AE-landUSE) model. *Environments* 11(5), 91.

Wang, G., Shi, R., Mi, L. and Hu, J. (2022) Agricultural eco-efficiency: Challenges and progress. *Sustainability* 14(3), 1051.

Wang, M., Liao, G. and Li, Y. (2021) The relationship between environmental regulation, pollution and corporate environmental responsibility. *International Journal of Environmental Research and Public Health* 18(15), 8018.

Weih, M., Mínguez, M.I. and Tavoletti, S. (2022) Intercropping systems for sustainable agriculture. *Agriculture* 12(2), 291.

Zenda, M. and Rudolph, M. (2024) A systematic review of agroecology strategies for adapting to climate change impacts on smallholder crop farmers' livelihoods in South Africa. *Climate* 12(3), 33.

Zhao, N., Zhang, J., Li, X., Ma, J., Cao, J. *et al.* (2024) Limited advantages of green manure planting on soil nutrients and productivity in intensive agriculture: A case study of wheat–maize–sunflower rotation in Hetao Irrigation District. *Agronomy* 14(1), 100.

Zhu, S.J., Mfuni, T.I. and Powell, B. (2024) Greater attention to wild foods and cultural knowledge supports increased nutrition outcomes associated with agroecology. *Sustainability* 16(10), 3890.

12 Bio-Inoculants as Biological Tools to Mitigate Water Deficit in Crops

Urley A. Pérez-Moncada[1], Pablo Cornejo[2,3], Christian Santander[4], Cledir Santos[4], Humberto Aponte[4], Sivagnanam Silambarasan[5] and Gladys Vidal[6,7]*

[1]Doctorado en Ciencias de Recursos Naturales, Universidad de La Frontera, Temuco, Chile; [2]Núcleo Científico Tecnológico en Biorecursos (BIOREN-UFRO), Universidad de La Frontera, Temuco 4811230, Chile; [3]Centro de Estudios Avanzados en Fruticultura (CEAF); Camino Las Parcelas N° 882, Sector los Choapinos, Rengo, Región de O'Higgins, Chile; [4]Departamento de Ciencias Químicas y Recursos Naturales, Universidad de La Frontera, Temuco, Chile; [5]Laboratorio de Ecología Microbiana del Suelo y Biogeoquímica, Instituto de Ciencias Agroalimentarias, Pecuarias y Ambientales (ICA3), Universidad de O'Higgins, San Fernando, Chile; [6]Grupo de Ingeniería y Biotecnología Ambiental (GIBA-UDEC), Facultad Ciencias Ambientales, Universidad de Concepción, Concepción, Chile; [7]Centro de Recursos Hídricos para la Agricultura y la Minería (CRHIAM), Centro ANID Fondap, Victoria 1295, Concepción, Chile

Abstract

The increasing global population and the concurrent rise in competition for limited water resources necessitate a substantial increase in food production by 2050. Climate change, characterized by altered rainfall patterns and frequent droughts, poses significant challenges to agricultural productivity. This chapter explores the potential of bio-inoculants, specifically plant growth-promoting microorganisms (PGPM) such as bacteria, free-living fungi, and arbuscular mycorrhizal fungi (AMF), in mitigating drought stress in crops. Bio-inoculants can enhance plant tolerance to water deficit through various mechanisms, including the production of growth-regulating substances, osmolytes, exopolysaccharides, and volatile organic compounds. These microorganisms also facilitate nutrient solubilization and uptake, which is crucial under drought conditions. Bacteria, particularly those capable of producing 1-aminocyclopropane-1-carboxylic acid (ACC) deaminase, play a pivotal role in reducing ethylene levels in plants under stress, thus promoting growth. The chapter emphasizes the need for further research into the mechanisms exerted by free-living fungi and the potential benefits of microbial consortia. The integration of bio-inoculants with conventional drought mitigation strategies, such as plant breeding and the exogenous application of substances, could provide holistic solutions to enhance crop resilience against drought in the context of climate change.

Keywords: bio-inoculants, plant growth-promoting microorganisms, arbuscular mycorrhizal fungi, agricultural productivity

*Corresponding author: glvidal@udec.cl

 Plant Biotechnology and Sustainable Agriculture (ed. Tariq Aftab)
DOI: 10.1079/9781800627659.0012

12.1 Introduction

It is expected that by 2050 the world population will have increased to 9 billion people, generating strong competition for the limited water resources worldwide (FAO, 2020). To supply the nutritional demand of the expected world population, it is necessary to increase food production by 70% (Lesk *et al.*, 2016); however, the current scenario of climate change (CC) may not allow this increase.

Global CC has been a critical constraint for agriculture during the last century since it has influenced several environmental characteristics such as the usual climatic conditions, which have changed drastically in recent years, including intensive changes in rainfall patterns, thus increasing the frequency of drought events (Compant *et al.*, 2010). Within the scenario of accelerated CC, drought is one of the abiotic factors with the greatest impact on plant growth and production, which causes many losses in crop yield. Globally, crop losses have been around 75% of the harvested area and, together with the heat waves, these losses have tripled (IPCC, 2022).

Sati *et al.* (2022) defined drought as a climate-specific event characterized by less than normal or no precipitation in a given time; however, salt accumulation in soil, extreme temperature fluctuations, and high light intensity can also cause drought stress (Farooqi *et al.*, 2020). In plants, the absence of water causes significant morphophysiological changes such as the loss and curling of leaves, modification of root architecture, closure of stomata, and with this, a decrease in the photosynthetic rate due to lack of CO_2 (Fang and Xiong, 2015). The reduction in relative water content and chlorophyll content (Alfaro-Corres *et al.*, 2023), the increase in antioxidant enzyme activity and nonenzymatic antioxidant compounds (Wang *et al.*, 2021), and the overproduction of reactive oxygen species (ROS) are some of the biochemical characteristics that cause water deficit in plants (Farooqi *et al.*, 2020). However, the regulation of these mechanisms will depend on the plant species or genotype, the stage of development, and other environmental factors (Demirevska *et al.*, 2009; Fang and Xiong, 2015).

Plants have developed different strategies to resist drought, such as avoidance, tolerance, escape, and recovery, avoidance and tolerance being the two main mechanisms (Fang and Xiong, 2015). Leaflet closure, stomatal closure, reduced transpiration, stomata arrangement, length, weight, volume, and density of root, and root/shoot ratio are indicators of avoidance that have been studied in plants under water deficit (Fang and Xiong, 2015; Feng *et al.*, 2022; Twala *et al.*, 2022). On the other hand, osmoprotectant production, ABA production, differential expression of genes and metabolites, transcription factors, and protein folding are indicators of tolerance to water deficit in the plants that have been studied (Vu *et al.*, 2022; Wassie *et al.*, 2023).

Constant drought events due to CC and increasingly limited access to water for irrigation are an emerging problem for water security; however, water security in many cases will not depend on CC but on socio-political factors (FAO, 2020; IPCC, 2022). Nevertheless, CC poses risks to water security in many parts of the world through either flooding or water scarcity (IPCC, 2022). For the latter, different strategies have been evaluated, ranging from the production of drought-resistant cultivars to the application of different abiotic (growth regulators, salts, polyethylene glycol, ketoconazole, and ethanol) (Fahad *et al.*, 2017; Vu *et al.*, 2022; Alfaro-Corres *et al.*, 2023) and biotic inducers (Hanaka *et al.*, 2021). Among the biotic inducers, the application of bio-inoculants belonging to the group of plant growth-promoting microorganisms (PGPM) such as bacteria, free-living fungi (rhizospheric or filamentous fungi and yeast), and arbuscular mycorrhizal fungi (AMF) has been used to mitigate drought effects on plants (Hanaka *et al.*, 2021).

Bacteria, free-living fungi, and AMF have the ability to promote drought stress tolerance and plant growth simultaneously (Sati *et al.*, 2022; Uzma *et al.*, 2022). Of these three groups of PGPM, the most studied in drought tolerance in plants is bacteria (Sati *et al.*, 2022; Uzma *et al.*, 2022), followed by AMF (Marro *et al.*, 2022; Wang *et al.*, 2023) with free-living fungi being the least studied. Although fungi can tolerate drought to a greater degree than bacteria (Treseder and Lennon, 2015), they have been little studied under these conditions because the focus has been on the biological control of plant pathogens (Tyśkiewicz *et al.*, 2022). However, within free-living fungi, the rhizospheric or filamentous fungi such as *Trichoderma* spp. (Mona *et al.*, 2017; Cornejo-Ríos *et al.*, 2021; Scudeletti *et al.*, 2021) and unicellular fungi such as yeasts (Silambarasan *et al.*, 2019b; Abdelaal *et al.*, 2021; Alzandi and Naguib, 2022) have gained considerable interest for drought tolerance in plants.

Bacteria and free-living fungi share similar mechanisms that allow them to tolerate drought. These mechanisms include the production of ACC deaminase (Silambarasan *et al.*, 2019b; Moreno-Galvan *et al.*, 2021; Chandwani and Amaresan, 2022; Tyśkiewicz *et al.*, 2022) and regulating hormones (Gowtham *et al.*, 2021), segregation of osmolytes such as trehalose, proline, choline, and some sugars (Sangiorgio *et al.*, 2020; Gamalero and Glick, 2022; Shao *et al.*, 2022), exopolysaccharide production (Silambarasan *et al.*, 2019a, b ; Khan *et al.*, 2020; Santillana Villanueva, 2021; Sati *et al.*, 2022), volatile organic compounds (VOCs) (Velázquez-Becerra *et al.*, 2011; Heenan-Daly *et al.*, 2019), enzyme activity (protease, urease, glucosidase, and phosphatase among others) (Bilek *et al.*, 2020; Alzandi and Naguib, 2022) and transport of nutrients and water through extraradical mycelial networks (Santander *et al.*, 2017; Zhang *et al.*, 2018; Leyva-Morales *et al.*, 2019).

To cope with the damage caused by drought in agriculture, it is necessary to know the biological mechanisms of PGPM, free-living fungi, and AMF as drought mitigators in plants in depth, in order to develop a new generation of bio-inoculants that might be applied either individually or in consortia to agricultural systems. This focus could provide economical biotechnological tools which are easy to apply, efficient and environmentally friendly, integrating them with other conventional strategies (e.g. plant breeding, exogenous application of substances, etc.) to counteract drought stress in plants (Fahad *et al.*, 2017; Ilyas *et al.*, 2021). Based on the above, with this comprehensive review we aim to discuss the biological basis of the effects caused by water shortage in plants, as well as the mechanisms exerted by PGPB, free-living fungi and AMF, the modifications they make in plants grown under drought stress conditions and the improvements expected of their use as a mitigation strategy in a CC scenario.

12.1.1 Drought Effects on Plants

Drought has become the most significant risk to global food security, causing greater crop losses than pathogen attacks (Gupta *et al.*, 2020; Ilyas *et al.*, 2021). Losses range from 10 to 65% in crops such as barley, corn, wheat, rice, chickpea, soybean, sunflower, and common beans (Sati *et al.*, 2022). Considering that 80–90% of the fresh weight in herbaceous plants and 50% in woody plants consist of water (Azcón-Bieto and Talón, 2003), water deficit caused by drought stress causes changes at molecular, biochemical, and morphophysiological levels in plants (Farooqi *et al.*, 2020), depending on the plant species and the intensity and duration of the drought.

As plants are sessile organisms, they cannot avoid stressful environmental situations. They have developed mechanisms to resist drought, among which avoidance and tolerance are the two main mechanisms (Azcón-Bieto and Talón, 2003; Fang and Xiong, 2015; Taiwo *et al.*, 2020). Changes occurring during avoidance include adjustment of morphological structures, such as root enlargement, leaf fall, and curling, premature flowering, closure and rearrangement of stomata, wax accumulation, and slowing of growth, among others (Aslam *et al.*, 2022b; Feng *et al.*, 2022). On the other hand, tolerance involves regulation at the molecular level of genes, proteins, and metabolic pathways to repair the damage caused by water deficit (Raza *et al.*, 2020; Taiwo *et al.*, 2020), especially the damage caused by overproduction of ROS that causes lipid peroxidation and DNA damage (Fang and Xiong, 2015).

Antioxidant enzyme activity, nonenzymatic antioxidant compounds, growth-regulating substances, and others are increased in order to maintain osmotic adjustment in the tissues and decrease water losses (Gupta *et al.*, 2020). Table 12.1 shows the most recent studies on avoidance and tolerance mechanisms that plants have employed to resist drought without the addition of any biotic or abiotic effect.

Table 12.1. Avoidance and tolerance mechanisms plants employ to resist drought.

Plant	Drought resistance mechanism	Drought related traits	Irrigation levels	References
Sorghum, Pegah and Speedfeed varieties	Avoidance	↓Plant height ↓Number of leaves per plant ↓Stem diameter	60% soil moisture	Rad *et al.* (2023)
Afrocarpus falcatus Podocarpus henkelii	Avoidance	↑Leaf shedding and senescence ↓Stem diameter Biomass, no effect Root:shoot ratio, no effect	0% soil moisture	Twala *et al.* (2022)
Wheat, genotypes inb, Dingxi24, Monkhead, 021–128, 92–46y Longchun8275	Avoidance, tolerance	↓Stomatal conductance ↑Leaf rolling ↑Leaf waxiness ↓Plant height and tiller number ↓ Leaf area ↑Proline ↑Soluble sugar ↑H_2O_2 ↑Superoxide dismutase (SOD), catalase (CAT), ascorbate peroxidase (APX), glutathione reductase (GR)	50% field water-holding capacity	Li *et al.* (2022a)
Quercus ilex	Tolerance	↑Secondary metabolites (71% of total metabolites)	1% soil moisture	Gargallo-Garriga *et al.* (2018)
Soybean, genotypes L11644, L13241, and EMBRAPA 48-C48	Avoidance, tolerance	↓Relative growth rate of the shoot ↓Leaf area ↓Root length Biomass Root:shoot ratio no effect ↑Proline ↑Glycine betaine ↑APX and GR ↑Metabolic profile ↓Gene expression such as rubisco ativase (RCA), cytochrome B6F complex subunit 7 (PetM cytochrome b6f, electron transport in chloroplasts) and light-harvesting complex II (LHBC1, energy absorption in antenna complex)	Leaf water potential at predawn of −1.5±0.2 MPa	Alves da Silva *et al.* (2022)

Continued

Table 12.1. Continued

Plant	Drought resistance mechanism	Drought related traits	Irrigation levels	References
Two *Triticum monococcum* diploid genotypes – 'MO1' and 'MO4' Two *T. dicoccum* tetraploid genotypes – 'DM22' and 'DM31' Two hexaploid modern genotypes – 'Longchun8275' (LC8275) and 'Monkhead' (MOH)	Avoidance	↑ Allocate biomass to the root system in diploid and tetraploid genotypes ↓ Allocate biomass to roots and ↑ biomass to shoots in hexaploid genotypes	50% field water-holding capacity	Li *et al.* (2022b)
Ipomoea batatas	Tolerance	↑ ABA-shoot ↓ ABA-tuber	Drip irrigation system, with 1.6 mm for control and 0.9 mm for water deficit. In total 77 mm of water for control and 54 mm for drought	Gouveia *et al.* (2021)
Sorghum genotypes DRT1019 (drought tolerant) and ICSV95022 (drought susceptible)	Avoidance, tolerance	Delayed leaf rolling in DRT1019 ↓ Stomatal conductance ↑ Epicuticular wax in DRT1019 ↓ Root length both genotypes ↑ Transcription factors Change in metabolites	30% soil moisture	Rajarajan *et al.* (2021)

Nowadays, seed priming with different abiotic inducers such as salts (sodium, potassium, calcium chlorides, or nitrates), polyethylene glycol, ascorbic acid, and putrescine, among others (Fahad *et al.*, 2017), is the most common strategy to help plants tolerate drought. In addition, growth regulators, osmoprotectants, ketoconazole, and ethanol have also been used as inducers of drought tolerance (Jaleel *et al.*, 2007; Fahad *et al.*, 2017; Vu *et al.*, 2022; Alfaro-Corres *et al.*, 2023). However, application of some of these substances should be studied in depth due to the effect they could have on the interaction of the microorganisms present in the plants, and their applicability could present some limitations.

The microbiome associated with plant roots plays a vital role in plant tolerance to different stresses, such as drought. It has been evidenced that drought changes the root metabolome (Gargallo-Garriga *et al.*, 2018) and that these changes could help select beneficial microbiomes to enhance plant growth under drought conditions (Aslam *et al.*, 2022a). Bacteria are the most studied group of microorganisms in drought stress tolerance; however, other groups of microorganisms, such as rhizospheric or filamentous fungi, and unicellular fungi, such as yeasts, have yet to be explored.

12.2 Mechanisms Exerted by PGPM Under Drought Conditions

In the rhizosphere, there are several interactions among different groups of microorganisms and plant roots, which can be beneficial, neutral or deleterious. Plants can select microorganisms to cope with any type of stress due to radical exudates (carbohydrates, amino acids, and organic acids) (Barnawal *et al.*, 2019). On the other hand, drought may have positive effects on the rhizosphere microbiome due to the greater proportion of photosynthates allocated to root exudates rather than going to biomass production (Williams and de Vries, 2020). Nevertheless, the quantity and quality of exudates will depend on the plant species and severity of drought (Aslam *et al.*, 2022a). Bio-inoculants based on PGPM, such as bacteria and fungi, can directly or indirectly modulate plant physiology through the production of compounds such as growth-regulating hormones, osmolyte production, induction of stress-responsive genes, solubilization of P and other nutrients (Santander *et al.*, 2017; Hanaka *et al.*, 2021; Kumar *et al.*, 2021; Gamalero and Glick, 2022; Sati *et al.*, 2022), and through induced systemic tolerance.

12.2.1 Bacteria

Bacteria influence plant behavior under drought stress conditions through different mechanisms, one of which is the production of growth-regulating substances such as auxins, mainly indole-3-acetic acid (IAA) (Moreno-Galvan *et al.*, 2021) (Fig. 12.1). There are multiple routes by which bacteria can synthesize IAA, which means that bacteria can use several alternatives for synthesizing this hormone (Duca and Glick, 2020). For instance, IAA biosynthesis within the bacterial cell can be tryptophan dependent or independent. However, in bacteria, the tryptophan concentration is low (Duca and Glick, 2020); therefore, bacteria use the tryptophan released by the root exudates to biosynthesize IAA and this can be absorbed by the roots to increase the endogenous concentration in the plant (Gamalero and Glick, 2022). IAA plays an important role in plant growth by controlling xylem and phloem differentiation, primary and secondary root development and elongation,

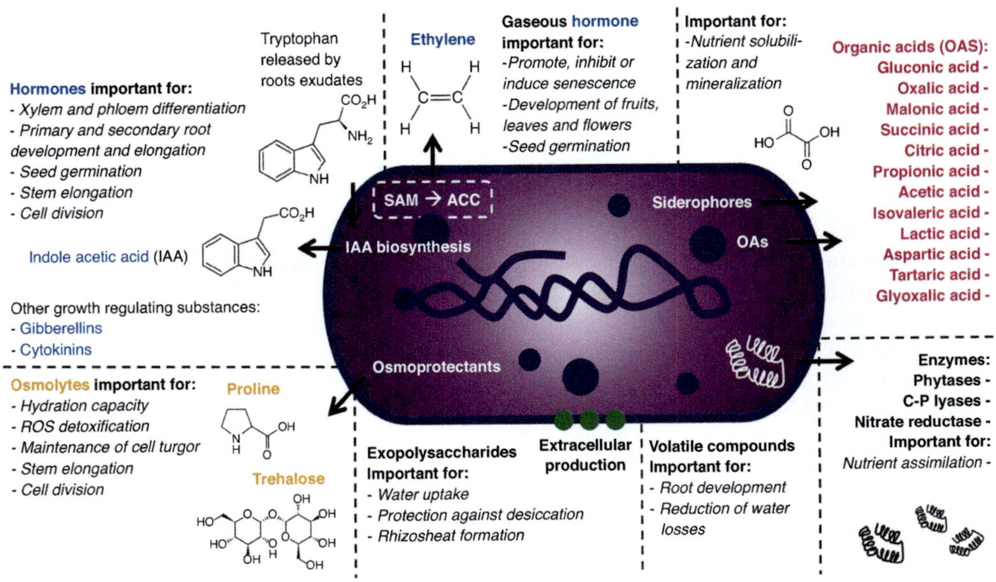

Fig. 12.1. Mechanisms exerted by plant growth-promoting bacteria (PGPB).

which are potentially important for tolerating drought stress (Ullah *et al.*, 2018; Gamalero and Glick, 2022; Sati *et al.*, 2022). Other growth-regulating substances, such as gibberellins and cytokinins, are also produced by bacteria (Sati *et al.*, 2022). These two hormones play a key role in seed germination, stem elongation, cell division in the plant, and acclimatization of plants to environmental stress (Ullah *et al.*, 2018; Sati *et al.*, 2022).

Bacteria can also produce osmolytes (osmoprotectants) that can interact with plant osmolytes to improve water potential, cell turgor, and membrane stability to prevent damage under drought conditions (Sangiorgio *et al.*, 2020) (see Fig. 12.1). Among the osmoprotectants produced by bacteria, proline and trehalose are highlighted. Proline is an amino acid that presents a hydrophobic side that can bind to proteins and a hydrophilic part that binds to water and gives it a great capacity for hydration, allowing proteins access to more water (Fang and Xiong, 2015). Proline also aids in the detoxification of ROS and the maintenance of cell turgor by contributing to the osmotic adjustment of the plant (Naseem *et al.*, 2018). On the other hand, trehalose is a nonreducing sugar composed of two glucose subunits linked by an $\alpha-1$ glycosidic bond disaccharide (α-d-glucopyranosyl-($1\rightarrow1$)-α-d-glucopyranoside) (Shao *et al.*, 2022) and is found in different groups of microorganisms and plants. Trehalose behaves as a xeroprotectant as it protects the cell against oxidative stress, maintains membrane integrity, and increases photosynthetic efficiency (Fang and Xiong, 2015; Gamalero and Glick, 2022; Shao *et al.*, 2022).

Ethylene is a gaseous hormone that regulates several physiological processes in the plant such as seed germination, fruit ripening, and response to different stresses (Ilyas *et al.*, 2021); however, the increase of this hormone under abiotic stress reduces growth and productivity in plants. Depending on concentrations, it can promote, inhibit or induce senescence and governs the development of fruits, leaves, and flowers (Pierik *et al.*, 2006). Ethylene is biosynthesized from S-adenosylmethionine (SAM), then SAM is converted by ACC synthetase to ACC which is oxidized by ACC oxidase to ethylene (Ilyas *et al.*, 2021; Moreno-Galvan *et al.*, 2021; Chandwani and Amaresan, 2022). Under drought conditions, ethylene concentration increases in plants but some bacteria synthesize ACC deaminase that degrades ACC into α-ketobutyrate and ammonia, leading to a decrease in ethylene stress (Chandwani and Amaresan, 2022).

Some microbes can produce exopolysaccharides (EPS) extracellularly. EPS are a heterogeneous mixture of polysaccharides, proteins, nucleic acids, and uranic acid and can represent between 40% and 95% of the bacterial weight, being important in the formation of biofilms (Naseem *et al.*, 2018; Khan and Bano, 2019). Under drought conditions, EPS produced by bacteria have many beneficial effects in the rhizosphere such as improving water uptake, protecting against desiccation and promoting the rhizosheath formation (Naseem *et al.*, 2018; Aslam *et al.*, 2022a) (see Fig. 12.1). The rhizosheath is formed from the mucilage produced by bacterial EPS, resulting in soil particles that adhere to root hairs, thus allowing the formation of soil micro- and macroaggregates, which increases water-holding capacity and nutrient flow, supporting plant–microbe and microbe–microbe interactions (Sandhya *et al.*, 2009).

Under drought conditions, the availability of water in the soil is scarce and this limits the availability of nutrients such as P, N, and K. Most of the P present in the soil is not available; however, some bacteria have the ability to solubilize and mineralize P through the production of organic acids such as gluconic, oxalic, malonic, succinic, citric, propionic, acetic, isovaleric, lactic, aspartic, tartaric and glyoxalic acids (Choudhary *et al.*, 2019) (see Fig. 12.1). In this sense, bacteria can produce siderophores, exopolysaccharides, and H_2S, and can lower the pH through direct release of protons (Rawat *et al.*, 2021). Additionally, bacteria can synthesize enzymes such as phytases, phosphatases, and C-P lyases, among others. P deficiencies decrease photosynthetic rate, water use efficiency, and plant yield under drought conditions (Olszewska, 2006).

On the other hand, N, a crucial element for drought tolerance, is involved in the processes of N assimilation, amino acid, and protein synthesis (Li *et al.*, 2020). Nitrate reductase is the first enzyme involved in N assimilation but under drought conditions, its activity substantially decreases (Fresneau *et al.*, 2007). Approximately

78% of N is found in the air and can be fixed by some groups of free-living and symbiotic bacteria (Bonilla-Buitrago *et al.*, 2021). The decrease in N assimilation due to drought affects biological N fixation via low nitrogenase activity, accumulation of ureides, and reduction of enzymatic activity. N-fixing bacteria can tolerate low soil moisture in drought stress through mechanisms such as the production of EPS, osmolytes, and ACC deaminase, among others (Santillana Villanueva, 2021).

The ability of some bacteria to produce VOCs has attracted attention in recent years. Although very few studies have been conducted regarding these compounds in abiotic stress tolerance, they have been reported to have a beneficial effect on drought stress (Heenan-Daly *et al.*, 2019) (see Fig. 12.1). VOCs can decrease primary root growth and increase lateral root formation, thus playing an important role in root development (Velázquez-Becerra *et al.*, 2011) and contributing to stomatal closure as well as reducing water losses. VOCs have also been shown to indirectly increase drought resistance in plants by mediating the production of exopolysaccharides. VOCs are small compounds with low molecular mass (<300 Da), low vapor pressure, and low boiling point (Heenan-Daly *et al.*, 2019), which facilitates their spread above and below the soil. These compounds can be organic (e.g. 2,3-butanediol) and inorganic (e.g. HCN), and include different chemical groups such as alcohols, alkenes, ketones, terpenes, benzenoids, pyrazines, esters, and acids (Piechulla and Degenhardt, 2014).

Additionally, some bacteria can naturally synthesize ABA analogues (xanthoxin-like compound) that have been shown to improve drought tolerance in plants, inducing stomatal closure as well as ABA. This compound can be absorbed by plants and remain in its original form or be converted to ABA, which provides the plant with a physiological adaptation to cope with drought stress (Gowtham *et al.*, 2021).

Finally, all bacterial-derived elicitors such as those described here act as inducers of induced systemic tolerance (IST). The ability of PGPB to generate responses to abiotic factors, such as drought, has been described as IST, and was proposed by Kim *et al.*

12.2.2 Free-living fungi

Fungi represent 55–89% of the total microbial biomass and can mediate P and N cycles by releasing extracellular enzymes (Treseder and Lennon, 2015). Some microorganisms that exhibit desiccation tolerance have shown the ability to protect plants against drought stress (Hanaka *et al.*, 2021). In fact, fungi may be more tolerant to drought conditions than bacteria (Treseder *et al.*, 2018) due to several characteristics that allow them to maintain their activity under extreme conditions (Fig. 12.2). These include the presence of β−1,3-glucan, trehalose, RNA helicase, melanin in the walls of fungi, and growth by budding (Treseder and Lennon, 2015) as in the case of yeasts.

Yeasts are generally found in very stressful conditions, such as very cold, dry, saline, or acidic environments (Treseder and Lennon, 2015). Yeasts are unicellular fungi whose beneficial effect on plants under drought conditions has been scarcely studied; however, it has been evidenced that, under drought conditions, yeasts can increase the activity of soil enzymes such as protease, urease, glucosidase, phosphatase, and catalase (Alzandi and Naguib, 2022), which improves the availability of nutrients such as P, zinc salts, and N (NH_3) (Bilek *et al.*, 2020). Yeasts have also been implicated in the production of EPS composed of mannose, glucose, and galactose (Silambarasan *et al.*, 2019a). Additionally, yeasts under drought conditions can promote IAA and siderophore production as well as ACC deaminase (Silambarasan *et al.*, 2019b), similarly to the above mentioned for bacteria (see Fig. 12.2).

Plant growth-promoting fungi (PGPF) can improve root and shoot growth in plants by increasing the contents of chlorophyll, proline, phenolic, and flavonoid compounds under drought conditions (Mona *et al.*, 2017). The latter are nonenzymatic compounds that allow the detoxification of ROS due to their antioxidant properties. In plants under drought stress, PGPF maintain high contents of growth-regulating substances such as IAA, gibberellic acid, and indole-3-butyric acid (Mona *et al.*, 2017). Indole-3-butyric acid is an auxin precursor that can be converted to IAA through β-oxidation peroxisomal activity (Damodaran

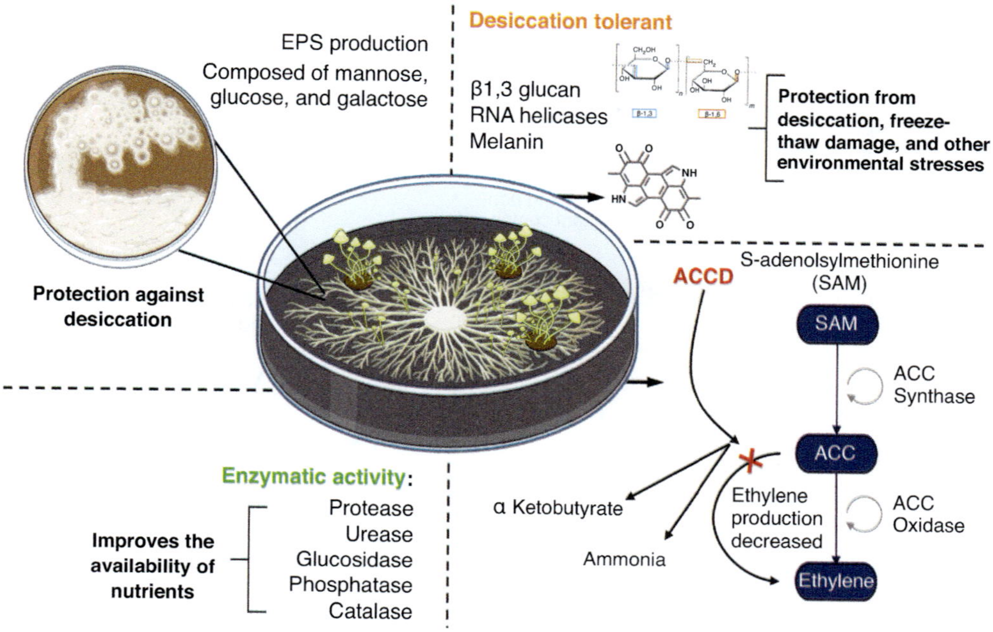

Fig. 12.2. Mechanisms exerted by free-living fungi.

and Strader, 2019), and is able to improve the root antioxidant system and adventitious root formation under abiotic stress (Li *et al.*, 2018). Additionally, it has been evidenced that PGPFs upregulate the expression of genes encoding antioxidant enzymes such as catalase (CAT), ascorbate peroxidase (APX), superoxide dismutase (SOD), and glutathione reductase (GR), thus preventing H_2O_2 overproduction under drought stress (Guler *et al.*, 2016).

12.3.3 Arbuscular mycorrhizal fungi

The oldest known plant–microbe interaction is the mutualistic symbiotic association between plant roots and AMF; this association dates back more than 460 million years (Redecker *et al.*, 2000) and may have played an important role in the transition of plants from an aquatic to a terrestrial system (Strullu-Derrien *et al.*, 2018). AMF exert drought stress tolerance in plants through direct mechanisms such as uptake and transport of water and nutrients (Santander *et al.*, 2017) (Fig. 12.3), as well as through indirect mechanisms that are more related to biochemical mechanisms in plants such as osmotic adjustment, higher production of enzymes and antioxidant compounds, production of plant growth-regulating hormones, glomalin production, and aquaporin-related mechanisms (Zhang *et al.*, 2018).

The absorption of water by AMF and transport to the plants occurs through their external mycelial network, which has the ability to penetrate soil pores and reach areas inaccessible to the root system (Santander *et al.*, 2017) (see Fig. 12.3), extending up to 100 mm into the soil and transporting up to 1973 mg H_2O/h/mm under drought conditions (Zhang *et al.*, 2018). Arbuscular mycorrhizal symbiosis can regulate the expression of genes encoding aquaporins in plants (Santander *et al.*, 2021), which are a family of integral pore- or channel-forming membrane proteins that regulate the movement of water through roots (Bahadur *et al.*, 2019), facilitating the passive transport of water and other low molecular weight molecules (Santander *et al.*, 2017).

Although drought conditions can affect the life cycle of AMF, such as spore production, germination and colonization on host plants,

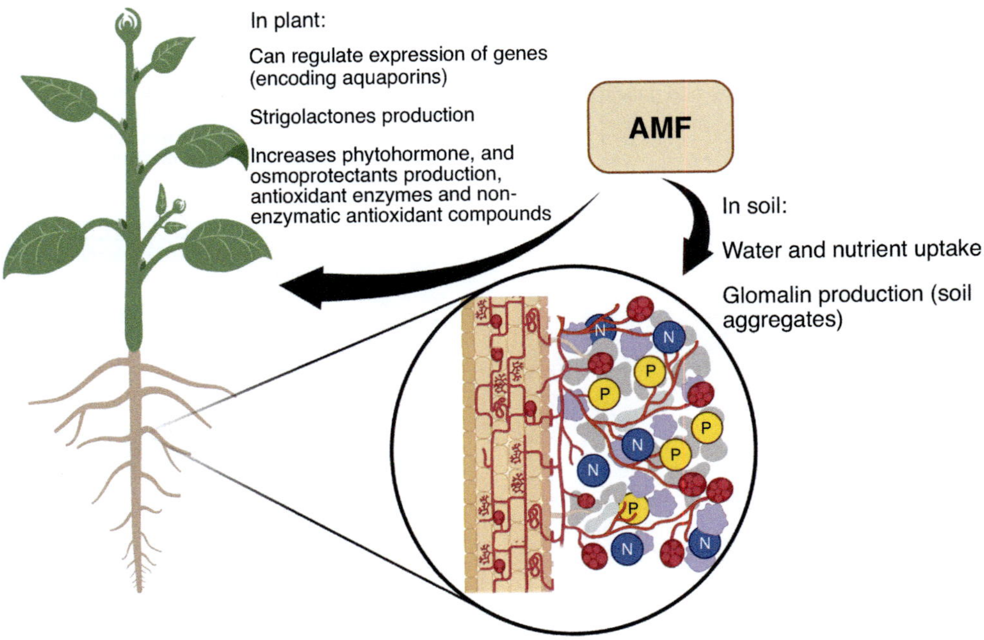

Fig. 12.3. Mechanisms exerted by arbuscular mycorrhizal fungi (AMF).

the hyphae can grow and survive in soil with extremely low water potentials (Allen, 2007). For instance, Leyva-Morales *et al.* (2019) evaluated how water starvation affects *Rhizophagus intraradices*, finding that the fungus improved its survival through (i) reducing the surface area exposed to desiccation by maintaining a large diameter hypha, and (ii) concentrating respiratory activity and resources in small sections that can act as propagules under more severe or prolonged drought conditions (Leyva-Morales *et al.*, 2019).

In addition to water transport, the external mycelium of AMF can absorb and transport soil nutrients, especially P, which quickly establish a depletion zone around the root due to its low mobility (Smith and Read, 2008). Plants have developed two main pathways for P absorption: (i) physiological strategies such as modification of root morphology, increase in root volume, exudation of organic anions that release P from metal complexes and production of enzymes such as phosphatases (Smith and Read, 2008; Kaur and Suseela, 2020), and (ii) biotic interactions with soil microorganisms (Choudhary *et al.*, 2019), highlighting the symbiotic association with AMF. P may possibly alter the stomatal response to environmental stress in two ways: one is by affecting the energy involved in guard cell parameters and the other may be by increasing the stiffness of the cell wall that governs stomatal movements (Augé, 2001). AM symbiosis can modify stomata behavior under drought or normal irrigation conditions, playing an important role in plant productivity, and promoting stomatal conductance (g_s) by up to 24% more when compared with non-AM colonized plants (Augé *et al.*, 2015).

Perhaps the most relevant indirect mechanism related to drought stress in AMF is glomalin production through the extraradical mycelium. Glomalin was described by Wright and Upadhyaya (1998) as a glycoprotein, operationally defined as glomalin-related soil protein (GRSP) (Rillig, 2004). The function of glomalin is to improve the structure of soil aggregates by enhancing infiltration and water retention in the rhizosphere (Santander *et al.*, 2017), increasing plant tolerance to drought stress (Gao *et al.*, 2019), and promoting soil organic carbon and heavy metal sequestration (Cornejo *et al.*, 2008), among others (see Fig. 12.3).

Another indirect mechanism of AM symbiosis in plants growing under drought stress is the increased production of phytohormones as strigolactones (Ruiz-Lozano *et al.*, 2016), as AM symbiosis increases the production of this and other phytohormones such as ABA and jasmonic acid by increasing hydraulic conductivity (Bahadur *et al.*, 2019). Additionally, under drought conditions AM symbiosis improves osmotic adjustment in the plant by increasing the production of free amino acids, proline, and sugars, protecting the structural and functional integrity of the membrane (Bahadur *et al.*, 2019; Begum *et al.*, 2021).

Finally, it has been shown that the AM association with plants under drought stress increases the activity of antioxidant enzymes such as CAT, APX, GR, SOD, and peroxide dismutase (POD) (Begum *et al.*, 2021; Tereucán *et al.*, 2022) increases phenolic compounds such as apigenin-C-pentoside-C-hexoside III (Nahuelcura *et al.*, 2022) and flavonoids (Begum *et al.*, 2021; Tereucán *et al.*, 2022), lipoxygenase (LOX) and phenylalanine ammonia lyase (PAL) activity (Begum *et al.*, 2021), the contents of a, b and total chlorophyll, carotenoids, and photosystem II (PSII) activity (Bahadur *et al.*, 2019; Begum *et al.*, 2021; Nahuelcura *et al.*, 2022; Tereucán *et al.*, 2022).

12.2.4 Growth Promotion and Drought Tolerance in Plants Through Mechanisms Exerted by PGPM

It is necessary to find solutions to mitigate the effects of drought on plants and help them to improve their productivity under water deficit conditions. Intensive agriculture and bad agricultural practices, such as the inefficient use of fertilizers and pesticides, among others, have reduced and changed the communities of microorganisms present in the rhizosphere which has led to significant changes in agroecosystems resulting in lower crop productivity (Shilev *et al.*, 2019). Faced with this situation, the agricultural sector is trying to find solutions that are more environmentally friendly and economically viable, such as the use of bio-inoculants to counteract the damage caused by drought.

Little attention has been paid to plant exudation under stressful environmental conditions such as drought, which has been shown to change the quality and quantity of plant exudates (Williams and de Vries, 2020). In the face of drought, plants make a 'cry for help' and change their metabolome in order to recruit microbes that allow them to tolerate drought events (Ali *et al.*, 2022). Therefore the use of microorganisms in agricultural systems in order to enhance positive microbe–microbe and plant–microbe interactions is necessary.

Bacteria and AMF have been proposed as potential bio-inoculants in drought tolerance in plants because they have been better studied. For bacteria, genera such as *Bacillus*, *Pseudomonas*, *Azospirillum*, *Herbaspirillum*, *Curtubacterium*, *Ochrobactrum*, *Paenibacillus*, *Nocardiopsis*, and *Streptomyces* have been the most studied (Saikia *et al.*, 2018; Ghosh *et al.*, 2019; Silambarasan *et al.*, 2019c, d, 2022; Carlson *et al.*, 2020; Kour *et al.*, 2020; Gowtham *et al.*, 2021; Mariotti *et al.*, 2021; Cortés-Patiño *et al.*, 2021; Yasmin *et al.*, 2021; Uzma *et al.*, 2022; Papadopoulou *et al.*, 2022). For the AMF, the genera *Funneliformis*, *Rhizophagus*, *Glomus*, *Claroideoglomus*, *Paraglomus*, and *Gigaspora* are those most reported in the literature (Zhang *et al.*, 2018; Ji *et al.*, 2019; Moradtalab *et al.*, 2019; Begum *et al.*, 2021; Li *et al.*, 2021; Fiorilli *et al.*, 2022; Hamedani *et al.*, 2022; Nahuelcura *et al.*, 2022; Tereucán *et al.*, 2022). However, other groups of microorganisms such as free-living fungi (rhizospheric or filamentous fungi and yeasts) are currently being explored, including genera such as *Trichoderma*, *Curvularia*, *Penicillium*, *Rhodotorula*, and bakery yeast being the most reported (Guler *et al.*, 2016; Mona *et al.*, 2017; Silambarasan *et al.*, 2019b; Poveda, 2020; Kour *et al.*, 2020; Abdelaal *et al.*, 2021; Scudeletti *et al.*, 2021; Cornejo-Ríos *et al.*, 2021; Adhikari *et al.*, 2022; Alzandi and Naguib, 2022).

Based on all the mechanisms exerted by bacteria, free-living fungi, and AMF, the application of these bio-inoculants has the potential to improve growth, nutrition, crop yield, and plant tolerance to drought stress, thus playing an important role as ecological actors to counteract the effects generated by drought on plant production. Although their responses vary according to the type of microorganisms, the plant species, and the intensity and duration of the drought, it has been found that these bio-inoculants can improve the biomass production of both the aerial and root parts of plants, leading them to better tolerate drought conditions.

Exopolysaccharide production is one of the most sought-after mechanisms in PGPM because EPS provide microbial cells with greater drought tolerance and better communication with other microbes and the plant, increasing water retention capacity around the root and giving it an advantage over those that do not produce water (Karimi *et al.*, 2022). However, to date, EPS production has only been documented in bacteria (Silambarasan *et al.*, 2019c; Cortés-Patiño *et al.*, 2021; Papadopoulou *et al.*, 2022). After treating wheat plants with high EPS-producing strains, Lafit *et al.* (2022) found increased drought tolerance in plants due to better adherence of bacterial cells to the root, improving colonization and allowing an increase in root diameter and surface area.

Another trait sought in PGPM to tolerate drought is ACC deaminase activity because this enzyme inhibits the high ethylene production that occurs under drought conditions (Sadeghi and Ghanaatiyan, 2017). This trait has been reported in bacteria and yeasts. Silambarasan *et al.* (2019b) reported that a potent isolate of *Rhodotolura mucilaginosa* CAM_4, obtained from the rhizospheric soil from *Rubus geoides* grown on acidic Andisols (southern Chile), with ACC deaminase activity had a greater effect on shoot and root length and biomass in lettuce plants under drought stress compared to the noninoculated stressed control.

The production of auxins, mainly IAA, and of organic acids and enzymes by PGPMs to solubilize and mineralize nutrients are essential traits in drought tolerance. Bacteria with high capacity to produce IAA (up to 116 μg/ml) and solubilize phosphates (3.33 mm) improved growth (shoot length) and yield in two varieties of *Vigna radiata* (Uzma *et al.*, 2022). In addition, the P-solubilizing fungus *Penicillium* sp. strain EU-DSF-10 was able to increase up to four and two times the contents of chlorophyll a and b, respectively, compared to the control treatment under drought stress in *Sorghum bicolor* plants (Kour *et al.*, 2020). Alzandi and Naguib (2022) inoculated bakery yeast in corn plants subjected to drought stress (40% soil moisture), reporting increases in enzymatic soil activity (protease, urease, glucosidase, phosphatase, and catalase), N, P, and K contents in the rhizosphere and in the roots.

Many studies have reported the benefits of bacteria and some free-living fungi in plant tolerance via mechanisms such as the production of several compounds mentioned above. However, there is another group of microorganisms, such as AMF, that play an important ecological role because of their coevolution with plants, allowing through their external mycelial network the capture and transport of water and nutrients in drought conditions (Smith and Read, 2008). Zhang *et al.* (2018) demonstrated that these fungi can transport water through their external mycelium, up to 2 mg H_2O/h/mm in trifoliate orange plants inoculated with *Funneliformis mosseae* and *Paraglomus occultum* subjected to drought stress (50% field capacity). The external mycelium of AMF can extend up to 100 cm into the soil (Zhang *et al.*, 2018), producing GRSP (Rillig, 2004) as an indirect mechanism of plant tolerance to drought stress. In one study, the external mycelium of *Gigaspora margarita* produced more easily extractable GRSP (EE-GRSP) and total GRSP (T-GRSP) under drought conditions than *F. mosseae*. Glomalin production by both AMF led to an increase in water-stable soil macroaggregates (>5.00 mm) with increases of 72 and 82% for *F. mosseae* and *G. margarita* respectively, improving plant tolerance to drought, and also increasing shoot dry weight, total root length, and root volume under drought stress (Ji *et al.*, 2019).

The benefits provided by PGPM-based bio-ionculants such as those mentioned in this review will depend on the range of drought to which the microorganisms are subjected. This review confirms the hypothesis that fungi are more resistant to drought than bacteria, since positive results were found in promoting plant growth up to 30% soil moisture or field capacity (Moradtalab *et al.*, 2019; Li *et al.*, 2021), while most of the bacteria had a range of action between 40 and 50% soil moisture, having almost no effect at lower percentages of soil moisture (Karimi *et al.*, 2022). For instance, Karimi *et al.* (2022) evaluated three strains of biofilm-forming bacteria (*Peribacillus simplex*-B1, *Bacillus pumilus*-B2, *P. simplex*-B3) on two wheat cultivars (Kohdasht and Chamran) in three water regimes at 80%, 50%, and 30%

soil moisture, finding that their effect on wheat plant tolerance decreased at 30% soil moisture compared to 50%. However, this could be enhanced with the use of microbial consortia, since it has been shown that coinoculations between different bacterial strains (Saikia *et al.*, 2018; Cortés-Patiño *et al.*, 2021) or between AFM and bacteria could increase drought tolerance in plants more than when they are inoculated individually (Ashwin *et al.*, 2022). It is also proposed that these PGPM should be used in drought-susceptible rather than in drought-tolerant cultivars, since the latter do not show a significant improvement in plant growth and drought tolerance (Karimi *et al.*, 2022; Nahuelcura *et al.*, 2022).

12.2.5 Prospects: From *In Vitro* to *In Situ* Studies

The global population is increasing and arable land is decreasing, with CC changing rainfall patterns, so droughts will continue to become more frequent. In this sense, biotechnological strategies are needed to cope with drought and meet the food needs of the world's growing population. As discussed, PGPM could be a viable and environmentally friendly alternative but these tools must be brought to the field. Numerous studies have been conducted in laboratory and greenhouse conditions, but almost nothing in the field. It is necessary to validate the results obtained in the greenhouse under real field conditions, with consortia of versatile microorganisms as far as possible, without endangering the health of plants, food, and humans. More research is needed in the scaling and development of bio-inoculants to be used in the field. Perhaps one reason for the lack of literature on work done in the field under drought conditions has been that under these conditions, a medium that can protect the microbial cells is needed to obtain consistent data.

Results expected in the field are difficult to predict since under these conditions there are a number of environmental variables that are not controlled for in the laboratory or greenhouse (Mitter *et al.*, 2021). Although many molecular mechanisms underlying plant–microbe interaction under stress conditions are known, it is necessary to use omics techniques (transcriptomics, proteomics, metabolomics) to cover more of these mechanisms, especially in microbial consortia where little is known. The integration of these data could allow a better understanding of the mutualistic relationships that could occur under these conditions for the development of new-generation bio-inoculants. (Vidal *et al.*, 2022). With the development of bio-inoculants to cope with drought in plants, it is necessary to integrate other mitigation strategies such as the application of abiotic effectors, in order to provide more holistic solutions to this stress.

Finally, we recommend understanding the water requirements of each crop and integrating this knowledge with evaluations of these types of bio-inoculants under field conditions, which will allow us to make efficient use of water.

12.2.6 Conclusion

This chapter discusses the effects that water scarcity in plants can have on the mechanisms by which plant growth-promoting microorganisms such as bacteria, free-living fungi, and arbuscular mycorrhizal fungi can help plants to cope with drought stress, which is of great importance in the current climate change scenario. The mechanistic analysis provided by recent studies has shown that these microorganisms play an important role as effective and environmentally friendly biological agents to be used as bio-inoculants. It is interesting to note that there are very few studies regarding the role of fungi (except for AMF) and their potential for improving drought tolerance in plants, as well as the production of compounds that allow them to resist drought. Therefore, further exploration of this group needs to be pursued.

Although this review did not emphasize the work done in consortia of bacteria, fungi, and AMF, studies reported to date are scarce, representing only 10% of studies on this topic. Based on the above, further studies on consortia of beneficial microorganisms, including different groups of microorganisms, could be of great interest in order to allow the development of multifunctional bio-inoculants for different environmental conditions such as soil type, temperature, pH and radiation, since these conditions could affect the success of bio-inoculants in field application.

It is still necessary to extend our knowledge of the mechanisms by which beneficial microorganisms could confer greater drought tolerance to plants, especially when consortia are considered in comparison with individual inoculations. The use of these bio-inoculants should be integrated into other drought tolerance practices such as the breeding of new plant genotypes through classical and molecular methods, exogenous application of abiotic effectors such as hormones, sugars, and amino acids, etc. Additionally, it was observed that seed priming could be an effective strategy for the inoculation of these microorganisms since it allows, from very early stages of plant growth and development, more chances to overcome stressful conditions due to the microbial load.

Finally, there are many reports of the application of microorganisms such as those covered here (bacteria, fungi, and AMF) in laboratory and greenhouse assays, but only a few have been carried out in field conditions. Therefore, we need more focused studies in realistic conditions, to determine the most suitable types of microorganisms and the mechanisms by which they exert tolerance, as a first step to integrate the use of advanced formulations of bioinoculants into large-scale agricultural systems to support other global initiatives that might enhance water and food security in the current context of climate change.

Acknowledgments

This work was funded by ANID/FONDAP/15130015 and ANID/FONDAP/1523A0001 grants.

References

Abdelaal, K., Attia, K.A., Niedbała, G., Wojciechowski, T., Hafez, Y. *et al.* (2021) Mitigation of drought damages by exogenous chitosan and yeast extract with modulating the photosynthetic pigments, antioxidant defense system and improving the productivity of garlic plants. *Horticulturae* 7, 510.

Adhikari, A., Khan, M.A., Imran, M., Lee, K.-E., Kang, S.-M. *et al.* (2022) The combined inoculation of *Curvularia lunata* AR11 and biochar stimulates synthetic silicon and potassium phosphate use efficiency, and mitigates salt and drought stresses in rice. *Frontiers in Plant Science* 13, 816858.

Alfaro-Corres, A.E., Lecona-Guzman, C.A., Gutiérrez-Miceli, F.A., Martínez-Esteves, M., Silverio-Gómez, M.C. *et al.* (2023) Effect of water stress induced by polyethylene glycol on growth, proline accumulation in *Agave americana* L. *Phyton-International Journal of Experimental Botany* 92(2), 629–643.

Ali, S., Tyagi, A., Park, S., Mir, R.A., Mushtaq, M. *et al.* (2022) Deciphering the plant microbiome to improve drought tolerance: Mechanisms and perspectives. *Environmental and Experimental Botany* 201, 104933.

Allen, M.F. (2007) Mycorrhizal fungi: Highways for water and nutrients in arid soils. *Vadose Zone Journal* 6(2), 291–297.

Alves da Silva, A., Oliveira Silva, C., Rosario Rosa, V., Silva Santos, M.F., Naomi Kuki, K. *et al.* (2022) Metabolic adjustment and regulation of gene expression are essential for increased resistance to severe water deficit and resilience post-stress in soybean. *PeerJ* 10, e13118.

Alzandi, A.A. and Naguib, D.M. (2022) Effect of yeast application on soil health and root metabolic status of corn seedlings under drought stress. *Archives of Microbiology* 204(4), 1–12.

Ashwin, R., Joseph, D. and Basavaiah, B. (2022) Dual inoculation with rhizobia and arbuscular mycorrhizal fungus improves water stress tolerance and productivity in soybean. *Plant Stress* 4, 100084.

Aslam, M.M., Okal, E.J., Idris, A.L., Qian, Z., Xu, W. *et al.* (2022a) Rhizosphere microbiomes can regulate plant drought tolerance. *Pedosphere* 32(1), 61–74.

Aslam, M.M., Rashid, M.A., Siddiqui, M.A., Khan, M.T., Farhat, F. *et al.* (2022b) Recent insights into signaling responses to cope drought stress in rice. *Rice Science* 29(2), 105–117.

Augé, R.M. (2001) Water relations, drought and vesicular-arbuscular mycorrhizal symbiosis. *Mycorrhiza* 11(1), 3–42.

Augé, R.M., Toler, H.D. and Saxton, A.M. (2015) Arbuscular mycorrhizal symbiosis alters stomatal conductance of host plants more under drought than under amply watered conditions: A meta-analysis. *Mycorrhiza* 25(1), 13–24.

Azcón-Bieto, J. and Talón, M. (2003) *Fundamentos de Fisiología Vegetal*. McGraw-Hill, New York.

Bahadur, A., Batool, A., Nasir, F., Jiang, S., Mingsen, Q. *et al.* (2019) Mechanistic insights into arbuscular mycorrhizal fungi-mediated drought stress tolerance in plants. *International Journal of Molecular Sciences* 20(17), 1–18.

Barnawal, D., Singh, R. and Singh, R.P. (2019) Role of plant growth promoting rhizobacteria in drought tolerance. In: Singh, A., Kumar, A. and Singh, P. (eds) *PGPR Amelioration in Sustainable Agriculture*. Elsevier, St Louis.

Begum, N., Akhtar, K., Ahanger, M.A., Iqbal, M., Wang, P. *et al.* (2021) Arbuscular mycorrhizal fungi improve growth, essential oil, secondary metabolism, and yield of tobacco (*Nicotiana tabacum* L.) under drought stress conditions. *Environmental Science and Pollution Research* 28(33), 45276–45295.

Bilek, F.N., Rezki, M.A., Grondin, C., Yahia, N. and Bekki, A. (2020) Plant growth promoting characteristics and stress tolerance of yeasts isolated from Algerian agricultural soils. *South Asian Journal of Experimental Biology* 10(6), 413–426.

Bonilla-Buitrago, R.R., Bashan, L.E., Pedraza, R.O. and Bonilla, E. (2021) G.a. In: Pardo Díaz, S. (ed.) *Bacterias Promotoras de Crecimiento Vegetal En Sistemas de Agricultura Sostenible*. Agrosavia, Cundinamarca.

Carlson, R., Tugizimana, F., Steenkamp, P.A., Dubery, I.A., Hassen, A.I. *et al.* (2020) Rhizobacteria-induced systemic tolerance against drought stress in *Sorghum bicolor* (L.) Moench. *Microbiological Research* 232, 126388.

Chandwani, S. and Amaresan, N. (2022) Role of ACC deaminase producing bacteria for abiotic stress management and sustainable agriculture production. *Environmental Science and Pollution Research International* 29(16), 22843–22859.

Choudhary, M., Meena, V.S., Yadav, R.P., Parihar, M., Pattanayak, A. *et al.* (2019) Does PGPR and mycorrhizae enhance nutrient use efficiency and efficacy in relation to crop productivity. In: Maheshwar, D. and Dheeman, S. (eds) *Field Crops: Sustainable Management by PGPR. Sustainable Development and Biodiversity*. Springer, Cham, pp. 45–68.

Compant, S., Heijden, M.G.A. and Sessitsch, A. (2010) Climate change effects on beneficial plant–microorganism interactions. *FEMS Microbiology Ecology* 73(2), 197–214.

Cornejo, P., Meier, S., Borie, G., Rillig, M.C. and Borie, F. (2008) Glomalin-related soil protein in a Mediterranean ecosystem affected by a copper smelter and its contribution to Cu and Zn sequestration. *Science of the Total Environment* 406(1–2), 154–160.

Cornejo-Ríos, K., Osorno-Suárez, M.D.P., Hernández-León, S., Reyes-Santamaría, M.I., Juárez-Díaz, J.A. *et al.* (2021) Impact of *Trichoderma asperellum* on chilling and drought stress in tomato (*Solanum lycopersicum*). *Horticulturae* 7(10), 385.

Cortés-Patiño, S., Vargas, C., Álvarez-Flórez, F., Bonilla, R. and Estrada-Bonilla, G. (2021) Potential of *Herbaspirillum* and *Azospirillum* consortium to promote growth of perennial ryegrass under water deficit. *Microorganisms* 9(1), 1–16.

Damodaran, S. and Strader, L.C. (2019) Indole 3-butyric acid metabolism and transport in *Arabidopsis thaliana*. *Frontiers in Plant Science* 10, 851.

Demirevska, K., Zasheva, D., Dimitrov, R., Simova-Stoilova, L., Stamenova, M. *et al.* (2009) Drought stress effects on rubisco in wheat: Changes in the rubisco large subunit. *Acta Physiologiae Plantarum* 31(6), 1129–1138.

Duca, D.R. and Glick, B.R. (2020) Indole-3-acetic acid biosynthesis and its regulation in plant-associated bacteria. *Applied Microbiology and Biotechnology* 104(20), 8607–8619.

Fahad, S., Bajwa, A.A., Nazir, U., Anjum, S.A., Farooq, A. *et al.* (2017) Crop production under drought and heat stress: Plant responses and management options. *Frontiers in Plant Science* 8, 1147.

Fang, Y. and Xiong, L. (2015) General mechanisms of drought response and their application in drought resistance improvement in plants. *Cellular and Molecular Life Sciences* 72(4), 673–689.

FAO (2020) The state of food and agriculture – overcoming water challenges in agriculture. Available at: h ttps://doi.org/10.4060/cb1447en (accessed 9 April 2025).

Farooqi, Z.U.R., Ayub, M.A., Rheman, M.Z., Sohail, M.I., Usma, M. *et al.* (2020) Regulation of drought stress in plants. In: Tripathi, D.K., Singh, V.P., Chauhan, D.K., Sharma, S. and Prasad, S.M. (eds)

Plant Life under Changing Environment: Responses and Management. Academic Press, New York, pp. 77–104.

Feng, X., Zhong, L., Tian, Q. and Zhao, W. (2022) Leaf water potential-dependent leaflet closure contributes to legume leaves cool down and drought avoidance under diurnal drought stress. *Tree Physiology* 42(11), 2239–2251.

Fiorilli, V., Maghrebi, M., Novero, M., Votta, C., Mazzarella, T. *et al.* (2022) Arbuscular mycorrhizal symbiosis differentially affects the nutritional status of two durum wheat genotypes under drought conditions. *Plants* 11(6), 804.

Fresneau, C., Ghashghaie, J. and Cornic, G. (2007) Drought effect on nitrate reductase and sucrose-phosphate synthase activities in wheat (*Triticum durum* L.): Role of leaf internal CO_2. *Journal of Experimental Botany* 58(11), 2983–2992.

Gamalero, E. and Glick, B.R. (2022) Recent advances in bacterial amelioration of plant drought and salt stress. *Biology* 11(3), 437.

Gao, W.Q., Wang, P. and Wu, Q.S. (2019) Functions and application of glomalin-related soil proteins: A review. *Sains Malaysiana* 48(1), 111–119.

Gargallo-Garriga, A., Preece, C., Sardans, J., Oravec, M., Urban, O. *et al.* (2018) Root exudate metabolomes change under drought and show limited capacity for recovery. *Scientific Reports* 8(1), 1–15.

Ghosh, D., Gupta, A. and Mohapatra, S. (2019) A comparative analysis of exopolysaccharide and phytohormone secretions by four drought-tolerant rhizobacterial strains and their impact on osmotic-stress mitigation in *Arabidopsis thaliana*. *World Journal of Microbiology & Biotechnology* 35(6), 1–15.

Gouveia, C.S.S., Ganança, J.F.T., Slaski, J.J., Lebot, V. and Carvalho, M.Â.A. (2021) Abscisic acid phytohormone estimation in tubers and shoots of *Ipomoea batatas* subjected to long drought stress using competitive immunological assay. *Physiologia Plantarum* 172(2), 419–430.

Gowtham, H.G., Duraivadivel, P., Ayusman, S., Sayani, D., Gholap, S.L. *et al.* (2021) ABA analogue produced by *Bacillus marisflavi* modulates the physiological response of *Brassica juncea* L. under drought stress. *Applied Soil Ecology* 159, 103845.

Guler, N.S., Pehlivan, N., Karaoglu, S.A., Guzel, S. and Bozdeveci, A. (2016) *Trichoderma atroviride* ID20G inoculation ameliorates drought stress-induced damages by improving antioxidant defence in maize seedlings. *Acta Physiologiae Plantarum* 38(6).

Gupta, A., Rico-Medina, A. and Caño-Delgado, A.I. (2020) The physiology of plant responses to drought. *Science* 368(6488), 266–269.

Hamedani, N.G., Gholamhoseini, M., Bazrafshan, F., Habibzadeh, F. and Amiri, B. (2022) Yield, irrigation water productivity and nutrient uptake of arbuscular mycorrhiza inoculated sesame under drought stress conditions. *Agricultural Water Management* 266, 107569.

Hanaka, A., Ozimek, E., Reszczyńska, E., Jaroszuk-Ściseł, J. and Stolarz, M. (2021) Plant tolerance to drought stress in the presence of supporting bacteria and fungi: An efficient strategy in horticulture. *Horticulturae* 7(10), 390.

Heenan-Daly, D., Velivelli, S.L.S. and Prestwich, B.D. (2019) The role of rhizobacterial volatile organic compounds in a second Green Revolution—the story so far. In: Maheshwari, D. and Dheeman, S. (eds) *Field Crops: Sustainable Management by PGPR. Sustainable Development and Biodiversity*, Vol. 23. Springer, Cham, pp. 191–220.

Ilyas, M., Nisar, M., Khan, N., Hazrat, A., Khan, A.H. *et al.* (2021) Drought tolerance strategies in plants: A mechanistic approach. *Journal of Plant Growth Regulation* 40(3), 926–944.

IPCC (2022) *Climate Change 2022: Impacts, Adaptation, and Vulnerability. Contribution of Working Group II to the Sixth Assessment Report of the Intergovernmental Panel on Climate Change*. Cambridge University Press, Cambridge.

Jaleel, C.A., Manivannan, P., Sankar, B., Kishorekumar, A., Gopi, R. *et al.* (2007) Induction of drought stress tolerance by ketoconazole in *Catharanthus roseus* is mediated by enhanced antioxidant potentials and secondary metabolite accumulation. *Colloids and Surfaces B: Biointerfaces* 60(2), 201–206.

Ji, L., Tan, W. and Chen, X. (2019) Arbuscular mycorrhizal mycelial networks and glomalin-related soil protein increase soil aggregation in calcaric regosol under well-watered and drought stress conditions. *Soil and Tillage Research* 185, 1–8.

Karimi, E., Aliasgharzad, N., Esfandiari, E., Hassanpouraghdam, M.B., Neu, T.R. *et al.* (2022) Biofilm forming rhizobacteria affect the physiological and biochemical responses of wheat to drought. *AMB Express* 12, 93.

Kaur, S. and Suseela, V. (2020) Unraveling arbuscular mycorrhiza-induced changes in plant primary and secondary metabolome. *Metabolites* 10(8), 1–30.

Khan, N. and Bano, A. (2019) Exopolysaccharide producing rhizobacteria and their impact on growth and drought tolerance of wheat grown under rainfed conditions. *PLoS One* 14(9), 1–9.

Khan, N., Ali, S., Tariq, H., Latif, S., Yasmin, H. *et al.* (2020) Water conservation and plant survival strategies of rhizobacteria under drought stress. *Agronomy* 10(11), 1–23.

Kour, D., Rana, K.L., Kaur, T., Sheikh, I., Yadav, A.N. *et al.* (2020) Microbe-mediated alleviation of drought stress and acquisition of phosphorus in great millet (*Sorghum bicolour* L.) by drought-adaptive and phosphorus-solubilizing microbes. *Biocatalysis and Agricultural Biotechnology* 23, 101501.

Kumar, M., Giri, V.P., Pandey, S., Gupta, A., Patel, M.K. *et al.* (2021) Plant-growth-promoting rhizobacteria emerging as an effective bioinoculant to improve the growth, production, and stress tolerance of vegetable crops. *International Journal of Molecular Sciences* 22(22), 12245.

Lesk, C., Rowhani, P. and Ramankutty, N. (2016) Influence of extreme weather disasters on global crop production. *Nature* 529(7584), 84–87.

Leyva-Morales, R., Gavito, M.E. and Carrillo-Saucedo, S.M. (2019) Morphological and physiological responses of the external mycelium of *Rhizophagus intraradices* to water stress. *Mycorrhiza* 29(2), 141–147.

Li, L., Zhang, H., Tang, M. and Chen, H. (2021) Nutrient uptake and distribution in mycorrhizal cuttings of *Populus* × *Canadensis* 'Neva' under drought stress. *Journal of Soil Science and Plant Nutrition* 21(3), 2310–2324.

Li, P.-F., Ma, B.-L., Palta, J.A., Ding, T.-T., Cheng, Z.-G. *et al.* (2022a) Distinct contributions of drought avoidance and drought tolerance to yield improvement in dryland wheat cropping. *Journal of Agronomy and Crop Science* 208(3), 265–282.

Li, P.-F., Ma, B.-L., Guo, S., Ding, T.-T. and Xiong, Y.-C. (2022b) Bottom-up redistribution of biomass optimizes energy allocation, water use and yield formation in dryland wheat improvement. *Journal of the Science of Food and Agriculture* 102, 3336–3349.

Li, S., Zhou, L., Addo-Danso, S.D., Ding, G., Sun, M. *et al.* (2020) Nitrogen supply enhances the physiological resistance of Chinese fir plantlets under Polyethylene Glycol (PEG)-induced drought stress. *Scientific Reports* 10(1), 1–8.

Li, S.W., Zeng, X.Y., Leng, Y., Feng, L. and Kang, X.H. (2018) Indole-3-butyric acid mediates antioxidative defense systems to promote adventitious rooting in mung bean seedlings under cadmium and drought stresses. *Ecotoxicology and Environmental Safety* 161, 332–341.

Mariotti, L., Scartazza, A., Curadi, M., Picciarelli, P. and Toffanin, A. (2021) *Azospirillum baldaniorum* Sp245 induces physiological Purple Basil. *Plants* 10(6), 1141.

Marro, N., Grilli, G., Soteras, F., Caccia, M., Longo, S. *et al.* (2022) The effects of arbuscular mycorrhizal fungal species and taxonomic groups on stressed and unstressed plants: A global meta-analysis. *Phytologist* 235, 320–332.

Mitter, E.K., Tosi, M., Obregón, D., Dunfield, K.E. and Germida, J.J. (2021) Rethinking crop nutrition in times of modern microbiology: Innovative biofertilizer technologies. *Frontiers in Sustainable Food Systems* 5, 606815.

Mona, S.A., Hashem, A., Abd-Allah, E.F., Alqarawi, A.A., Soliman, D.W.K. *et al.* (2017) Increased resistance of drought by *Trichoderma harzianum* fungal treatment correlates with increased secondary metabolites and proline content. *Journal of Integrative Agriculture* 16(8), 1751–1757.

Moradtalab, N., Hajiboland, R., Aliasgharzad, N., Hartmann, T.E. and Neumann, G. (2019) Silicon and the association with an arbuscular-mycorrhizal fungus (*Rhizophagus clarus*) mitigate the adverse effects of drought stress on strawberry. *Agronomy* 9(1), 1–22.

Moreno-Galvan, A.E., Cortés, S.L., Mendoza, J.A. and Bécquer, C.J. (2021) Aplicación de bacterias promotoras de crecimiento vegetal en la mitigación de estrés. In: Bonilla-Buitrago, R.R., Bashan, L.E., Pedraza, R.O., Estrada Bonilla, G.A. and Pardo Díaz, S. (eds) *Bacterias Promotoras de Crecimiento Vegetal En Sistemas de Agricultura Sostenible*. Agrosavia, Cundinamarca, p. 372.

Nahuelcura, J., Ruiz, A., Gomez, F. and Cornejo, P. (2022) The effect of arbuscular mycorrhizal fungi on the phenolic compounds profile, antioxidant activity and grain yields in wheat cultivars growing under hydric stress. *Journal of the Science of Food and Agriculture* 102(1), 407–416.

Naseem, H., Ahsan, M., Shahid, M.A. and Khan, N. (2018) Exopolysaccharides producing rhizobacteria and their role in plant growth and drought tolerance. *Journal of Basic Microbiology* 58(12), 1009–1022.

Olszewska, M. (2006) Effect of water stress on physiological processes, leaf greenness (SPAD index) and dry matter yield of *Lolium perenne* and *Dactylis glomerata*. *Polish Journal of Natural Sciences* 21(2), 533–562.

Papadopoulou, A., Matsi, T., Kamou, N., Avdouli, D., Mellidou, I. *et al.* (2022) Decoding the potential of a new *Pseudomonas putida* strain for inducing drought tolerance of tomato (*Solanum lycopersicum*) plants through seed biopriming. *Journal of Plant Physiology* 271, 153658.

Piechulla, B. and Degenhardt, J. (2014) The emerging importance of microbial volatile organic compounds. *Plant, Cell & Environment* 37(4), 811–812.

Pierik, R., Tholen, D., Poorter, H., Visser, E.J.W. and Voesenek, L.A.C.J. (2006) The Janus face of ethylene: Growth inhibition and stimulation. *Trends in Plant Science* 11(4), 176–183.

Poveda, J. (2020) *Trichoderma parareesei* favors the tolerance of rapeseed (*Brassica napus* L.) to salinity and drought due to a chorismate mutase. *Agronomy* 10(1), 118.

Rad, R.D., Sharifabad, H.H., Torabi, M., Azizinejad1, R., Salemi, H. *et al.* (2023) Drought stress tolerance based on selection indices of resistant crops variety. *Global Journal of Environmental Science and Management* 9(2), 1–12.

Rajarajan, K., Ganesamurthy, K., Raveendran, M., Jeyakumar, P., Yuvaraja, A. *et al.* (2021) Differential responses of sorghum genotypes to drought stress revealed by physio-chemical and transcriptional analysis. *Molecular Biology Reports* 48(3), 2453–2462.

Rawat, P., Das, S., Shankhdhar, D. and Shankhdhar, S.C. (2021) Phosphate-solubilizing microorganisms: Mechanism and their role in phosphate solubilization and uptake. *Journal of Soil Science and Plant Nutrition* 21(1), 49–68.

Raza, A., Ashraf, F., Zou, X., Zhang, X. and Tosif, H. (2020) Plant adaptation and tolerance to environmental stresses: mechanisms and perspectives. In: Hasanuzzaman, M. (ed.) *Plant Ecophysiology and Adaptation under Climate Change: Mechanisms and Perspectives I*. Springer, Singapore, pp. 117–145.

Redecker, D., Kodner, R. and Graham, L.E. (2000) Glomalean fungi from the Ordovician. *Science* 289(5486), 1920–1921.

Rillig, M.C. (2004) Arbuscular mycorrhizae, glomalin, and soil aggregation. *Canadian Journal of Soil Science* 84(4), 355–363.

Ruiz-Lozano, J.M., Aroca, R., Zamarreño, Á.M., Molina, S., Andreo-Jiménez, B. *et al.* (2016) Arbuscular mycorrhizal symbiosis induces strigolactone biosynthesis under drought and improves drought tolerance in lettuce and tomato. *Plant, Cell & Environment* 39(2), 441–452.

Sadeghi, H. and Ghanaatiyan, K. (2017) Probing the responses of four chicory ecotypes by ethylene accumulation and growth characteristics under drought stress. *Italian Journal of Agronomy* 12(3), 177–182.

Saikia, J., Sarma, R.K., Dhandia, R., Yadav, A., Bharali, R. *et al.* (2018) Alleviation of drought stress in pulse crops with ACC deaminase producing rhizobacteria isolated from acidic soil of Northeast India. *Scientific Reports* 8(1), 1–16.

Sandhya, V., Ali SK, Z., Grover, M., Reddy, G. and Venkateswarlu, B. (2009) Alleviation of drought stress effects in sunflower seedlings by the exopolysaccharides producing *Pseudomonas putida* strain GAP-p45. *Biology and Fertility of Soils* 46(1), 17–26.

Sangiorgio, D., Cellini, A., Donati, I., Pastore, C., Onofrietti, C. *et al.* (2020) Facing climate change: Application of microbial biostimulants to mitigate stress in horticultural crops. *Agronomy* 10(6), 794.

Santander, C., Aroca, R., Ruiz-Lozano, J.M., Olave, J., Cartes, P. *et al.* (2017) Arbuscular mycorrhiza effects on plant performance under osmotic stress. *Mycorrhiza* 27(7), 639–657.

Santander, C., Aroca, R., Cartes, P., Vidal, G. and Cornejo, P. (2021) Aquaporins and cation transporters are differentially regulated by two arbuscular mycorrhizal fungi strains in lettuce cultivars growing under salinity conditions. *Plant Physiology and Biochemistry* 158, 396–409.

Santillana Villanueva, N. (2021) Mecanismos de inducción de rizobios para reducir el estrés por sequía en las leguminosas. *Revista de Investigaciones Altoandinas - Journal of High Andean Research* 23(4), 258–265.

Sati, D., Pande, V., Pandey, S.C. and Samant, M. (2022) Recent advances in PGPR and molecular mechanisms involved in drought stress resistance. *Journal of Soil Science and Plant Nutrition* 23, 106–124.

Scudeletti, D., Crusciol, C.A.C., Bossolani, J.W., Moretti, L.G., Momesso, L. *et al.* (2021) *Trichoderma asperellum* inoculation as a tool for attenuating drought stress in sugarcane. *Frontiers in Plant Science* 12, 645542.

Shao, J., Wu, W., Rasul, F., Munir, H., Huang, K. *et al.* (2022) Trehalose induced drought tolerance in plants: Physiological and molecular responses. *Notulae Botanicae Horti Agrobotanici Cluj-Napoca* 50(1), 12584.

Shilev, S., Azaizeh, H., Vassilev, N., Georgiev, D. and Babrikova, I. (2019) Interactions in soil-microbe-plant system: adaptation to stressed agriculture. In: Singh, D., Gupta, V. and Prabha, R. (eds) *Microbial Interventions in Agriculture and Environment*. Springer, Singapore, pp. 131–171.

Silambarasan, S., Logeswari, P., Cornejo, P. and Kannan, V.R. (2019a) Evaluation of the production of exopolysaccharide by plant growth promoting yeast *Rhodotorula* sp. strain CAH2 under abiotic stress conditions. *International Journal of Biological Macromolecules* 121, 55–62.

Silambarasan, S., Logeswari, P., Cornejo, P., Abraham, J. and Valentine, A. (2019b) Simultaneous mitigation of aluminum, salinity and drought stress in *Lactuca sativa* growth via formulated plant growth promoting *Rhodotorula mucilaginosa* CAM4. *Ecotoxicology and Environmental Safety* 180, 63–72.

Silambarasan, S., Logeswari, P., Cornejo, P. and Kannan, V.R. (2019c) Role of plant growth-promoting rhizobacterial consortium in improving the *Vigna radiata* growth and alleviation of aluminum and drought stresses. *Environmental Science and Pollution Research International* 26(27), 27647–27659.

Silambarasan, S., Logeswari, P., Valentine, A. and Cornejo, P. (2019d) Role of *Curtobacterium herbarum* strain CAH5 on aluminum bioaccumulation and enhancement of *Lactuca sativa* growth under aluminum and drought stresses. *Ecotoxicology and Environmental Safety* 183, 109573.

Silambarasan, S., Logeswari, P., Vangnai, A.S., Kamaraj, B. and Cornejo, P. (2022) Plant growth-promoting actinobacterial inoculant assisted phytoremediation increases cadmium uptake in *Sorghum bicolor* under drought and heat stresses. *Environmental Pollution* 307, 119489.

Smith, S.E. and Read, D. (2008) Mineral nutrition, toxic element accumulation and water relations of arbuscular mycorrhizal plants. In: Smith, S.E. and Read, D. (eds) *Mycorrhizal Symbiosis*, 3rd edn. Academic Press, New York, pp. 145–VI.

Strullu-Derrien, C., Selosse, M.A., Kenrick, P. and Martin, F.M. (2018) The origin and evolution of mycorrhizal symbioses: From palaeomycology to phylogenomics. *New Phytologist* 220(4), 1012–1030.

Taiwo, A.F., Daramola, O., Sow, M. and Semwal, V.K. (2020) Ecophysiology and responses of plants under drought. In: Hasanuzzaman, M. (ed.) *Plant Ecophysiology and Adaptation under Climate Change: Mechanisms and Perspectives I*. Springer, Singapore, pp. 231–268.

Tereucán, G., Ruiz, A., Nahuelcura, J., Oyarzún, P., Santander, C. *et al.* (2022) Shifts in biochemical and physiological responses by the inoculation of arbuscular mycorrhizal fungi in *Triticum aestivum* growing under drought conditions. *Journal of the Science of Food and Agriculture* 102(5), 1927–1938.

Treseder, K.K. and Lennon, J.T. (2015) Fungal traits that drive ecosystem dynamics on land. *Microbiology and Molecular Biology Reviews* 79(2), 243–262.

Treseder, K.K., Berlemont, R., Allison, S.D. and Martiny, A.C. (2018) Drought increases the frequencies of fungal functional genes related to carbon and nitrogen acquisition. *PLoS One* 13(11), 1–17.

Twala, T.C., Witkowski, E.T.F. and Fisher, J.T. (2022) The effects of heat and drought stress on the ecophysiological responses and growth of *Afrocarpus falcatus* and *Podocarpus henkelii* seedlings. *South African Journal of Botany* 149, 258–268.

Tyśkiewicz, R., Nowak, A., Ozimek, E. and Jaroszuk-Ściseł, J. (2022) *Trichoderma*: The current status of its application in agriculture for the biocontrol of fungal phytopathogens and stimulation of plant growth. *International Journal of Molecular Sciences* 23(4), 2329.

Ullah, A., Manghwar, H., Shaban, M., Khan, A.H., Akbar, A. *et al.* (2018) Phytohormones enhanced drought tolerance in plants: A coping strategy. *Environmental Science and Pollution Research* 25(33), 33103–33118.

Uzma, M., Iqbal, A. and Hasnain, S. (2022) Drought tolerance induction and growth promotion by indole acetic acid producing *Pseudomonas aeruginosa* in *Vigna radiata*. *PLoS One* 17, e0262932.

Velázquez-Becerra, C., Macías-Rodríguez, L.I., López-Bucio, J., Altamirano-Hernández, J., Flores-Cortez, I. *et al.* (2011) A volatile organic compound analysis from *Arthrobacter agilis* identifies dimethylhexadecylamine, an amino-containing lipid modulating bacterial growth and *Medicago sativa* morphogenesis *in vitro*. *Plant and Soil* 339(1), 329–340.

Vidal, C., González, F., Santander, C., Pérez, R., Gallardo, V. *et al.* (2022) Management of rhizosphere microbiota and plant production under drought stress: A comprehensive review. *Plants* 11, 2437.

Vu, A.T., Utsumi, Y., Utsumi, C., Tanaka, M., Takahashi, S. *et al.* (2022) Ethanol treatment enhances drought stress avoidance in cassava (*Manihot esculenta* Crantz). *Plant Molecular Biology* 110(3), 269–285.

Wang, X., Li, Q., Xie, J., Huang, M., Cai, J. *et al.* (2021) Abscisic acid and jasmonic acid are involved in drought priming-induced tolerance to drought in wheat. *Crop Journal* 9(1), 120–132.

Wang, Y., Zou, Y.-N., Shu, B. and Wu, Q.-S. (2023) Deciphering molecular mechanisms regarding enhanced drought tolerance in plants by arbuscular mycorrhizal fungi. *Scientia Horticulturae* 308, 111591.

Wassie, M., Song, S., Cao, L. and Chen, L. (2023) A *Medicago truncatula* calcineurin B-like protein, MtCBL13 confers drought sensitivity in *Arabidopsis* through ABA-dependent pathway. *Environmental and Experimental Botany* 206, 105141.

Williams, A. and de Vries, F.T. (2020) Plant root exudation under drought: Implications for ecosystem functioning. *New Phytologist* 225(5), 1899–1905.

Wright, S.F. and Upadhyaya, A. (1998) A survey of soils for aggregate stability and glomalin, a glycoprotein produced by hyphae of arbuscular mycorrhizal fungi. *Plant and Soil* 198(1), 97–107.

Yasmin, H., Rashid, U., Hassan, M.N., Nosheen, A., Naz, R. *et al.* (2021) Volatile organic compounds produced by *Pseudomonas pseudoalcaligenes* alleviated drought stress by modulating defense system in maize (*Zea mays* L). *Physiologia Plantarum* 172(2), 896–911.

Zhang, F., Zou, Y.N. and Wu, Q.S. (2018) Quantitative estimation of water uptake by mycorrhizal extraradical hyphae in citrus under drought stress. *Scientia Horticulturae* 229, 132–136.

13 Genetic Engineering of Plants for Remediation of Toxic Metals and Metalloids

Syeda Gul Nisa[1], Maria Hameed[2], Ishmal Eman[1], Noor Fatima[1], Shahab Sadiq[1], Sawaira Rafique[1], Zeeshan Khan[1] and Tariq Shah[3]*

[1]Department of Plant Biotechnology, Atta-ur-Rahman School of Applied Biosciences, National University of Sciences and Technology, Islamabad, Pakistan; [2]Department of Botany, Faculty of Sciences, PMAS Arid Agriculture University, Rawalpindi, Pakistan; [3]Department of Agronomy, Faculty of Crop Production Sciences, University of Agriculture, Peshawar, Pakistan

Abstract

Genetic engineering in plants has transformed agricultural methods, providing novel options to increase crop output and resilience. This chapter explores the many approaches and uses of genetic engineering in plant science. The development of genetically modified organisms (GMOs) to increase stress tolerance, the development of genome editing technology, i.e., CRISPR-Cas9, and the use of genetic engineering in phytoremediation are important areas of study. The chapter focuses on how genetic changes might improve plant development and productivity in challenging environments by conferring resistance to both abiotic and biotic environmental stresses. Furthermore, it delves into the possibilities of genetically modified plants in sustainable agriculture, specifically with heavy metal removal and nutrient control. This chapter offers a thorough summary of the most recent developments and potential directions in plant genetic engineering by fusing research findings from the present with useful applications. These development implications for ensuring sustainable environment and global food security are also explored, highlighting the pivotal role that genetic engineering plays in resolving some of the most important issues facing modern agriculture.

Keywords: remediation, environmental stresses, CRISPR-Cas9, modern agriculture

13.1 Introduction

Water and soil pollution due to metalloids and metals is becoming more detrimental worldwide (Sharma *et al.*, 2021). Agriculture, metallurgy, mining, faulty waste disposal, military operations, and fossil fuel combustion are important sources for the release of massive quantities of toxic heavy metals/metalloids into the environment with serious health effects (Kotrba *et al.*, 2009). Heavy metals (HMs) are naturally occurring in nondepleted layers of the Earth's crust and pose a serious risk to agricultural production in fertile soil across the world by affecting yield, development, and crop growth due to rising levels above the permissible limit (Noor *et al.*, 2022).

*Corresponding author: tariq.shah94@gmail.com

DOI: 10.1079/9781800627659.0013

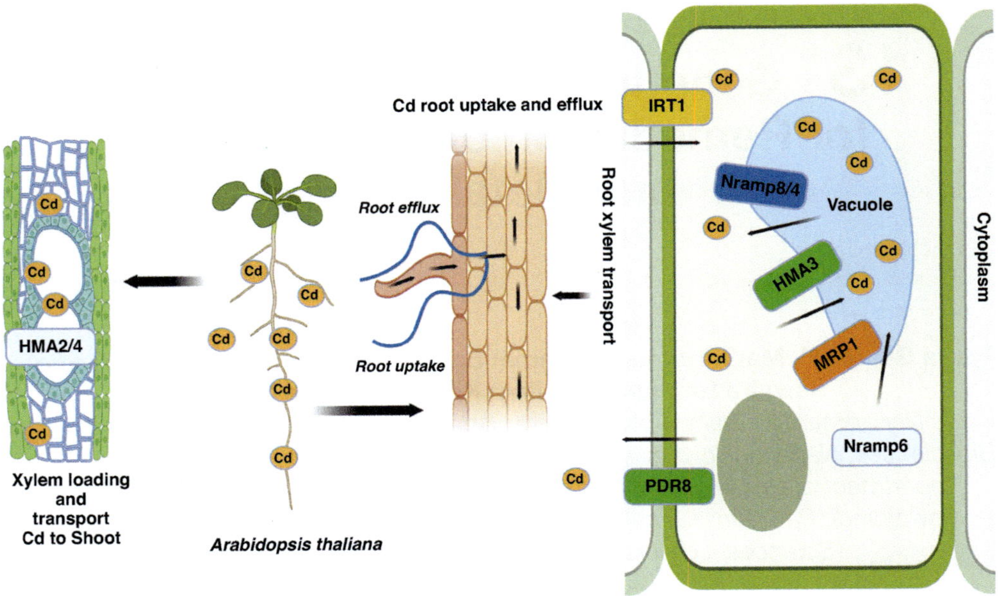

Fig. 13.1. The absorption and translocation of Cd by metal/metalloid transporter genes.

The term 'metals' or 'metalloids' refers to 51 elements in the periodic table with densities greater than 5000 kg/m^3 and atomic masses greater than 20 (Raffa *et al.*, 2021). Soils become more contaminated with HMs as a result of significant discharges of contaminated substances into the atmosphere from both natural and artificial sources (Morkunas *et al.*, 2018). Metals accumulate and percolate through the soil to contaminate groundwater. Since HMs are persistent and nonbiodegradable, they cannot be readily eradicated from the environment. Some HMs are mobile and can be taken up by plants via metal transporters, endocytosis, or diffusion, but others are stationary and cannot be removed from their accumulation sites (Noor *et al.*, 2022), as illustrated in Fig. 13.1. Inorganic elements with biological functions such as Mo, Fe, Mn, Ni, Zn, Co, and Cu are considered as heavy metals if present in excess amounts. Other elements such as As, Sb, Cd, Hg, Pb, and Ag with no known biological functions are more devastating even in low concentrations (Fasani *et al.*, 2018).

Complete elimination of HMs is not feasible, but they can be neutralized or transformed into a less toxic form that minimizes their injurious effects on the surrounding environment (Pratush *et al.*, 2018). Green approaches are urgently required to explore feasible and affordable ways to eradicate heavy metal pollution. Phytoremediation is an environmentally friendly, inexpensive, novel and efficient green approach to eliminate heavy metals (Ojha *et al.*, 2023). It is a relatively new and still evolving scientific field that uses a wide range of naturally occurring or genetically engineered plants to remove pollutants from contaminated waters and soils (Fasani *et al.*, 2018). The ability of plants to eliminate HMs from the environment through phytoremediation is greatly influenced by modern biotechnology. By employing genetic engineering techniques, it is possible to regulate certain resistance genes that are responsible for the influx and efflux of different substances as well as their transportation in transgenic plants (Noor *et al.*, 2022).

Genetically modified plants are better able to survive in soil or waters with elevated levels of metals without exhibiting unusual symptoms. Genetic engineering and nanotechnology are opening new avenues to counteract and treat heavy metal contaminants (Yadav *et al.*, 2022). Phytoremediation stands out as the most affordable and promising method for the remediation of heavy metals (Razzaq, 2017). The success of phytoremediation is dependent on different mechanisms such as metal uptake, accumulation, tolerance, and detoxification. Metalloid metabolism in plants includes absorption by the roots, translocation, and then transformation into less hazardous molecules that are then released into the environment through the process of phytovolatilization. For instance, when selenium (Se) is taken up by plants, different enzymes involved in Se metabolism, assimilation, and volatilization transform it into dimethyl selenide. Then, this is released into the atmosphere as a gas that is less harmful than inorganic Se (Gupta and Gupta, 2017).

13.2 Engineering Metal/Metalloid Transporter Genes and Uptake System

Much success has been achieved in the last few years in engineering plant uptake systems and metal/metalloid transporter genes to increase phytoremediation efficiency and tolerance to heavy metal/metalloid stress (Raza *et al.*, 2021). A list of important metal transporter genes that have been overexpressed in plants includes the following. The ZIP family (ZRT, IRT-like protein) is responsible for the uptake of Zn, Fe, Mn, and Cd in major crops and edible plants (Lira-Morales *et al.*, 2019; Mohammed *et al.*, 2023). The HMA family (heavy metal ATPase) is involved in mediating xylem loading and translocation of Zn, Cd, Cu, and Pb from roots to shoots in plants (Yang *et al.*, 2022). NRAMP (natural resistance-associated macrophage protein) transports Cu, Mn, and Fe from soil to plants (Mani and Sankaranarayanan, 2022). The role of the CDF family (cation diffusion facilitator) is to sequestrate Zn, Fe, Mn, and Cd into vacuoles for storage (Tiwari and Lata, 2022). The ABC transporters (ATP-binding cassette) are mostly involved in transportation of chelated metal (Dahuja *et al.*, 2021).

The metal transporter genes could be the mechanism of enhancement in plant phytoremediation. The upregulation of metal transporter genes can augment the plant's capacity to assimilate HMs residing in soil via the root system. Additionally, it can make it easier for metals to move from the roots to the shoots. These transporters can encapsulate the metals within vacuoles or cell walls of plants and can endure elevated metal concentrations in tissues. This improves the efficiency of plants' phytoremediation and their ability to thrive in contaminated soils (Yang *et al.*, 2022; Koźmińska *et al.*, 2018), depicted in Fig. 13.2. Examples of genes actively involved in phytoremediation include the overexpression of *OsHMA3* in rice which decreases the accumulation of Cd in rice grains and *OsNRAMP5* in rice which decreases Mn and Cd uptake (Shao *et al.*, 2017). The overexpression of *AtHMA4* in *Arabidopsis thaliana* enhances Cd and Zn tolerance (Verret *et al.*, 2004).

13.2.1 Genes encoding metal/metalloid transporters and their overexpression

Several silicon (Si) uptake genes are reported in plants for encoding metal/metalloid. Lsi1 (*OsNIP2;1*) is documented as a silicon influx transporter in rice, while Lsi2 as silicon efflux transporter is also reported (Zhao *et al.*, 2010). Expression of HvLsi1 in barley reportedly increased silicon uptake.

Genes responsible for selenium uptake, reduction, sequestration, and translocation can be overexpressed to enhance Se accumulation (Abou Seeda *et al.*, 2022), some of the genes which are responsible for the uptake of HMs are mentioned in Table 13.1. Expressing bacterial ars genes (part of the bacterial ars operon; arsenic resistance system) in plants can convert toxic selenate to less toxic selenite and enhance Se tolerance (Gupta and Gupta, 2017). The knockdown of the *OsPT4* gene in rice decreases arsenate uptake and inorganic As. Expression of *PvACR3;1* and *ScACR3* arsenate transporters genes in rice lowers the inorganic As content in the plant (Shri *et al.*, 2019). Some aquaporin channel proteins, e.g., *OsNIP3;3* and *OsNIP1;1*, are known to facilitate diffusion of As $(OH)_3$ across membranes of rice plants (Bienert *et al.*, 2008). Metallothioneins (MT) in plants are classified as Class II and can be divided into four groups, MT1 to MT4, and have been reported to moderate metal(loid) toxicity. For example, functional identification of an MT2 gene (*SsMT2*) and subsequent gene overexpression boosted $CdCl_2$ tolerance in *Arabidopsis* because of its raised Cd content (Sunitha *et al.*, 2013). *OsPRX38* overexpression in *A. thaliana* decreased

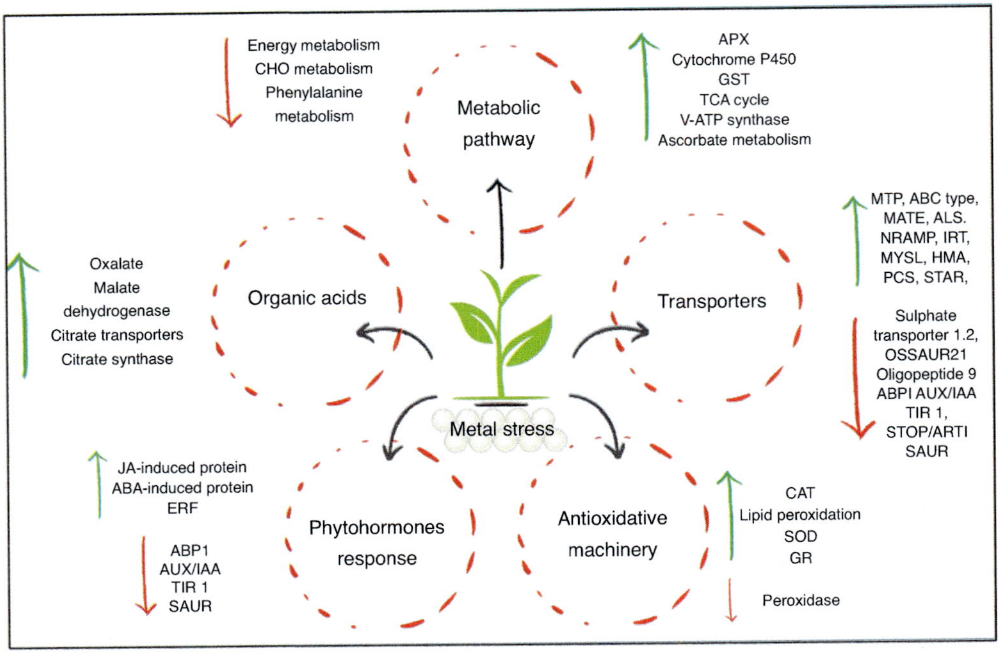

Fig. 13.2. Overview of the molecular mechanisms underlying metal and metalloid toxicity in plants.

As content by promoting apoplastic lignification, antioxidant enzymes SOD, PRX, and GST, decreasing H_2O_2, and MDA contents. Methylated bands were denser in *Noccaea caerulescens* (Ni hyperaccumulator) compared to *A. thaliana* (Ni sensitive) plants when grown in Ni-contaminated soil. Another study described complex changes in DNA methylation and its link to adaptive responses of *A. thaliana* treated with Cd, with increased methylation status and decreased expression of the *ROS1*/*DML2*/*DML3* DNA demethylase genes. DNA methylation induced by Cr accompanied by the increase in antioxidant potential helped regulation of plant genes to the exposure of Cr (Singh *et al.*, 2015).

13.2.2 Genes encoding metal/metalloids chelators and their overexpression

Plants absorb a broad range of trace metalloids/metals from the soil. Few of these are essential for plants but can be detrimental at higher volumes (Ni, Co, Zn, Co) while some are nonessential and even minute amounts of them are toxic for plants (Cd, As, Cr, Hg, Ag). Trace metals are natural constituents of the Earth's crust but in low concentrations. These trace metals are also known as heavy metals. Anthropogenic practices such as metal mining and smelting, fossil fuel burning, sewage sludges, application of fertilizers and pesticides, and irrigation with polluted water largely increase their concentration in soil (Tang *et al.*, 2023). To counter the detrimental effects of HMs, plants are equipped with detoxification mechanisms, as highlighted in Fig. 13.3. When HMs enter the cell, they are sequestered by specific metal-binding ligands, organic acids, glutathione (GSH), or amino acids (Hossain *et al.*, 2012).

Plants synthesize phytochelatins (PCs) in their cytoplasm. PCs are thiol-reactive, cysteine-rich peptides, which sequester metals by chelating them. PCs form a complex with the metals/metalloids which is then taken into a vacuole via ATP-binding cassette (ABC) transporters such as *YCF1* found in baker's yeast (Brunetti *et al.*, 2011). PCs will accumulate greater amounts of these metals and chelate to inactivate them and ABC transporters will pump them into a vacuole to store them (Hong-Bo *et al.*, 2010). A transgenic *Arabidopsis thaliana* plant was developed

Table 13.1. Genes overexpressed in plants that encode metal/metalloids transporters.

Genes	Plant	Role	Reference
OsIRT1, OsIRT2	*Oryza sativa*	Overexpression enhanced Cd uptake by roots	(Yang *et al.*, 2016)
OsZIP5, OsZIP9	*Oryza sativa*	Overexpression increased Cd uptake	(Liu *et al.*, 2019)
OsZIP7	*Oryza sativa*	Overexpression enhanced Cd translocation from roots to shoots	(Li *et al.*, 2024)
MsYSL1	*Miscanthus*	Overexpression increased Cd translocation in stems	(Chen *et al.*, 2018)
SnYSL3	*Solanum nigrum*	Overexpression enhanced Cd translocation in roots and stems	(Feng *et al.*, 2017)
BjYSL7	*Brassica juncea*	Overexpression increased Cd translocation in stems	(Wang *et al.*, 2013)
OsNIP2;1	*Oryza sativa*	Overexpression increased silicon and selenite uptake	(Liang *et al.*, 2019)
OsNIP1;1	*Oryza sativa*	Facilitate diffusion of $As(OH)_3$ across membranes	(Bienert *et al.*, 2008)
AtBOR1	*Arabidopsis*	Overexpression enhanced boron uptake and translocation	(Uraguchi *et al.*, 2014)
OsBOR1	*Oryza sativa*	Overexpression increased boron uptake	(Tanaka *et al.*, 2013)
OsPT8	*Oryza sativa*	Knockdown reduced arsenate uptake and inorganic As in rice grains	(Wang *et al.*, 2016)
PvACR3;1, ScACR3	*Oryza sativa*	Expressing decreased inorganic As in grains	(Duan *et al.*, 2012)
HvLsi1	Barley	Expressing increased silicon uptake	(Liang *et al.*, 2015)
OsHMA3	*Oryza sativa*	Overexpression reduced Cd accumulation in grains	(Zhang *et al.*, 2020)
OsNRAMP5	*Oryza sativa*	Overexpression decreased Mn and Cd uptake	(Chang *et al.*, 2020)
AtHMA4	*Arabidopsis*	Overexpression enhanced Zn and Cd tolerance	(Barabasz *et al.*, 2016)
γ-ECS	*Brassica indica*	Overexpression increased GSH and Cd tolerance	(Reisinger *et al.*, 2008)
PCS	*Phytochelatin synthase*	Expressing bacterial gene enhanced Cd, Cu, and Zn tolerance	(Li *et al.*, 2015)
Gly I, Gly II	*Brassica indica*	Overexpressing bacterial genes enhanced Ni and Cd tolerance	(Zaid *et al.*, 2019)

by overexpressing garlic's PC gene *AsPCS1* and *S. cerevisiae* vacuolar compartmentalization gene *YCF1*. When compared to wild-type (WT) and single-gene transgenic lines, double-gene transgenic lines were observed with the longest roots and greater As and Cd accumulation (Guo *et al.*, 2012).

Sulfur-containing metabolites like cysteine (Cys) have a pivotal role in the plant life cycle. Cysteine is involved in the biosynthesis of methionine, GSH, proteins, PCs, and other sulfur-containing compounds which are known for mitigating HM stress. The production of cysteine is catalyzed by Cys synthase (CS) [O-acetylserine

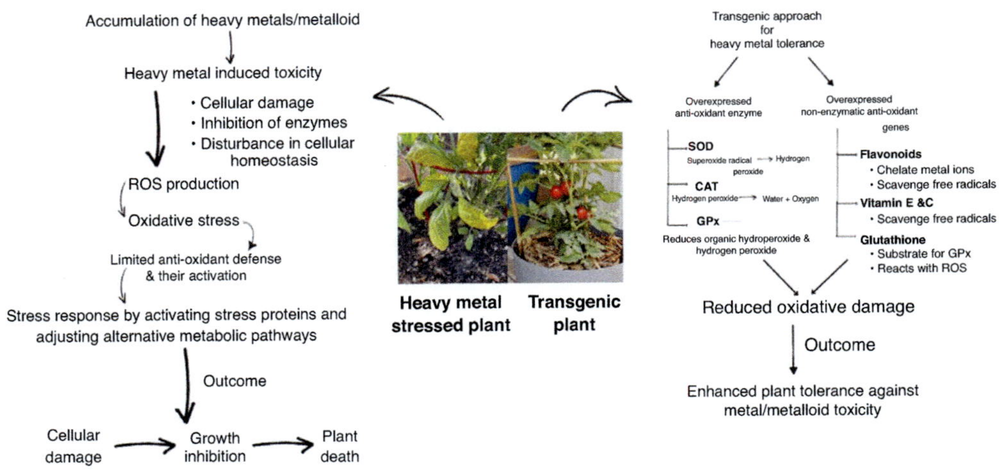

Fig. 13.3. Comparison of antioxidant defense mechanisms in a heavy metal stressed plant and transgenic plant.

(thiol)-lyase] and serine acetyltransferase (SAT) (Hell and Wirtz, 2011). Transgenic tobacco plants were developed by overexpressing both CS and SAT genes simultaneously. The transgenic plants overexpressing dual genes showed more resistance to Cd stress compared to the WT and single-gene transgenic plants. Moreover, dual transgenic plants demonstrated higher production of PCs which can chelate Cd (Nakamura *et al.*, 2014).

Metallothioneins (MTs) are low molecular weight cysteine-rich polypeptides that bind to both essential and nonessential metals to detoxify them and maintain cellular homeostasis. They are essential in distribution of HMs because they use the thiol group of the cysteine residue to form complexes with metal ions (Hegelund *et al.*, 2012). Transcriptome analysis of *Phytolacca americana* showed that MT genes, especially type III, were upregulated under Cd stress (Chen *et al.*, 2017). The *P. americana* MT gene *PaMT3-1* was inserted into a tobacco plant and the transgenic tobacco plant was treated with $CdCl_2$ for 7 days. The roots of the transgenic plants and the WT had similar Cd content but the Cd level of transgenic leaves was about 66.28–78.70% of the WT (Zhi *et al.*, 2020). Many reports have suggested that it is possible to mitigate HM toxicity by overexpressing the genes responsible for HM absorption, sequestration, and translocation. MTs and PCs are key components in moderating HM stress (Sunitha *et al.*, 2013). Transgenic plants overexpressing metallothionein and phytochelatin synthase genes have been widely reported (Table 13.2).

13.2.3 Genes encoding antioxidant defense machinery and their overexpression

The soil contaminated with heavy metal has devastating effects on agricultural crops. These metallic elements (Hg, Cr, As, Cd, and Pb) can trigger the production of reactive oxygen species (ROS) such as hydroxyl radicals (•OH), singlet oxygen ($^1O_2$), hydrogen peroxide (H_2O_2), and superoxide anion (O^2 •−). Long-term exposure of heavy metals can elevate ROS production which changes the natural redox state of cells and destroys DNA, lipids, proteins, and other components of the cell (Kumar *et al.*, 2020). Plants react to environmental stressors at the molecular level by turning on several kinds of stress resistance genes and proteins. To counteract HM stress, plants upregulate genes that cause overexpression of antioxidant enzymes such as ascorbate peroxidase (APX), superoxide dismutase (SOD), glutathione peroxidase (GPx), and catalase (CAT) (Alaraidh *et al.*, 2018).

In one study, fenugreek seedlings were tested for their antioxidant activity against metallic stress caused by Cr, Pb, and Cd. Quantitative real-time PCR was used to examine

Table 13.2. Genes encoding MTs and PCs are overexpressed in transgenic plants.

Genes	Plants	Effects	Reference
CUP1	Sunflower	Cd tolerance	(Watanabe *et al.*, 2005)
AtPCS1/CePCS1	Tobacco	As tolerance and accumulation	(Wojas *et al.*, 2010)
AtPCS1	Mustard	As and Cd tolerance	(Gasic and Korban, 2007)
AtPCS1	Tobacco	Cd tolerance and accumulation	(Pomponi *et al.*, 2006)
AtPCS1	*Arabidopsis*	Cd tolerance and accumulation	(Lee *et al.*, 2003)
		Cd hypersensitivity and As tolerance	(Li *et al.*, 2004)
		Cd tolerance	(Brunetti *et al.*, 2011)
AtMT2a	Broad bean	Cd tolerance	(Lee *et al.*, 2004)
Yeast*MT (CUP1)*	*Brassica oleraceae*	Cd accumulation	(Krämer and Chardonnens, 2001)
Mammalian *MT*	Tobacco	Reduced root–shoot Cd transport	(Krämer and Chardonnens, 2001)
TaPCS1	Poplar	Accumulation of Zn and Pb	(Couselo *et al.*, 2010)
TaPCS1	Rice	Cd hypersensitivity	(Wang *et al.*, 2012)
CdPCS1	Tobacco	Accumulation of Cd and As	(Shukla *et al.*, 2013)
Tc MT2a	*Arabidopsis*	Cd and Zn hypersensitivity	(Schiller *et al.*, 2014)
		No change in tolerance	(Hassinen *et al.*, 2009)
TcPCS1	Tobacco	Accumulation of Cd	(Liu *et al.*, 2011)
NnPCS1	*Arabidopsis*	Accumulation of Cd	(Liu *et al.*, 2012)
Tc MT3	*Arabidopsis*	No change in tolerance	(Hassinen *et al.*, 2009)
CcMT1	*Arabidopsis*	Cd and Cu tolerance	(Sekhar *et al.*, 2011)
NnMT2a	*Arabidopsis*	Increased seed germination vigor	(Zhou *et al.*, 2012)
Mouse *MT*	*Ralstonia*	Enhanced Cd immobilizing ability	(Valls *et al.*, 2000)
AtMT3	Broad bean	Cd tolerance	(Lee *et al.*, 2004)
PtPCS1	Poplar	Zn accumulation	(Adams *et al.*, 2011)

the transcription profiles of the genes (POD, CAT, and APX) encoding antioxidant enzymes. When compared to untreated seedlings, seedlings treated with Cd, Cr, and Pb metal ions showed a significant rise in the transcript level for all the genes that encode for antioxidant enzymes (Alaraidh *et al.*, 2018). In a hydroponics system, ryegrass seedlings were subjected to Pb stress for 7 days to investigate antioxidant enzyme transcript levels. Within 24 hours of Pb treatment, an upregulation was observed in RNA transcript level of eight genes that encode for antioxidant enzymes (Li *et al.*, 2012). Similar results were produced in cotton treated with Pb as higher transcript levels of antioxidant genes, i.e., APX, CAT, SOD, and guaiacol peroxidase, were observed (Bharwana *et al.*, 2013).

Plants enriched with anthocyanin have high antioxidant content and are more adaptable to biotic and abiotic stresses (Dehghan *et al.*, 2014). This increased resilience due to flavonoid accumulation may be linked to enhanced ROS scavenging. Overexpression of the transcription factor *IbMYB1* responsible for anthocyanin biosynthesis conferred greater tolerance against abiotic stresses in tobacco plant by producing antioxidants (Naing *et al.*, 2017). The transcription factor *RsMYB1* is a positive regulator to transcriptionally activate anthocyanin bioproduction and ultimately antioxidant production in *Arabidopsis* and radish plants (Lim *et al.*, 2016).

Glutaredoxins (Grxs) are ubiquitous, cysteine-rich, heat-stable, low molecular weight (10–15 kDa) proteins which can scavenge ROS by elevating antioxidant activity. A study was conducted on the chickpea glutaredoxin (CaGrx) gene (*LOC101493651*) in relation to different heavy metals, including As(V), As(III), Cd, and Cr(VI). Results showed that the higher levels of CaGrx gene in chickpea plants mitigate metal stress. The CaGrx gene from chickpea was inserted and overexpressed in *Arabidopsis thaliana* to study various physiological and biochemical parameters under the influence of metal stressors. Transgenic plants showed higher transcript levels of CaGrx gene along with higher survival efficiency, seed germination, and longer root in response to each of the metals under study. Significantly higher antioxidant activity was observed in transgenic *Arabidopsis* plants to counter metal stress (Kumar *et al.*, 2020).

Spermidine (Spd) is a polyamine that serves multiple regulatory purposes in cells such as cell division, DNA replication, transcription and translation, and membrane stability. Besides this, evidence suggests that it also has a role in mitigation of oxidative stress supported by its antioxidant potential. Spermidine synthase (SPDS) is an enzyme involved in biosynthesis of spermidine (Wen *et al.*, 2008). To confirm whether spermidine can confer HM tolerance, the apple SPDS gene (*MdSPDS1*) was overexpressed in European pear (*Pyrus communis* L. Line # 32). $PbCl_2$, $ZnCl_2$, and $CdCl_2$ were used as HM stressors. Both Spd and SPDS levels increased in transgenic plants but when the GSH content was observed in stress conditions, Line # 32 showed a noticeably lower level of GSH, although not quite as much as the WT (Wen *et al.*, 2010).

Metallothioneins are widely distributed, cysteine-rich, low molecular weight metal-chelating proteins that can also scavenge ROS efficiently under HM stress by regulating antioxidant activity (Hassinen *et al.*, 2011). The chickpea MT1 gene was overexpressed in *A. thaliana* and this transgenic plant was tested against Cd, Cr(VI), As (As(III) and As(V)) toxicity. Under stress conditions, MT1 overexpressing lines showed enhanced antioxidant activities such as glutathione reductase (GRX), APX, and GPX (Dubey *et al.*, 2021). Another study showed that tobacco plants that overexpressed the SaMT2 gene had elevated antioxidant activity under Cd and Zn stress (Zhang *et al.*, 2014).

13.2.4 Chloroplast genome transformation and its role in phytoremediation

The chloroplast is a vital organelle of plant cells which is distinct from proplastids and is essential for photosynthesis. In this process, plants convert light energy from sunlight into biochemical energy which they use to sustain themselves (Bobik and Burch-Smith, 2015). Besides its significant vital role in the photosynthesis process, the chloroplast is also associated with other essential cellular activities in plants, including synthesis of secondary metabolites, generation of amino acids, biosynthesis of fatty acids, and nitrogen metabolism (Rolland *et al.*, 2012). The estimated 100–200 kb gene sizes of the currently recognized chloroplast suggest that its genome size is smaller than that of its progenitor (Jensen and Leister, 2014).

Chloroplast transformation

Research on chloroplast transformation has been carried out since its discovery almost 20 years ago, and findings show that it is a good substitute for traditional nuclear transformation techniques (Kumar *et al.*, 2021). Engineering of chloroplast is an eco-friendly tool and is preferable to nuclear engineering in certain crops. Chloroplast transformation entails genome modification or a foreign gene introduction into the chloroplast. This procedure involves designing the DNA construct, which incorporates selectable markers, and delivering this construct into the chloroplast. The design of the selectable marker and vector is crucial for placing the desired gene into the DNA construct, facilitating successful identification of transformed chloroplasts based on their intended use (Liu *et al.*, 2023).

The plastid genome is highly polyploid, permitting the incorporation of multiple inserts of foreign genes into plant cells. Transformation of chloroplasts has several inherent advantages, such as the capacity to manipulate several genes simultaneously in one transformation event, including enhanced expression of foreign genes, avoidance of gene silencing and positional effects, reduced risk of transgene outcrossing, and precise containment and regulation of foreign proteins. Recent advancements in chloroplast genome engineering techniques have enhanced plant resistance to herbicides, insect pests, and drought, as well as the production of biopharmaceuticals such as vaccine antigens (Verma, 2019). Continued advancements in chloroplast transformation are expected to facilitate genetic modifications and enhancements in plants, enabling cost-effective and environmentally friendly production of pharmaceutical components (Verma, 2019).

Approaches used for chloroplast transformation

Chloroplast transformation has been recognized as an ideal method for transgene expression, as mentioned in Table 13.3. In angiosperm plants, chloroplast genomes are predominantly inherited maternally, which limits the dissemination of transgenes. This feature makes chloroplast genomes more useful in developing genetically altered plants. Chloroplast transformation typically results in higher levels of proteins than nuclear genome transformation. For instance, studies have shown that approximately 46% of the protein produced originates from the inserted transgene. Furthermore, proteins produced through chloroplast transformation

Table 13.3. Comparison of chloroplast transformation methods.

Methods	Description	Advantages	Disadvantages	References
Biolistic method	DNA-coated particles (typically tungsten or gold) are shot into plant cells via gene gun at high velocity	Suitable for several plant species. Efficient delivery of large DNA constructs. Minimal requirement for plant regeneration protocols	High cost of equipment and consumables. Potential for cell damage due to high pressure. Risk of gene silencing	(Purton, 2007)
PEG-mediated transformation	Disrupts protoplast membrane, facilitating DNA uptake	Simple and cost-effective. High transformation efficiency	Understanding of long-term effects is limited. Optimization required for different plant species. Potential toxicity concerns	(Zienkiewicz *et al.*, 2019)
UV-laser microbeam	Beam creates transient pores in cell membranes for DNA delivery	Precise spatial control over gene delivery. Direct DNA uptake without complex equipment	High equipment costs, cell damage, limited application in chloroplast transformation	(Saba *et al.*, 2023)
Agrobacterium-mediated transformation	Delivers DNA into chloroplasts via T-DNA transfer	Mimics natural plant transformation. Efficient integration of DNA into chloroplast	Specific strains of *Agrobacterium*, limited to certain plant species and cell types	(Kumar and Ling, 2021)

are free from viral contamination or human pathogens (Wang *et al.*, 2009).

Chloroplast transformation has been achieved using methods such as *Agrobacterium*-mediated delivery, biolistic delivery, or polyethylene glycol (PEG)-mediated delivery. These techniques necessitate the regeneration of genetically modified offspring plants, a process that is often difficult and time-consuming. However, successful transformation protocols have been developed for a limited number of plants to date (Kumar and Ling, 2021). Previous researchers discovered that single-walled carbon nanotubes and artificially manufactured nanoparticles can readily pass through the stiff walls and membranes of plant cells as well as the double lipid bilayers (outer and inner membranes) of chloroplasts smoothly (Giraldo *et al.*, 2014). In addition, research on employing nanomaterials to carry foreign DNA into plant chloroplasts has made significant advances. The single-walled carbon nanotubes were designed and screened to successfully transfer DNA into chloroplasts isolated from various plant species without biolistics or any additional chemicals. This nanosized DNA delivery tool has significant advantages compared to classic delivery techniques and is considered as a potential transformation method with high efficacy in plants (Kwak *et al.*, 2019).

BIOLISTIC METHOD. The biolistic method, also called the gene gun technique, was established in the late 1980s and continues to be an extensively used approach in plant transformation because of its flexibility and effectiveness (Matsumoto and Gonsalves, 2012). This process entails coating DNA constructs on purified particles, usually tungsten or gold, and delivering them into plant cells by electric discharge or high-pressure helium gas. The method successfully integrates DNA into chloroplasts by avoiding cell surface obstacles, which can result in modifications (Carter and Shieh, 2015). For example, investigations on potatoes revealed improved protein yields, and studies with *Artemisia annua* indicated enhanced synthesis of artemisinin (Kaushal *et al.*, 2020). The technique can change plastid genomes and has wide applications, but it also has disadvantages including high costs, the possibility of damaging cells, and the possibility of gene silencing when several genes are added. Notwithstanding these difficulties, the biolistic approach is a useful tool in genetic engineering, particularly in situations where other approaches are inappropriate (Jinturkar *et al.*, 2011; Bhatia *et al.*, 2015).

POLYETHYLENE GLYCOL-MEDIATED TRANSFORMATION. Transformation using PEG is widely practiced in plant bioengineering, especially for chloroplast transformation in protoplasts, due to its simple protocol and low cost (Kumar and Ling, 2021). This technique involves increasing cell membrane permeability with PEG, which allows the desired DNA segment to move through in vesicles. The process begins with the isolation and purification of protoplasts and chloroplasts, followed by insertion of the desired DNA construct using a 60% PEG solution, and concludes with the regeneration of transformed chloroplasts (Wang *et al.*, 2022). PEG-mediated transformation frequently results in high transformation efficiency and can absorb huge DNA constructs without rupturing the cell membrane, although it is difficult to generate a significant number of active protoplasts; it also frequently generates many temporary transformants (Kumar and Ling, 2021).

CARBON NANOTUBE CARRIERS. Carbon nanotube carriers (CNT), developed by Sumio Iijima in 1991, are emerging as a promising tool to deliver biomolecules and drugs because of their unique structure and physical properties (Irshad *et al.*, 2020). These nanoparticles are positively charged, typically 10–20 nanometers, and can effectively introduce pDNA into chloroplasts by a process known as lipid exchange envelope penetration (LEEP), which allows them to move through the chloroplast envelope (Newkirk *et al.*, 2021). Although delivery using CNTs has shown promising results in *A. thaliana* mesophyll protoplasts with more than 75% effectiveness, further development is necessary to ensure that CNTs are specifically engineered to uptake and transport pDNA selectively into chloroplasts (Newkirk *et al.*, 2021).

This method is economical and does not require sophisticated equipment. Conversely, UV laser microbeam-mediated gene delivery, which employs a UV laser to generate temporary openings in cell membranes for DNA intake, provides precise spatial control but is less prevalent due to expensive equipment and the risk of cell damage. Although it can be effective, such as with *Brassica napus* protoplasts, further research is necessary to determine its overall effectiveness and wider uses (Ihien Katche and Mason, 2023).

UV LASER MICROBEAM. Gene delivery using UV laser microbeam is an attractive approach because it can precisely control the laser beam generated by the optical fiber, enabling spatial precision. This approach includes four different types of laser-assisted gene delivery: (i) transfection via laser-induced stress waves, (ii) optoinjection, (iii) selective cell targeting using light-absorbing particles, and (iv) photochemical internalization. The UV laser beam induces cell perforations, creating holes in the cell membrane (Kumar and Ling, 2021). The beam leaves a hole about 0.5 μm in size, but it heals itself and usually closes in 5 s. Exogenous DNA or pDNA may enter the cells during this brief opening, aiding in the transformation process. Fluorescence microscopy has been utilized to confirm incorporation of the desired DNA into chloroplasts of *Brassica napus* using a UV laser microbeam technique. However, because it is expensive to purchase the equipment to produce a laser beam with the necessary 100 nm dimensions, this method is not commonly used. UV laser light also poses a risk of damaging cells and chloroplasts. As there are not many studies on chloroplast transformation using this strategy at the moment, more investigation is necessary to fully appreciate its efficacy (Rivera *et al.*, 2012).

AGROBACTERIUM-MEDIATED TRANSFORMATION. Transformation using *A. tumefaciens* is important as it mimics the natural process of plant transformation. This method is highly regarded for its nonintrusive gene delivery capabilities. In this method, the plant is infected by a bacterium which then inserts the desired gene into the chloroplast. The process involves five main steps: (i) recognition of signals, (ii) processing of T-DNA (transfer DNA), (iii) movement of T-DNA inside the host cell, (iv) incorporation of T-DNA in the host genome, and (v) expression of the transferred DNA (T-DNA) (Kumar and Ling, 2021). When *A. tumefaciens* attaches to plant cells, the desired gene contained within the pathogenic bacterium is shifted to the T-DNA located in the Ti (tumor-inducing) plasmid. Through nonhomologous recombination, this T-DNA is inserted into the chloroplast's genome. Following insertion of the desired gene, the chloroplast undergoes regeneration and multiplication (Pratiwi and Surya, 2020).

Chloroplast transformation and phytoremediation

The main reason why plants are preferred as better bioreactors is that they can precisely carry out posttranslational modifications. These include disulfide bond production, amidation, phosphorylation, appropriate protein folding, and the proper assembly of complex multi-subunit proteins. (Řepková, 2010). Phytoremediation is widely considered a safe and economically viable approach for remediating polluted environments by employing plants. Among the most hazardous form of mercury are organomercurial compounds, which primarily target the chloroplast genome in plants. This makes the chloroplast an ideal target for engineering resistance and detoxification mechanisms against organomercurials and metallic mercury.

By genetically modifying the chloroplasts of plants to synthesize metal chelators, their ability to uptake metals has been significantly enhanced (Hussein *et al.*, 2007; Occhialini *et al.*, 2016). Two bacterial genes that encode enzymes, organomercurial lyase (merB) and mercuric ion reductase (merA), were overexpressed together as an operon in transgenic tobacco chloroplasts. This resulted in the accumulation of mercury in roots, reaching concentrations exceeding those found in the soil, up to 200 μg/g. Importantly, this accumulation did not cause any harmful effects. Additionally, the transgenic plants could accumulate mercury in leaves at levels 100 times higher than nontransformed plants (Occhialini *et al.*, 2016).

For remediation of toxic mercury, chloroplasts of tobacco plants were engineered with metallothionein enzyme (Agrawal *et al.*, 2011). Expressing the MT1 gene in chloroplasts enhances plant resistance to mercury, permitting high mercury accumulation without harmful effects. Transgenic plants resisted mercury levels up to 20 µm, compared to 10 µm resistance in nuclear transgenic plants with different chelator genes. Our study used mercury-enriched media to demonstrate that transplastomic seeds can sprout and grow in the presence of mercury. These plants grew better and accumulated more mercury than untransformed plants, confirming that engineering chloroplasts with MT genes can phytoremediate mercury. This method prevents mercury volatilization and allows for industrial mercury recycling, enhancing phytoremediation's cost-effectiveness. Additionally, to quantify transcript abundance, homoplasmy, and transgene copy number, real-time PCR was used as a better quantitative tool than traditional Southern blots for molecular identification of transplastomic lines (Ruiz *et al.*, 2011).

The research shows that in transgenic *Arabidopsis* plants, targeting the *Brassica rapa* type 1 MT gene (BrMT1) to chloroplasts improves resistance to cadmium and ROS. Metallothioneins are known to bind HMs, such as exposure to cadmium. BrMT1, which was first discovered to give cadmium resistance in yeast, was successfully introduced into *Arabidopsis* chloroplasts using a RbsS transit peptide and a constitutive CaMV35S promoter. This focused localization successfully reduced H_2O_2 and cadmium stressors. Specifically, paraquat-induced leaf chlorosis and H_2O_2 buildup were greatly inhibited by BrMT1, which targeted chloroplasts. This work is the first study on chloroplast-targeted type 1 MT1, indicating that it may be useful in creating plants that are more resilient to environmental stresses (Nikolić and Tomašević, 2021; Kumar *et al.*, 2021).

Apart from these chelators, two genes encoding enzymes, organomercurial lyase and mercuric ion reductase, can convert toxic mercury contaminants into nontoxic forms. These two enzymes along with metal chelators were inserted into the chloroplast genome and expressed together as an operon. This approach has been effectively applied in genetically engineered tobacco and brassica plants, demonstrating successful conversion of mercury contaminants, as evidenced by several studies (Meagher *et al.*, 2020). Tobacco (*Nicotiana tabacum*) plastids were engineered using a refined genetic construct to generate the biodegradable and renewable material polyhydroxybutyrate (PHB). This was accomplished by using the extended operon approach (Gokul *et al.*, 2024).

13.3 Role of Transgenic Plants in Remediation of Metals/Metalloids: Case Studies

Phytoremediation is an environmentally friendly method which has gained a lot of attention. The scientific achievement of recent decades suggests that transgenic plants are advantageous in remediation of toxic metals/metalloids. Physicochemical methods and methods which involve the immobilization of metal ions pose hazardous risks to soil microbiota (Kotrba *et al.*, 2009). Phytoextraction of metals and metalloids using transgenic plants results in aboveground accumulation of these metals. Accumulation of metal biomass on aboveground soil leads to the production of chelators and sulfur metabolism enzymes (de Mello-Farias *et al.*, 2011).

Populus angustifolia, *Silene cucubalis*, and *Nicotiana tabacum* are among the plant species that have been altered genetically to express certain enzymes that provide enhanced metal accumulation in soil compared to the WT of these plants (Fulekar *et al.*, 2009). Plants with high biomass are best suited for genetic engineering and Indian mustard (*Brassica juncea*), yellow poplar (*Liriodendron tulipifera*), and sunflower (*Helianthus annuus*) are plants for which protocols have been developed already (de Mello-Farias *et al.*, 2011). Therefore, transgenic plant varieties have provided a powerful means to remediate heavy metals and environmental pollutants (Gunarathne *et al.*, 2019).

Brassica juncea was genetically engineered by introducing the *gshl* gene from *Escherichia coli*. It was observed that the transgenic version of *B. juncea* showed a higher amount of phytochelatins, glutathione, and nonprotein thiols. In addition, in the presence of Cd this transgenic plant showed enhanced tolerance compared to the wild type (Yang *et al.*, 2005). It also showed enhanced sensitivity to As both in As-hypertolerant and nonadapted plants (Tong *et al.*, 2004). Similarly, the *A. thaliana* gene *CAX-2* was introduced into tobacco plants to enhance tolerance to HMs. The transgenic tobacco plant showed enhanced tolerance to these heavy metals (Ow, 1996).

Another case study shows that transgenic *A. thaliana* plants were made by inserting glutathione-S-transferase from tobacco plants. This transgenic plant demonstrated more tolerance to Al, Na, and Cu. Similarly, the ferretin gene from soybean was inserted into tobacco and rice and transgenic plants showed enhanced iron accumulation compared to the wild species. The *MerA* gene was isolated from bacteria and inserted into *A. thaliana* to make it resistant to metals like Hg (Pulford and Watson, 2003). The Indian mustard plant demonstrated tolerance to heavy metals like Se when glutathione reductase was inserted (Ow, 1996).

From these case studies, it is evident that transgenic plant varieties play a significant role in enhancing tolerance against HMs.

13.4 Engineering Metal/Metalloid Smart Plants Using Gene and Genome Editing by CRISPR-Cas9

Heavy metal pollution is hazardous to human health, farmland, and food security as HMs residing in water and soil can easily be absorbed and accumulated in plants, ultimately polluting the food chain. HMs are not biodegradable and therefore persist in soil (Pal *et al.*, 2023; Wu *et al.*, 2022). High levels of heavy metals are harmful to plants (Riyazuddin *et al.*, 2021). Other remediation methods are not eco-friendly and are time-consuming and expensive (Rajendran *et al.*, 2022). To address this issue, the use of green technologies like phytoremediation is timely. Genetic engineering is another green approach to enhance phytoremediation capacity by modifying genes that can improve plants' capacity to mobilize, stabilize, and accumulate metals (Fasani *et al.*, 2018; Ozyigit *et al.*, 2021).

CRISPR-Cas9 is a powerful tool which has overshadowed other genome editing techniques because it is pocket friendly, simple, easy to design and implement, and has a high rate of success (Fig. 13.4) (Borrelli *et al.*, 2018). Using CRISPR-Cas9, smart plants can be engineered with improved plant immunity systems and heavy metal remediation traits (Ghosh and Dey, 2022).

The CRISPR-Cas9 gene editing tool provides opportunities for engineering smart plants with enhanced phytoremediation ability because it can target and modify genes that are responsible for the uptake, translocation, and sequestration of metalloids in plant tissues. To achieve this, guide RNAs (gRNA) are designed to target genes that encode chelators, metal/metalloid transporters, and sequestration proteins. Furthermore, it can be used for the overexpression of positive regulators and knockout of negative regulators, for example knockout of metal transporter (*OsNRAMP5*) in rice root cells, resulting in reduction of Cd

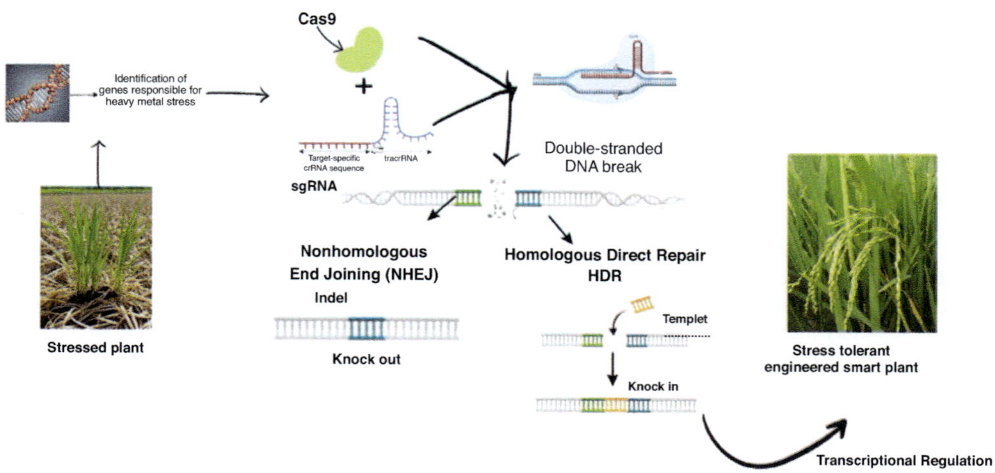

Fig. 13.4. Gene editing in plants by CRISPR-Cas9.

accumulation without affecting rice yield (Basharat *et al.*, 2018; Sathee *et al.*, 2022).

CRISPR/Cas9 can be used to engineer smart plants with enhanced tolerance to heavy metal/metalloid toxicity by focusing on regulatory, cis-regulatory, and structural genes that are associated with plant stress responses, i.e., metal chelation, antioxidant defense, and stress response pathways (Zafar *et al.*, 2020; Jain, 2015). In *Arabidopsis*, overexpression of *AsPCS1* and *YCF1* genes increases accumulation of Cd and As and improves plant stress tolerance (Guo *et al.*, 2012).

13.5 Conclusion

Genetic engineering has much potential for improving plant science and solving environmental issues. The many transformation methods covered in this chapter – biolistic, PEG mediated, UV laser microbeam, and *Agrobacterium* mediated – all have their own pros and cons. Through the creative application of the CRISPR-Cas9 system, these techniques have made it possible to develop transgenic plants with improved stress tolerance and phytoremediation capabilities. It is critical to keep improving these methods as research develops to increase productivity, lower expenses, and minimize dangers. The effective application of genetically modified plants for environmental remediation emphasizes how these technologies may support environmentally friendly farming practices and environmental preservation. To fully realize the benefits of this transformational science, future efforts should concentrate on overcoming regulatory and ethical issues, promoting multidisciplinary collaboration, and broadening the breadth of applications for genetic engineering.

References

Abou Seeda, M., Abou El-Nour, E.Z., Abdallah, M. and El-Bassiouny, H. (2022) Impacts of metal, metalloid and their effects in plant physiology: A review. *Middle East Journal of Agricultural Research* 11(3), 838–931.

Adams, J.P., Adeli, A., Hsu, C.-Y., Harkess, R.L., Page, G.P. *et al.* (2011) Poplar maintains zinc homeostasis with heavy metal genes HMA4 and PCS1. *Journal of Experimental Botany* 62(11), 3737–3752.

Agrawal, P., Verma, D. and Daniell, H. (2011) Expression of *Trichoderma reesei* β-mannanase in tobacco chloroplasts and its utilization in lignocellulosic woody biomass hydrolysis. *PLoS One* 6(12), e29302.

Alaraidh, I.A., Alsahli, A.A. and Abdel Razik, E.S. (2018) Alteration of antioxidant gene expression in response to heavy metal stress in *Trigonella foenum-graecum* L. *South African Journal of Botany* 115, 90–93.

Barabasz, A., Klimecka, M., Kendziorek, M., Weremczuk, A. and Ruszczyńska, A. (2016) The ratio of Zn to Cd supply as a determinant of metal-homeostasis gene expression in tobacco and its modulation by overexpressing the metal exporter AtHMA4. *Journal of Experimental Botany* 67, 6201–6214.

Basharat, Z., Novo, L.A. and Yasmin, A. (2018) Genome editing weds CRISPR: What is in it for phytoremediation? *Plants* 7(3), 51.

Bharwana, S.A., Ali, S., Farooq, M.A., Iqbal, N., Abbas, F. *et al.* (2013) Alleviation of lead toxicity by silicon is related to elevated photosynthesis, antioxidant enzymes suppressed lead uptake and oxidative stress in cotton. *Journal of Bioremediation and Biodegradation* 4(4), 10–4172.

Bhatia, S., Bera, T., Dahiya, R., Bera, T., Bhatia, S. *et al.* (2015) Classical and nonclassical techniques for secondary metabolite production in plant cell culture. In: Bhatia, S., Sharma, K., Dahiya, R. and Bera, T. (eds) *Modern Applications of Plant Biotechnology in Pharmaceutical Sciences*. Elsevier, St Louis, pp. 231–291.

Bienert, G.P., Thorsen, M., Schüssler, M.D., Nilsson, H.R., Wagner, A. *et al.* (2008) A subgroup of plant aquaporins facilitate the bi-directional diffusion of As $(OH)_3$ and Sb $(OH)_3$ across membranes. *BMC Biology* 6, 1–15.

Bobik, K. and Burch-Smith, T.M. (2015) Chloroplast signaling within, between and beyond cells. *Frontiers in Plant Science* 6, 781.

Borrelli, V.M., Brambilla, V., Rogowsky, P., Marocco, A. and Lanubile, A. (2018) The enhancement of plant disease resistance using CRISPR/Cas9 technology. *Frontiers in Plant Science* 9, 1245.

Brunetti, P., Zanella, L., Proia, A., Paolis, A., Falasca, G. *et al.* (2011) Cadmium tolerance and phytochelatin content of *Arabidopsis* seedlings over-expressing the phytochelatin synthase gene AtPCS1. *Journal of Experimental Botany* 62(15), 5509–5519.

Carter, M. and Shieh, J. (eds) (2015) Cell culture techniques. In: *Guide to Research Techniques in Neuroscience*. Academic Press, New York, pp. 295–310.

Chang, J.-D., Huang, S., Konishi, N., Wang, P., Chen, J. *et al.* (2020) Overexpression of the manganese/cadmium transporter OsNRAMP5 reduces cadmium accumulation in rice grain. *Journal of Experimental Botany* 71(18), 5705–5715.

Chen, H., Zhang, C., Guo, H., Hu, Y., He, Y. *et al.* (2018) Overexpression of a *Miscanthus sacchariflorus* yellow stripe-like transporter MsYSL1 enhances resistance of *Arabidopsis* to cadmium by mediating metal ion reallocation. *Plant Growth Regulation* 85(1), 101–111.

Chen, Y., Zhi, J., Zhang, H., Li, J., Zhao, Q. *et al.* (2017) Transcriptome analysis of *Phytolacca americana* L. in response to cadmium stress. *PLoS One* 12(9), e0184681.

Couselo, J.L., Navarro-Avñó, J. and Ballester, A. (2010) Expression of the phytochelatin synthase TaPCS1 in transgenic aspen, insight into the problems and qualities in phytoremediation of Pb. *International Journal of Phytoremediation* 12(4), 358–370.

Dahuja, A., Kumar, R.R., Sakhare, A., Watts, A., Singh, B. *et al.* (2021) Role of ATP-binding cassette transporters in maintaining plant homeostasis under abiotic and biotic stresses. *Physiologia Plantarum* 171(4), 785–801.

de Mello-Farias, P.C., Chaves, A.L.S. and Lencina, C.L. (2011) Transgenic plants for enhanced phytoremediation–physiological studies. In: Alvarez, M. (ed.) *Genetic Transformation*. IntechOpen, London, pp. 305–328.

Dehghan, S., Sadeghi, M., Pöppel, A., Fischer, R., Lakes-Harlan, R. *et al.* (2014) Differential inductions of phenylalanine ammonia-lyase and chalcone synthase during wounding, salicylic acid treatment, and salinity stress in safflower, *Carthamus tinctorius*. *Bioscience Reports* 34(3), e00114.

Duan, G., Kamiya, T., Ishikawa, S., Arao, T. and Fujiwara, T. (2012) Expressing ScACR3 in rice enhanced arsenite efflux and reduced arsenic accumulation in rice grains. *Plant and Cell Physiology* 53(1), 154–163.

Dubey, A.K., Kumar, A., Kumar, N., Kumar, S., Meenakshi, L. *et al.* (2021) Over-expression of chickpea metallothionein 1 gene confers tolerance against major toxic heavy metal stress in *Arabidopsis*. *Physiology and Molecular Biology of Plants* 27(12), 2665–2678.

Fasani, E., Manara, A., Martini, F., Furini, A. and DalCorso, G. (2018) The potential of genetic engineering of plants for the remediation of soils contaminated with heavy metals. *Plant, Cell & Environment* 41(5), 1201–1232.

Feng, S., Tan, J., Zhang, Y., Liang, S., Xiang, S. *et al.* (2017) Isolation and characterization of a novel cadmium-regulated yellow stripe-like transporter (SnYSL3) in *Solanum nigrum*. *Plant Cell Reports* 36(2), 281–296.

Fulekar, M., Singh, A. and Bhaduri, A.M. (2009) Genetic engineering strategies for enhancing phytoremediation of heavy metals. *African Journal of Biotechnology* 8(4).

Gasic, K. and Korban, S.S. (2007) Expression of Arabidopsis phytochelatin synthase in Indian mustard (*Brassica juncea*) plants enhances tolerance for Cd and Zn. *Planta* 225(5), 1277–1285.

Ghosh, S. and Dey, G. (2022) Biotic and abiotic stress tolerance through CRISPR-Cas mediated genome editing. *Journal of Plant Biochemistry and Biotechnology* 31(2), 227–238.

Giraldo, J.P., Landry, M.P., Faltermeier, S.M., McNicholas, T.P., Iverson, N.M. *et al.* (2014) Plant nanobionics approach to augment photosynthesis and biochemical sensing. *Nature Materials* 13(4), 400–408.

Gokul, G.S., Pradeepkumar, M., Prarthana, S.T.D. and Christi, V.I. (2024) Phytoplastics based on (PHB) producing transgenic plants as an alternative to plastics: A review. In: *ICSETPSD 2023: Proceedings of the First International Conference on Science, Engineering and Technology Practices for Sustainable Development, ICSETPSD 2023*, European Alliance for Innovation, Coimbatore, Tamilnadu, India, pp. 17–18.

Gunarathne, V., Mayakaduwa, S., Ashiq, A., Weerakoon, S.R., Biswas, J.K. *et al.* (2019) Transgenic plants: Benefits, applications, and potential risks in phytoremediation. In: Prasad, M. (ed.) *Transgenic Plant Technology for Remediation of Toxic Metals and Metalloids*. Elsevier, St Louis, pp. 89–102.

Guo, J., Xu, W. and Ma, M. (2012) The assembly of metals chelation by thiols and vacuolar compartmentalization conferred increased tolerance to and accumulation of cadmium and arsenic in transgenic *Arabidopsis thaliana*. *Journal of Hazardous Materials* 199, 309–313.

Gupta, M. and Gupta, S. (2017) An overview of selenium uptake, metabolism, and toxicity in plants. *Frontiers in Plant Science* 7, 2074.

Hassinen, V.H., Tervahauta, A.I., Schat, H. and Kärenlampi, S.O. (2011) Plant metallothioneins – metal chelators with ROS scavenging activity? *Plant Biology* 13(2), 225–232.

Hassinen, V.H., Tuomainen, M., Peräniemi, S., Schat, H., Kärenlampi, S.O. *et al.* (2009) Metallothioneins 2 and 3 contribute to the metal-adapted phenotype but are not directly linked to Zn accumulation in the metal hyperaccumulator, *Thlaspi caerulescens*. *Journal of Experimental Botany* 60(1), 187–196.

Hegelund, J.N., Schiller, M., Kichey, T., Hansen, T.H., Pedas, P. *et al.* (2012) Barley metallothioneins: MT3 and MT4 are localized in the grain aleurone layer and show differential zinc binding. *Plant Physiology* 159(3), 1125–1137.

Hell, R. and Wirtz, M. (2011) Molecular biology, biochemistry and cellular physiology of cysteine metabolism in *Arabidopsis thaliana*. *Arabidopsis Book* 9, e0154.

Hong-Bo, S., Li-Ye, C., Cheng-Jiang, R., Hua, L., Dong-Gang, G. *et al.* (2010) Understanding molecular mechanisms for improving phytoremediation of heavy metal-contaminated soils. *Critical Reviews in Biotechnology* 30(1), 23–30.

Hossain, M.A., Piyatida, P., Silva, J.A.T. and Fujita, M. (2012) Molecular mechanism of heavy metal toxicity and tolerance in plants: Central role of glutathione in detoxification of reactive oxygen species and methylglyoxal and in heavy metal chelation. *Journal of Botany* 2012(1), 1–37.

Hussein, H.S., Ruiz, O.N., Terry, N. and Daniell, H. (2007) Phytoremediation of mercury and organomercurials in chloroplast transgenic plants: Enhanced root uptake, translocation to shoots, and volatilization. *Environmental Science & Technology* 41(24), 8439–8446.

Ihien Katche, E. and Mason, A.S. (2023) Resynthesized rapeseed (*Brassica napus*): Breeding and genomics. *Critical Reviews in Plant Sciences* 42(2), 65–92.

Irshad, A., Sarwar, N., Zahid, M. and Irshad, I. (2020) Interaction of carbon nanotubes with rhizosphere microbial communities. In: Abd-Elsalam, K. (ed.) *Carbon Nanomaterials for Agri-Food and Environmental Applications*. Elsevier, St Louis, pp. 487–504.

Jain, M. (2015) Function genomics of abiotic stress tolerance in plants: A CRISPR approach. *Frontiers in Plant Science* 6, 375.

Jensen, P.E. and Leister, D.J.F. (2014) Chloroplast evolution, structure and functions. *F1000 Prime Reports* 6, 40.

Jinturkar, K.A., Rathi, M.N. and Misra, A. (2011) Gene delivery using physical methods. In: Misra, A. (ed.) *Challenges in Delivery of Therapeutic Genomics and Proteomics*. Elsevier, St Louis, pp. 83–126.

Kaushal, C., Abdin, M.Z. and Kumar, S. (2020) Chloroplast genome transformation of medicinal plant *Artemisia annua*. *Plant Biotechnology Journal* 18(11), 2155–2157.

Kotrba, P., Najmanova, J., Macek, T., Ruml, T. and Mackova, M. (2009) Genetically modified plants in phytoremediation of heavy metal and metalloid soil and sediment pollution. *Biotechnology Advances* 27(6), 799–810.

Koźmińska, A., Wiszniewska, A., Hanus-Fajerska, E. and Muszyńska, E. (2018) Recent strategies of increasing metal tolerance and phytoremediation potential using genetic transformation of plants. *Plant Biotechnology Reports* 12(1), 1–14.

Krämer, U. and Chardonnens, A. (2001) The use of transgenic plants in the bioremediation of soils contaminated with trace elements. *Applied Microbiology and Biotechnology* 55(6), 661–672.

Kumar, A.U. and Ling, A.P.K. (2021) Gene introduction approaches in chloroplast transformation and its applications. *Journal of Genetic Engineering & Biotechnology* 19, 1–10.

Kumar, A., Dubey, A.K., Kumar, V., Ansari, M.A., Narayan, S. *et al.* (2020) Over-expression of chickpea glutaredoxin (CaGRx) provides tolerance to heavy metals by reducing metal accumulation and improved physiological and antioxidant defence system. *Ecotoxicology and Environmental Safety* 192, 110252.

Kumar, S., Yadav, A., Kumar, A., Verma, R., Lal, S. *et al.* (2021) Plant metallothioneins as regulators of environmental stress responses. *International Journal of Plant and Environment* 7(1), 27–38.

Kwak, S.-Y., Lew, T.T.S., Sweeney, C.J., Koman, V.B., Wong, M.H. *et al.* (2019) Chloroplast-selective gene delivery and expression *in planta* using chitosan-complexed single-walled carbon nanotube carriers. *Nature Nanotechnology* 14(5), 447–455.

Lee, J., Shim, D., Song, W.-Y., Hwang, I. and Lee, Y. (2004) *Arabidopsis* metallothioneins 2a and 3 enhance resistance to cadmium when expressed in *Vicia faba* guard cells. *Plant Molecular Biology* 54(6), 805–815.

Lee, S., Moon, J.S., Ko, T.-S., Petros, D., Goldsbrough, P.B. *et al.* (2003) Overexpression of *Arabidopsis* phytochelatin synthase paradoxically leads to hypersensitivity to cadmium stress. *Plant Physiology* 131(2), 656–663.

Li, H., Cong, Y., Lin, J. and Chang, Y. (2015) Enhanced tolerance and accumulation of heavy metal ions by engineered *Escherichia coli* expressing *Pyrus calleryana* phytochelatin synthase. *Journal of Basic Microbiology* 55(3), 398–405.

Li, H., Luo, H., Li, D., Hu, T. and Fu, J. (2012) Antioxidant enzyme activity and gene expression in response to lead stress in perennial ryegrass. *Journal of the American Society for Horticultural Science* 137(2), 80–85.

Li, M., Hu, D., Liu, X., Zhang, R., Liu, H. *et al.* (2024) The OsZIP2 transporter is involved in root-to-shoot translocation and intervascular transfer of cadmium in rice. *Plant, Cell & Environment* 47(10), 3865–3881.

Li, Y., Dhankher, O.P., Carreira, L., Lee, D., Chen, A. *et al.* (2004) Overexpression of phytochelatin synthase in arabidopsis leads to enhanced arsenic tolerance and cadmium hypersensitivity. *Plant & Cell Physiology* 45(12), 1787–1797.

Liang, Y., Nikolic, M., Bélanger, R., Gong, H. and Song, A. (2015) Silicon uptake and transport in plants: Physiological and molecular aspects. In: Liang, Y., Nikolic, M., Belanger, R., Gong, H. and Song, A. (eds) *Silicon in Agriculture: From Theory to Practice*. Springer, Cham, pp. 69–82.

Liang, Y., Su, Y., Li, L., Huang, X., Panhwar, F.H. *et al.* (2019) Quick selenium accumulation in the selenium-rich rice and its physiological responses in changing selenium environments. *BMC Plant Biology* 19(1), 11.

Lim, S.-H., Song, J.-H., Kim, D.-H., Kim, J.K., Lee, J.-Y. *et al.* (2016) Activation of anthocyanin biosynthesis by expression of the radish R2R3-MYB transcription factor gene RsMYB1. *Plant Cell Reports* 35(3), 641–653.

Lira-Morales, J.D., Varela-Bojórquez, N., Montoya-Rojo, M.B. and Sañudo-Barajas, J.A. (2019) The role of ZIP proteins in zinc assimilation and distribution in plants: Current challenges. *Czech Journal of Genetics and Plant Breeding* 55(2), 45–54.

Liu, G.-Y., Zhang, Y.-X. and Chai, T.-Y. (2011) Phytochelatin synthase of *Thlaspi caerulescens* enhanced tolerance and accumulation of heavy metals when expressed in yeast and tobacco. *Plant Cell Reports* 30(6), 1067–1076.

Liu, X.S., Feng, S.J., Zhang, B.Q., Wang, M.Q., Cao, H.W. *et al.* (2019) OsZIP1 functions as a metal efflux transporter limiting excess zinc, copper and cadmium accumulation in rice. *BMC Plant Biology* 19, 1–16.

Liu, Y., Li, F., Gao, L., Tu, Z., Zhou, F. *et al.* (2023) Advancing approach and toolbox in optimization of chloroplast genetic transformation technology. *Journal of Integrative Agriculture* 22(7), 1951–1966.

Liu, Z., Gu, C., Chen, F., Yang, D., Wu, K. *et al.* (2012) Heterologous expression of a *Nelumbo nucifera* phytochelatin synthase gene enhances cadmium tolerance in *Arabidopsis thaliana*. *Applied Biochemistry and Biotechnology* 166(3), 722–734.

Mani, A. and Sankaranarayanan, K. (2022) Natural resistance-associated macrophage proteins (NRAMPs): Functional significance of metal transport in plants. In: Kumar, K. and Srivastava, S. (eds) *Plant Metal and Metalloid Transporters*. Springer, Cham, pp. 91–107.

Matsumoto, T.K. and Gonsalves, D. (2012) Biolistic and other non-agrobacterium technologies of plant transformation. In: Altman, A. and Hasegawa, P. (eds) *Plant Biotechnology and Agriculture*. Elsevier, New York, pp. 117–129.

Meagher, R.B., Rugh, C.L., Kandasamy, M.K., Gragson, G. and Wang, N.J. (2020) Engineered phytoremediation of mercury pollution in soil and water using bacterial genes. In: Terry, N. and Banuelos, G. (eds) *Phytoremediation of Contaminated Soil and Water*. CRC Press, Boca Raton, pp. 201–219.

Mohammed, K.F.A., Kaul, T., Agrawal, P.K., Thangaraj, A., Kaul, R. *et al.* (2023) Function identification and characterization of *Oryza sativa* ZRT and IRT-like proteins computationally for nutrition and biofortification in rice. *Journal of Biomolecular Structure & Dynamics* 41(15), 7490–7510.

Morkunas, I., Woźniak, A., Mai, V.C., Rucińska-Sobkowiak, R. and Jeandet, P. (2018) The role of heavy metals in plant response to biotic stress. *Molecules* 23(9), 2320.

Naing, A.H., Park, K.I., Ai, T.N., Chung, M.Y., Han, J.S. *et al.* (2017) Overexpression of snapdragon *Delila* (*Del*) gene in tobacco enhances anthocyanin accumulation and abiotic stress tolerance. *BMC Plant Biology* 17(1), 65.

Nakamura, M., Ochiai, T., Noji, M., Ogura, Y. and Suzuki, K.T. (2014) An improved tolerance to cadmium by overexpression of two genes for cysteine synthesis in tobacco. *Plant Biotechnology* 31(2), 141–147.

Newkirk, G.M., Allende, P., Jinkerson, R.E. and Giraldo, J.P. (2021) Nanotechnology approaches for chloroplast biotechnology advancements. *Frontiers in Plant Science* 12, 691295.

Nikolić, M. and Tomašević, V. (2021) Implication of the plant species belonging to the Brassicaceae family in the metabolization of heavy metal pollutants in urban settings. *Polish Journal of Environmental Studies* 30(1), 523–534.

Noor, I., Sohail, H., Sun, J., Nawaz, M.A., Li, G. *et al.* (2022) Heavy metal and metalloid toxicity in horticultural plants: Tolerance mechanism and remediation strategies. *Chemosphere* 303, 135196.

Occhialini, A., Lin, M.T., Andralojc, P.J., Hanson, M.R. and Parry, M.A.J. (2016) Transgenic tobacco plants with improved cyanobacterial Rubisco expression but no extra assembly factors grow at near wild-type rates if provided with elevated CO_2. *Plant Journal* 85(1), 148–160.

Ojha, A., Jaiswal, S., Thakur, P. and Mishra, S.K. (2023) Bioremediation techniques for heavy metal and metalloid removal from polluted lands: A review. *International Journal of Environmental Science and Technology* 20(9), 10591–10612.

Ow, D.W. (1996) Heavy metal tolerance genes: Prospective tools for bioremediation. *Resources, Conservation and Recycling* 18(1–4), 135–149.

Ozyigit, I.I., Can, H. and Dogan, I. (2021) Phytoremediation using genetically engineered plants to remove metals: A review. *Environmental Chemistry Letters* 19(1), 669–698.

Pal, S., Patel, N., Malik, A., Sharma, A., Pal, U. *et al.* (2023) Eco-friendly treatment of wastewater and its impact on soil and vegetables using flood and micro-irrigation. *Agricultural Water Management* 275, 108025.

Pomponi, M., Censi, V., Girolamo, V., Paolis, A., Toppi, L.S. *et al.* (2006) Overexpression of *Arabidopsis* phytochelatin synthase in tobacco plants enhances Cd(2+) tolerance and accumulation but not translocation to the shoot. *Planta* 223(2), 180–190.

Pratiwi, R.A. and Surya, M.I. (2020) Agrobacterium-mediated transformation. In: To, K. (ed.) *Genetic Transformation in Crops*. IntechOpen, London, pp. 1–14.

Pratush, A., Kumar, A. and Hu, Z. (2018) Adverse effect of heavy metals (As, Pb, Hg, and Cr) on health and their bioremediation strategies: A review. *International Microbiology* 21(3), 97–106.

Pulford, I.D. and Watson, C. (2003) Phytoremediation of heavy metal-contaminated land by trees—A review. *Environment International* 29(4), 529–540.

Purton, S. (2007) Tools and techniques for chloroplast transformation of *Chlamydomonas*. *Advances in Experimental Medicine and Biology* 616, 34–45.

Raffa, C.M., Chiampo, F. and Shanthakumar, S. (2021) Remediation of metal/metalloid-polluted soils: A short review. *Applied Sciences* 11(9), 4134.

Rajendran, S., Priya, T.A.K., Khoo, K.S., Hoang, T.K.A., Ng, H.-S. *et al.* (2022) A critical review on various remediation approaches for heavy metal contaminants removal from contaminated soils. *Chemosphere* 287, 132369.

Raza, A., Habib, M., Charagh, S. and Kakavand, S.N. (2021) Genetic engineering of plants to tolerate toxic metals and metalloids. In: Hasanuzzaman, M. and Prasad, M. (eds) *Handbook of Bioremediation*. Elsevier, New York, pp. 411–436.

Razzaq, R. (2017) Phytoremediation: An environmental friendly technique – A review. *Journal of Environmental Analytical Chemistry* 4(2).

Reisinger, S., Schiavon, M., Terry, N. and Pilon-Smits, E.A. (2008) Heavy metal tolerance and accumulation in Indian mustard (*Brassica juncea* L.) expressing bacterial γ-glutamylcysteine synthetase or glutathione synthetase. *International Journal of Phytoremediation* 10(5), 440–454.

Řepková, J. (2010) Potential of chloroplast genome in plant breeding. *Czech Journal of Genetics and Plant Breeding* 46(3), 103–113.

Rivera, A.L., Gómez-Lim, M., Fernández, F. and Loske, A.M. (2012) Physical methods for genetic plant transformation. *Physics of Life Reviews* 9(3), 308–345.

Riyazuddin, R., Nisha, N., Ejaz, B., Khan, M.I.R., Kumar, M. *et al.* (2021) A comprehensive review on the heavy metal toxicity and sequestration in plants. *Biomolecules* 12(1), 43.

Rolland, N., Curien, G., Finazzi, G., Kuntz, M., Maréchal, E. *et al.* (2012) The biosynthetic capacities of the plastids and integration between cytoplasmic and chloroplast processes. *Annual Review of Genetics* 46(1), 233–264.

Ruiz, O.N., Alvarez, D., Torres, C., Roman, L. and Daniell, H. (2011) Metallothionein expression in chloroplasts enhances mercury accumulation and phytoremediation capability. *Plant Biotechnology Journal* 9(5), 609–617.

Saba, K., Ijaz, F., Malik, M.S., Batool, N., Lössl, A.G. *et al.* (2023) Targeting chloroplasts for plant molecular farming. In: Kole, C., Chaurasia, A., Hefferon, K. and Panigrahi, J. (eds) *Tools & Techniques of Plant Molecular Farming*. Springer, Dordrecht, pp. 251–271.

Sathee, L., Jagadhesan, B., Pandesha, P.H., Barman, D. and Adavi, B. (2022) Genome editing targets for improving nutrient use efficiency and nutrient stress adaptation. *Frontiers in Genetics* 13, 900897.

Schiller, M., Hegelund, J.N., Pedas, P., Kichey, T., Laursen, K.H. *et al.* (2014) Barley metallothioneins differ in ontogenetic pattern and response to metals. *Plant, Cell & Environment* 37(2), 353–367.

Sekhar, K., Priyanka, B., Reddy, V.D. and Rao, K.V. (2011) Metallothionein 1 (CcMT1) of pigeonpea (*Cajanus cajan*, L.) confers enhanced tolerance to copper and cadmium in *Escherichia coli* and *Arabidopsis thaliana*. *Environmental and Experimental Botany* 72(2), 131–139.

Shao, J.F., Fujii-Kashino, M., Yamaji, N., Fukuoka, S., Shen, R.F. *et al.* (2017) Isolation and characterization of a rice line with high Cd accumulation for potential use in phytoremediation. *Plant and Soil* 410, 357–368.

Sharma, P., Sirohi, R., Tong, Y.W., Kim, S.H. and Pandey, A. (2021) Metal and metal(loids) removal efficiency using genetically engineered microbes: Applications and challenges. *Journal of Hazardous Materials* 416, 125855.

Shri, M., Singh, P.K., Kidwai, M., Gautam, N., Dubey, S. *et al.* (2019) Recent advances in arsenic metabolism in plants: Current status, challenges and highlighted biotechnological intervention to reduce grain arsenic in rice. *Metallomics* 11(3), 519–532.

Shukla, D., Kesari, R., Tiwari, M., Dwivedi, S., Tripathi, R.D. *et al.* (2013) Expression of *Ceratophyllum demersum* phytochelatin synthase, CdPCS1, in *Escherichia coli* and *Arabidopsis* enhances heavy metal(loid)s accumulation. *Protoplasma* 250(6), 1263–1272.

Singh, S., Parihar, P., Singh, R., Singh, V.P. and Prasad, S.M. (2015) Heavy metal tolerance in plants: Role of transcriptomics, proteomics, metabolomics, and ionomics. *Frontiers in Plant Science* 6, 1143.

Sunitha, M.S., Prashant, S., Kumar, S.A., Rao, S.R., Narasu, M.L. *et al.* (2013) Cellular and molecular mechanisms of heavy metal tolerance in plants: A brief overview of transgenic plants overexpressing phytochelatin synthase and metallothionein genes. *Plant Cell Biotechnology and Molecular Biology* 14(2), 33–48.

Tanaka, N., Uraguchi, S., Saito, A., Kajikawa, M., Kasai, K. *et al.* (2013) Roles of pollen-specific boron efflux transporter, OsBOR4, in the rice fertilization process. *Plant and Cell Physiology* 54(12), 2011–2019.

Tang, Z., Wang, H.-Q., Chen, J., Chang, J.-D. and Zhao, F.-J. (2023) Molecular mechanisms underlying the toxicity and detoxification of trace metals and metalloids in plants. *Journal of Integrative Plant Biology* 65(2), 570–593.

Tiwari, S. and Lata, C. (2022) Cation diffusion facilitator family of transporters in plants. In: *Cation Transporters in Plants*. Elsevier, New York, pp. 375–384.

Tong, Y.-P., Kneer, R. and Zhu, Y.-G. (2004) Vacuolar compartmentalization: A second-generation approach to engineering plants for phytoremediation. *Trends in Plant Science* 9(1), 7–9.

Uraguchi, S., Kato, Y., Hanaoka, H., Miwa, K. and Fujiwara, T. (2014) Generation of boron-deficiency-tolerant tomato by overexpressing an *Arabidopsis thaliana* borate transporter *AtBOR1*. *Frontiers in Plant Science* 5, 125.

Valls, M., Atrian, S., Lorenzo, V. and Fernández, L.A. (2000) Engineering a mouse metallothionein on the cell surface of Ralstonia eutropha CH34 for immobilization of heavy metals in soil. *Nature Biotechnology* 18(6), 661–665.

Verma, A. (ed.) (2019) Chloroplast genetic engineering: Concept and industrial applications. In: *Industrial Biotechnology*. De Gruyter, Berlin, pp. 173–204.

Verret, F., Gravot, A., Auroy, P., Leonhardt, N., David, P. *et al.* (2004) Overexpression of AtHMA4 enhances root-to-shoot translocation of zinc and cadmium and plant metal tolerance. *FEBS Letters* 576(3), 306–312.

Wang, F., Wang, Z. and Zhu, C. (2012) Heteroexpression of the wheat phytochelatin synthase gene (TaPCS1) in rice enhances cadmium sensitivity. *Acta Biochimica et Biophysica Sinica* 44(10), 886–893.

Wang, H.-H., Yin, W.-B. and Hu, Z.-M. (2009) Advances in chloroplast engineering. *Journal of Genetics and Genomics* 36(7), 387–398.

Wang, J.-W., Li, Y., Zhang, Y.-X. and Chai, T.-Y. (2013) Molecular cloning and characterization of a *Brassica juncea* yellow stripe-like gene, BjYSL7, whose overexpression increases heavy metal tolerance of tobacco. *Plant Cell Reports* 32(5), 651–662.

Wang, P., Zhang, W., Mao, C., Xu, G. and Zhao, F.-J. (2016) The role of OsPT8 in arsenate uptake and varietal difference in arsenate tolerance in rice. *Journal of Experimental Botany* 67(21), 6051–6059.

Wang, Y., Zhang, Y., Dong, Y., Li, D., Shi, S. *et al.* (2022) A highly efficient mesophyll protoplast isolation and PEG-mediated transient expression system in eggplant. *Scientia Horticulturae* 304, 111303.

Watanabe, M., Shinmachi, F., Noguchi, A. and Hasegawa, I. (2005) Introduction of yeast metallothionein gene (*CUP1*) into plant and evaluation of heavy metal tolerance of transgenic plant at the callus stage. *Soil Science and Plant Nutrition* 51(1), 129–133.

Wen, X.-P., Pang, X.-M., Matsuda, N., Kita, M., Inoue, H. *et al.* (2008) Over-expression of the apple spermidine synthase gene in pear confers multiple abiotic stress tolerance by altering polyamine titers. *Transgenic Research* 17(2), 251–263.

Wen, X.-P., Ban, Y., Inoue, H., Matsuda, N. and Moriguchi, T. (2010) Spermidine levels are implicated in heavy metal tolerance in a spermidine synthase overexpressing transgenic European pear by exerting antioxidant activities. *Transgenic Research* 19(1), 91–103.

Wojas, S., Clemens, S., Skłodowska, A. and Maria Antosiewicz, D. (2010) Arsenic response of AtPCS1- and CePCS-expressing plants – effects of external as (V) concentration on As-accumulation pattern and NPT metabolism. *Journal of Plant Physiology* 167(3), 169–175.

Wu, J., Liang, J., Björn, L.O., Li, J., Shu, W. *et al.* (2022) Phosphorus-arsenic interaction in the 'soil-plant-microbe' system and its influence on arsenic pollution. *Science of the Total Environment* 802, 149796.

Yadav, M., Singh, M., Nanda, K. and Singh, S.P. (2022) Genetically engineered plants for phytoremediation of heavy metals. In: Sharma, P., Pandey, A., Tong, Y. and Ngo, H. (eds) *Current Developments in Biotechnology and Bioengineering*. Elsevier, New York, pp. 223–239.

Yang, X.-E., Jin, X.-F., Feng, Y. and Islam, E. (2005) Molecular mechanisms and genetic basis of heavy metal tolerance/hyperaccumulation in plants. *Journal of Integrative Plant Biology* 47(9), 1025–1035.

Yang, Y., Xiong, J., Chen, R., Fu, G., Chen, T. *et al.* (2016) Excessive nitrate enhances cadmium (Cd) uptake by up-regulating the expression of OsIRT1 in rice (*Oryza sativa*). *Environmental and Experimental Botany* 122, 141–149.

Yang, Z., Yang, F., Liu, J.-L., Wu, H.-T., Yang, H. *et al.* (2022) Heavy metal transporters: Functional mechanisms, regulation, and application in phytoremediation. *Science of the Total Environment* 809, 151099.

Zafar, S.A., Zaidi, S.S.-E.-A., Gaba, Y., Singla-Pareek, S.L., Dhankher, O.P. *et al.* (2020) Engineering abiotic stress tolerance via CRISPR/ Cas-mediated genome editing. *Journal of Experimental Botany* 71(2), 470–479.

Zaid, A., Mohammad, F., Wani, S.H. and Siddique, K.M. (2019) Salicylic acid enhances nickel stress tolerance by up-regulating antioxidant defense and glyoxalase systems in mustard plants. *Ecotoxicology and Environmental Safety* 180, 575–587.

Zhang, J., Zhang, M., Tian, S., Lu, L., Shohag, M.J.I. *et al.* (2014) Metallothionein 2 (SaMT2) from *Sedum alfredii* Hance confers increased Cd tolerance and accumulation in yeast and tobacco. *PLoS One* 9(7), e102750.

Zhang, L., Gao, C., Chen, C., Zhang, W., Huang, X.-Y. *et al.* (2020) Overexpression of rice OsHMA3 in wheat greatly decreases cadmium accumulation in wheat grains. *Environmental Science & Technology* 54(16), 10100–10108.

Zhao, X.Q., Mitani, N., Yamaji, N., Shen, R.F. and Ma, J.F. (2010) Involvement of silicon influx transporter OsNIP2;1 in selenite uptake in rice. *Plant Physiology* 153(4), 1871–1877.

Zhi, J., Liu, X., Yin, P., Yang, R., Liu, J. *et al.* (2020) Overexpression of the metallothionein gene *PaMT3-1* from *Phytolacca americana* enhances plant tolerance to cadmium. *Plant Cell, Tissue and Organ Culture* 143(1), 211–218.

Zhou, Y., Chu, P., Chen, H., Li, Y., Liu, J. *et al.* (2012) Overexpression of *Nelumbo nucifera* metallothioneins 2a and 3 enhances seed germination vigor in *Arabidopsis*. *Planta* 235(3), 523–537.

Zienkiewicz, M., Krupnik, T., Drożak, A. and Kania, K. (2019) PEG-mediated, stable, nuclear and chloroplast transformation of *Cyanidioschizon merolae*. *Bio-Protocol* 9(17), e3355.

14 Oxidative Stress Generation and Antioxidant Defense Response in Plants During Abiotic Stress

Faamiya Shajar, Azha Ufaq Nabi, Aamina Manzoor, Seerat Saleem, Naveed Ul Mushtaq and Reiaz Ul Rehman*

Department of Bioresources, School of Biological Sciences, University of Kashmir, Srinagar, India

Abstract

Plants are constantly faced with a barrage of environmental insults, which result in an overproduction of reactive oxygen species (ROS). The ROS arise within distinct cellular compartments like chloroplasts and mitochondria. They exhibit high reactivity due to their ability to engage with different cellular molecules and metabolites, resulting in irreversible metabolic disturbances and eventual death. Despite their damaging effects, ROS also act as signaling molecules involved in cellular responses to stress. Plants have robust antioxidant mechanisms to neutralize excess ROS, including various enzymes along with nonenzymatic compounds. This chapter explores the complexities of ROS formation and their adverse effects on various cellular targets. It highlights how plants deploy sophisticated antioxidant strategies to counteract ROS production, showcasing their adaptability to diverse stress conditions.

Keywords: abiotic stresses, reactive oxygen species, superoxide dismutase, peroxidase, catalase, ascorbate peroxidase

14.1 Introduction

Plants face a multitude of stresses on a daily basis, encompassing both abiotic stressors like salinity, heavy metals (HM), drought, extreme heat or cold, waterlogging, mineral imbalances, and radiation, and biotic stresses stemming from damage inflicted by various living organisms including viruses, fungi, bacteria, insects, parasites, weeds, etc. (Pardo, 2010). When plants are exposed to these stressors, it induces oxidative stress leading to the excessive generation and buildup of reactive oxygen species (ROS).

Such harsh conditions cause a notable decrease in crop yield. Due to their reactive nature, ROS are capable of damaging plant organelles, metabolites, and molecules, hampering multiple metabolic processes and ultimately causing cell death. ROS include superoxide radicals (O2•−) and hydroxyl radicals (•OH), hydrogen peroxide (H_2O_2), singlet oxygen ($^1O_2$), etc.

While ROS have detrimental impacts, they also serve crucial roles as signaling molecules in a number of cellular activities, aiding plants in coping with environmental stresses. The dual role of ROS, whether as signaling agents or instigators

*Corresponding author: rreiazbiores@gmail.com

DOI: 10.1079/9781800627659.0014

of damage, hinges on the intricate balance between their formation and elimination (Sharma *et al.*, 2012). Effective elimination of excess ROS under stress calls for the concerted effort of non-enzymatic and enzymatic antioxidants within the tissues (Hasanuzzaman *et al.*, 2021). Enzymatic antioxidants include superoxide dismutase (SOD), catalase (CAT), ascorbate peroxidase (APX), glutathione peroxidase (GPX), peroxidase (POD), glutathione-S-transferase (GST), mono-dehydroascorbate reductase (MDHAR), and glutathione reductase (GR), while the nonenzymatic antioxidants include flavonoids, carotenoids, ascorbate (AsA), glutathione and osmolytes like proline, sugars, etc. The ability of plants to mount a robust antioxidant defense has been associated with enhanced resilience against diverse environmental pressures (Zaefyzadeh *et al.*, 2009; Chen *et al.*, 2012).

14.2 Reactive Oxygen Species

14.2.1 Generation of ROS

Reactive oxygen species constitute a category of reactive molecules that arise from molecular oxygen (O_2). Research findings suggest that nearly 1% of the O_2 utilized by plants is allocated to ROS production (Sharma *et al.*, 2012). Elevated ROS levels can harm biomolecules, whereas lower concentrations function as secondary messengers involved in intracellular signaling pathways, coordinating a multitude of responses within plant cells. ROS essentially are atoms or groups of atoms with at least one unpaired electron. This configuration is extremely volatile, prompting rapid reactions with other molecules and generating additional free radicals. This phenomenon occurs because electrons typically seek pairing to form stable two-electron bonds.

The initiation of ROS formation occurs through molecular reduction of O_2, giving rise to various types of ROS like O_2•−, H_2O_2, •OH, and $^1O_2$. These ROS are formed through diverse means and are found in various cellular structures including the apoplast, mitochondria, chloroplasts, endoplasmic reticulum (ER), and peroxisomes (Fig. 14.1) (Ashraf, 2009).

Organelles characterized by high oxidizing metabolic activities or accelerated electron flow, including mitochondria, chloroplasts, and peroxisomes, serve as significant sources of ROS in plants. Within chloroplasts, O_2 is continually generated during photosynthetic electron transport while simultaneously being reduced and assimilated (Tripathy and Oelmüller, 2012).

14.2.2 Chloroplasts

Reactive oxygen species within chloroplasts originate from multiple sites, with the electron transport chains (ETCs) in photosystem I (PSI) and photosystem II (PSII) representing the major sources (Chen *et al.*, 2012). Conditions that limit CO_2 fixation, such as drought, salinity, and temperature fluctuations, along with excessive light stress, exacerbate ROS production from these sources. Under normal circumstances, the flow of electrons from PS to nicotinamide adenine dinucleotide phosphate ($NADP^+$), which is subsequently converted to NADPH, facilitates entry into the C3 cycle where the ultimate electron acceptor, CO_2, is reduced. However, during periods of ETC overload caused by reduced NADP availability due to stress, electron leakage from ferredoxin to oxygen (O_2) occurs, resulting in its reduction to superoxide (O_2•−) (Sharma *et al.*, 2012), a process often referred to as the Mehler reaction. In addition, electron leakage to O_2 can arise from the [2Fe-2S] and [4Fe-4S] clusters within the ETC of PSI. Within PSII, the electron acceptor site of the ETC encompasses QA and QB. Electron discharge from this side to O_2 results in the formation of superoxide (O_2•−). The production of O_2•− through O_2 reduction represents a crucial stage. Upon formation, superoxide forms additional ROS with heightened reactivity. Superoxide can undergo protonation to form hydroperoxyl radicals (HO_2•) on the inner membrane surface, termed the 'lumen.' Alternatively, it can be enzymatically disproportionated by SOD or concurrently converted to H_2O_2 on the 'stromal' side. In the presence of Fe-S centers containing available Fe^{2+}, hydrogen peroxide may undergo the Fenton reaction, triggering the production of significantly more hazardous OH• radicals (Sachdev *et al.*, 2021).

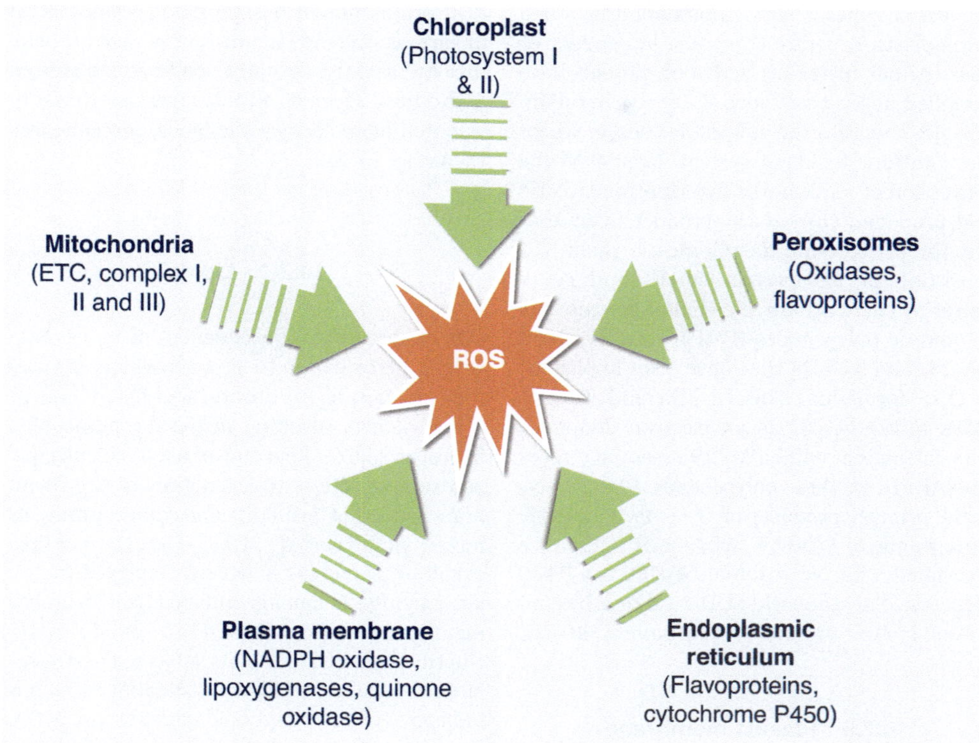

Fig. 14.1. Major sites of ROS generation in plants.

14.2.3 Mitochondria

Mitochondria represent an additional site where the formation of O_2•−, H_2O_2, and OH• can occur. Within mitochondria, the conversion of oxygen (O_2) to superoxide (O_2•−) takes place at the complex I (Arora *et al.*, 2002). In instances where NAD+-associated substrates for complex I are scarce, electron transfer may proceed from complex II to complex I, a phenomenon termed reverse electron flow. This mechanism is known to elevate ROS formation at complex I and is modulated by the breakdown of ATP (Turrens, 2003). Furthermore, in the ubiquinone-cytochrome site, or complex III, of the ETC, oxygen is converted to superoxide (O2•−). It is suggested that completely reduced ubiquinone transfers an electron to Cyt C1, while also generating a highly volatile ubisemiquinone radical. This radical promotes electron discharge to O_2, bringing about the production of O_2•− (Murphy, 2009). Under typical aerobic conditions in plants, the ETC and ATP synthesis processes operate in close coordination. However, various stress factors can hamper and modify their components, resulting in the overreduction of electron carriers and subsequent ROS production (Noctor *et al.*, 2007; Blokhina and Fagerstedt, 2010).

14.2.4 Peroxisomes

Another significant site of intracellular H_2O_2 generation is the peroxisome. This is primarily due to its oxidative metabolism (Nyathi and Baker, 2006). Various metabolic processes within peroxisomes contribute to the generation of H_2O_2, including glycolate oxidase activity, fatty acid β-oxidation, reactions involving flavin oxidases, and the dismutation of O_2•−. In the context of photorespiration, glycolate oxidase-driven oxidation of glycolate represents a significant source of H_2O_2 formation (Noctor *et al.*, 2002).

Peroxisomes, like mitochondria and chloroplasts, generate O_2•− as a by-product of their regular metabolic activities. Studies have identified at least two loci of O_2•− generation. One site is within the subcellular space, where the xanthine oxidase system facilitates the conversion of xanthine or hypoxanthine to uric acid, producing O_2•− as a by-product. In another site, the peroxisomal membranes, a small ETC consisting of flavoprotein NADH and cyt b operates. There are three essential peroxisomal membrane polypeptides (PMPs) roughly about 18, 29, and 32 kDa that have been implicated in O_2•− formation. The 18 kDa and 32 kDa PMPs utilize NADH as an electron donor for O_2•− formation, while PMP29 essentially relies on NADPH. Of these polypeptides, PMP18 acts as the primary producer of O_2•−. PMP32 likely corresponds to MDHAR, while PMP29 may be associated with peroxisomal NADPH:cyt P450 reductase. Subsequently, SOD catalyzes the conversion of O_2•− into H_2O_2 (Sharma *et al.*, 2012).

14.2.5 Plasma membrane

The NADPH oxidases present within the cell membranes are primary catalysts implicated in ROS production (Janků *et al.*, 2019).

O_2•− formation in plasmalemma has been linked with the activity of two main enzymes – NADPH oxidase and, if menadione is present, a quinone reductase (Heyno *et al.*, 2011). NADPH oxidase facilitates the movement of electrons from cytoplasmic NADPH to O_2, leading to the generation of O_2•−. Subsequently, O_2•− is either spontaneously disproportionate to H_2O_2 or through the activity of the enzyme SOD. NADPH oxidase is proposed to be central in the generation and excess buildup of ROS in plants during stressful conditions (Torres *et al.*, 2002; Apel and Hirt, 2004).

14.2.6 Biomolecules targeted by ROS

Oxidative stress manifests when the levels of ROS outstrip the protective potential of the cell. Strict control of both the synthesis and detoxification of ROS is crucial to keep a check on oxidative stress. The fine equilibrium between ROS formation and its elimination is jeopardized in various stressful circumstances such as metal toxicity, salinity, drought, pathogen invasion, and others. Elevated ROS is a serious threat to essential biomolecules like lipids, proteins, and DNA (Fig. 14.2).

14.2.7 Lipids

One of the major consequences of excess ROS is lipid peroxidation, a process which involves the oxidation of an unsaturated lipid chain by a free radical, resulting in the formation of a hydroperoxidized lipid and an alkyl radical. Lipid peroxidation alters the structure of the membrane, affecting its fluidity and compromising its integrity (Yadav *et al.*, 2019). The extent of lipid oxidation serves as a widely employed metric for gauging the damage inflicted by ROS on cell membranes during stress (Taïbi *et al.*, 2016; Gharibi *et al.*, 2016). Plants subjected to stresses often show elevated lipid peroxidation, which is indicative of increased lipid degradation. MDA, the endproduct of lipid oxidation, acts as a marker to gauge membrane damage (Sun *et al.*, 2010).

The entire process of lipid peroxidation is fundamentally divided into three phases: it begins with **initiation**, followed by **propagation**, and ends with **termination**. Initiation is marked by the activation of O_2. ROS such as O_2•− and •OH can interact with the methylene groups present in polyunsaturated fatty acids (PUFAs), leading to the creation of conjugated dienes, lipid peroxy radicals, and hydroperoxides (Anjum *et al.*, 2015). MDA formation represents the next step, known as propagation. Eventually, lipid radicals combine to form lipid dimers (Das and Roychoudhury, 2014). Lipid peroxidation contributes to membrane destabilization by affecting permeability, causing electrolyte leakage, enzyme and receptor deactivation, and promoting nucleic acid and protein oxidation (Sachdev *et al.*, 2021). Lipid radicals generated as a consequence of lipid peroxidation experience either enzymatic or nonenzymatic breakdown, producing reactive carbonyl species (RCS) (Slade and Radman, 2011). These RCS specifically interact with proteins via lipoxidation reactions, leading to impaired protein function.

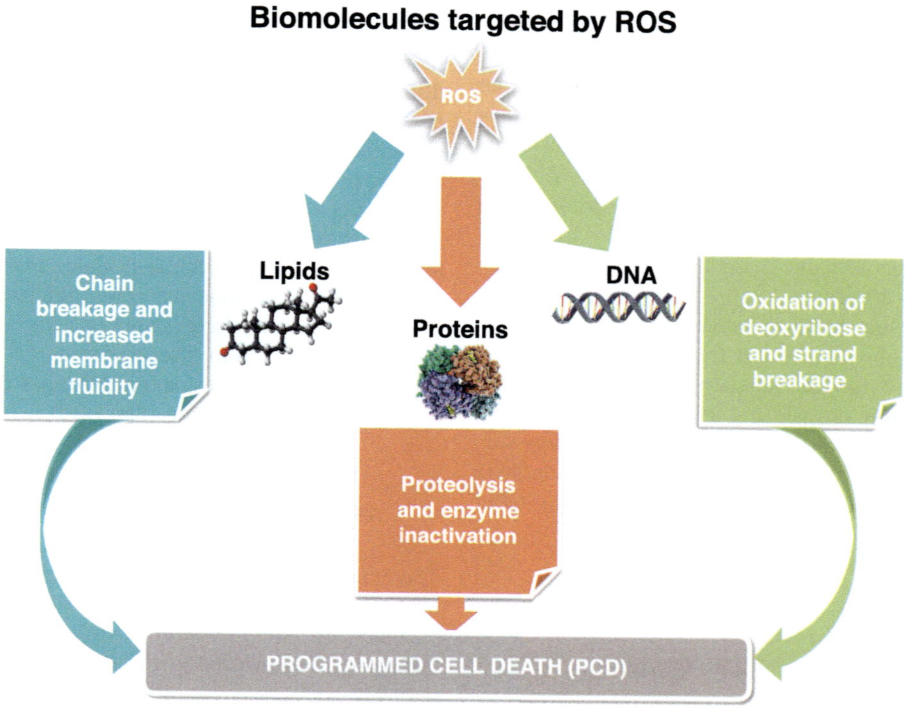

Fig. 14.2. The various biomolecules targeted by ROS.

14.2.8 Proteins

Reactive oxygen species have the ability to alter proteins through both direct and indirect mechanisms. Direct alteration entails modifying the protein activity via processes like nitrosylation, carbonylation, disulfide bond formation, and glutathionylation. Indirect alteration occurs when proteins conjugate with the by-products formed as a result of fatty acid (particularly PUFA) peroxidation (Yamauchi *et al.*, 2008). Overproduction of ROS results in modification of amino acids (AA), fragmentation of peptide chains, clustering of cross-linked reaction derivatives, changes in electrostatic charge, and heightened vulnerability of proteins to degradation by proteases. Tissues affected by oxidative stress typically exhibit elevated levels of carbonylated proteins, usual indicators of protein oxidation (Møller and Kristensen, 2004). Higher protein changes have been documented in plants subjected to stressors like drought (Sharma and Dubey, 2005), heavy metal (Maheshwari and Dubey, 2009), and salinity (Tanou *et al.*, 2009).

Different amino acids within a peptide exhibit different degrees of sensitivity to ROS. For instance, thiol groups and AA containing sulfur are especially prone to attack by ROS. ROS can remove an H-atom from cysteine residues, resulting in the creation of a thiyl radical that can subsequently undergo cross-linking with another thiyl radical to form a disulfide bond. HM such as cadmium, lead, and mercury reduce protein-bound thiol groups. Additionally, methionine can be oxidized to methionine sulfoxide, a redox biomarker. Moreover, ROS cause tyrosine to rapidly undergo cross-linking to produce bityrosine (Sharma *et al.*, 2012).

The oxidation of Fe-S centers by O_2•− results in irreversible enzyme deactivation. In these instances, the metal (Fe) interacts with the protein in a site-specific manner. This interaction triggers the Fenton reaction which leads to the generation of an •OH radical. This radical then promptly oxidizes specific AA residue situated close to the cation-binding region of the protein (Sharma *et al.*, 2012). Some have proposed that oxidation may render a protein more prone to ubiquitination, thereby rendering it a prime candidate for proteasomal breakdown (Cabiscol *et al.*, 2000). A study revealed that oxidized proteins degraded more effectively, leading to a notable rise in proteolytic activity due to Cd stress (Romero-Puertas *et al.*, 2002).

Numerous investigations have indicated that once a specific threshold of damage is crossed, further oxidation leads to the formation of highly cross-linked and aggregated reaction products. These products are ill-suited as substrates for degradation and have the potential to hinder proteases from further degrading other oxidized proteins (Sharma *et al.*, 2012).

14.2.9 DNA

Reactive oxygen species have the capacity to significantly harm DNA. This damage extends to nuclear, mitochondrial, and chloroplastic DNA. Because of the lack of histone proteins, chloroplastic and mitochondrial DNA show heightened sensitivity to ROS (Das and Roychoudhury, 2014).

Reactive oxygen species-induced DNA damage encompasses a spectrum of alterations, including deoxyribose sugar residue oxidation, strand breaks, nucleotide modifications, and DNA–protein cross-links. Moreover, it can cause dysfunction or total inactivation of encoded proteins.

The OH• radical targets the sugar backbone by removing hydrogen atoms. In addition, it interacts with the double bond's nitrogenous bases (Evans *et al.*, 2004). Base oxidation can lead to formation of various products, notably 8-hydroxyquinine, and others such as hydroxylmethylurea, dehydro-2′-deoxyguanosine, thymine glycol, etc. The activity of OH• is particularly known for generating DNA–protein conjugates. These cross-links pose challenges for repair mechanisms and could potentially be fatal to plant cells if not addressed promptly before replication or transcription commences.

14.3 Antioxidant Defense in Plants

Different organelles within the plant cells house distinct systems for both production and removal of ROS. These include chloroplasts, mitochondria, and peroxisomes. The pathways for neutralizing ROS across various compartments within the cells are intricately synchronized (Pang and Wang, 2008). As already mentioned, sudden spikes in intracellular ROS due to stress can cause substantial oxidative damage to biomolecules. Plants have advanced antioxidant defense mechanisms in place to counteract this damage. These mechanisms comprise enzymatic and nonenzymatic elements (Fig. 14.3).

14.3.1 Enzymatic antioxidants

The enzymatic elements of the plant antioxidant defense machinery encompass various enzymes, including SOD, POD, CAT, GPX, and other enzymes involved in the Halliwell-Asada pathway such as APX, MDHAR, DHAR, and GR. These enzymes function in distinct regions within the cells and collectively react whenever cells encounter oxidative stress.

Superoxide dismutase

The metalloenzyme SOD is pivotal in protecting aerobes from oxidative stress. SOD serves as the primary defense against harmful ROS (Gill and Tuteja, 2010). It essentially converts $O_2^{\bullet-}$ to O_2 and H_2O_2. Chemically:

$$O_2^{\bullet-} + O_2^{\bullet-} + 2H^+ \rightarrow 2H_2O_2 + O_2$$

Superoxide dismutase is distributed across various subcellular compartments where activated $O_2$is generated. Plants possess three main SOD isozymes: Cu/Zn-SOD primarily distributed in the cytosol, peroxisomes, and chloroplasts; Mn-SOD, specifically found within the mitochondria; and Fe-SOD, located predominantly in the chloroplasts (Mittler, 2002).

Plants experiencing environmental challenges often show elevated SOD activity (Sharma and Dubey, 2005; Mishra *et al.*, 2011). This heightened SOD activity is commonly linked with improved plant stress resilience (Xie *et al.*, 2019).

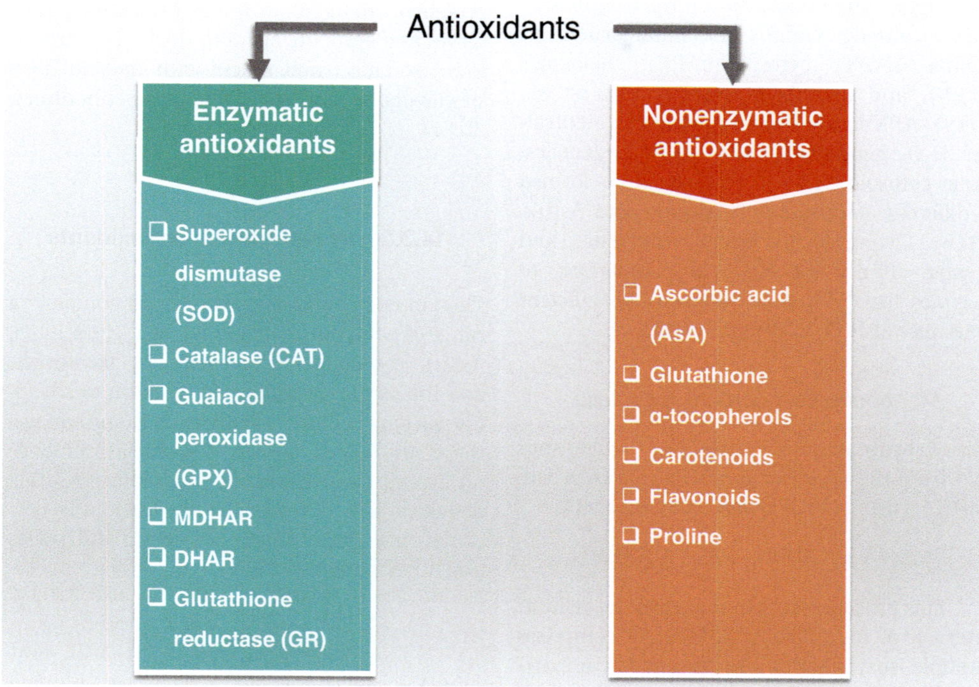

Fig. 14.3. Different antioxidants found in plants.

Catalase

This enzyme facilitates the conversion of H_2O_2 into H_2O and O_2:

$$H_2O_2 \rightarrow H_2O + (1/2)O_2$$

It is predominantly located within the peroxisomes, cytoplasmic matrix, chloroplasts, and mitochondria (Mhamdi *et al.*, 2010). During periods of cellular energy stress, where rapid H_2O_2 production occurs through catabolic processes, CAT) efficiently degrades H_2O_2 in an energetically favorable fashion (Mallick and Mohn, 2000). Stress can either amplify or deplete CAT activity, depending upon factors like severity, timespan, and the type of stress (Weisany *et al.*, 2012; Lum *et al.*, 2014). In CAT-deficient tobacco plants, an elevation of oxidized glutathione (GSSG) and a decline in AsA were recorded, underscoring the critical role of CAT in maintaining redox equilibrium during oxidative stress (Sharma *et al.*, 2012).

Guaiacol peroxidase

Guaiacol peroxidase oxidizes aromatic compounds like guaiacol and pyrogallol, utilizing H_2O_2 in the process (Mushtaq *et al.*, 2022). GPXs are commonly recognized as stress-responsive enzymes. They serve as powerful scavengers of reactive intermediate forms of O_2 and peroxy radicals during stress. GPX activity increases under conditions such as salinity (Jebara *et al.*, 2005), HM stress (Shah *et al.*, 2001) and drought (Sharma and Dubey, 2005).

$$H_2O_2 + GSH \rightarrow H_2O + GSSG$$

Ascorbate peroxidase

The enzyme ascorbate peroxidase is pivotal in the Halliwell-Asada cycle, which is indispensable for modulating intracellular levels of ROS. It utilizes AsA to convert H_2O_2 to H_2O with concurrent emergence of dehydroascorbate (DHA).

$$H_2O_2 + AsA \rightarrow 2H_2O + DHA$$

Higher plants possess several isozymes of APX, localized in various subcellular structures: stroma (sAPX), cytosol (cAPX), thylakoids (tAPX), and microbody (mAPX) (Liu *et al.*, 2019). APX located within organelles neutralizes H_2O_2 generated within those organelles, while cytosolic APX neutralizes H_2O_2 formed in different compartments (Goraya and Asthir, 2016). Owing to its broad occurrence and greater affinity for H_2O_2 in comparison to enzymes like CAT, APX serves as an efficient scavenger of H_2O_2 in stressed plants.

Monodehydroascorbate reductase

Monodehydroascorbate reductase replenishes AsA from the transient monodehydroascorbate (MDHA), using NADPH as a reducing agent.

$$MDHA + NADPH \rightarrow AsA + NADP^+$$

This process serves to replenish the cellular reservoir of AsA. Like APX, MDHAR comprises multiple isozymes that are present in compartments, such as chloroplasts, mitochondria, peroxisomes, cytosol, and glyoxysomes (Das and Roychoudhury, 2014).

Dehydroascorbate reductase

Dehydroascorbate reductase (DHAR) facilitates the conversion of DHA to ascorbic acid (AsA) by utilizing reduced glutathione (GSH) as an electron donor (Eltayeb *et al.*, 2007).

Alongside MDHAR, this enzyme aids in restoring the cellular AsA reservoir. DHAR is vital to controlling the magnitude of the AsA reservoir in symplasts as well as apoplasts, thereby preserving the redox balance within the cell.

$$DHA + 2GSH \rightarrow AsA + GSSG$$

Glutathione reductase

Glutathione reductase is a flavoprotein that utilizes NADPH to convert GSSG to GSH. GSH is consumed to regenerate AsA from MDHA and DHA, leading to its oxidation to GSSG. GR is a vital enzyme in the ASC-GSH cycle that brings about the formation of a disulfide bond in GSSG to uphold a high cellular GSH/GSSG ratio. It is primarily located in chloroplasts, with traces present in mitochondria and the cytosol. GSH serves as a reducing agent to shield thiol groups from oxidation and reacts with harmful ROS species like $^1O_2$ and OH• (Das and Roychoudhury, 2014).

14.3.2 Nonenzymatic antioxidants

Thenon-enzymatic antioxidants encompass a range of compounds including AsA, glutathione (GSH), α-tocopherol, carotenoids, flavonoids, and the osmolyte proline. In addition to shielding various cellular constituents from damage, these substances also exert a crucial influence on overall plant health by modulating routine activities like mitosis, cell elongation, senescence, and programmed cell death (PCD) (Das and Roychoudhury, 2014).

Ascorbic acid

Ascorbic acid stands out as the most prevalent and extensively studied antioxidant. Within plant cells, the Smirnoff-Wheeler pathway is predominantly responsible for synthesizing AsA, facilitated by L-galactano-γ-lactone dehydrogenase in the mitochondria, supplemented by its generation from D-galacturonic acid. Ascorbate acid undergoes oxidation in two stages. Initially, it is oxidized to MDHA in the presence of H_2O_2 and APX. This MDHA, if not promptly reverted to AsA, subsequently dismutates to DHA. It effectively scavenges H_2O_2, and OH• and aids in renewing α-tocopherol from tocopheroxyl radicals, thus safeguarding membranes from oxidative harm. Moreover, AsA serves an important function of preserving the functions of certain metal-binding enzymes. Additionally, it has been implicated in thwarting photooxidation by modulating PSII activity and downregulating it via pH-mediated mechanisms, alongside promoting zeaxanthin formation.

Research has revealed that levels of AsA fluctuate under stress (Guha *et al.*, 2012; Tuna *et al.*, 2013). The levels of AsA under stress are contingent upon the equilibrium between AsA synthesis and turnover rates associated with antioxidant requirement (Chaves *et al.*, 2002).

Elevated expression of enzymes responsible for AsA biogenesis improves plant resilience to abiotic stress (Upadhyaya *et al.*, 2009; Zhang *et al.*, 2011).

Glutathione

Glutathione, a low molecular weight tripeptide, is a strong antioxidant present in most living cells. It performs a vital function in safeguarding cells against oxidative stress and damage induced by ROS. Additionally, it participates in different metabolic and cellular processes, including detoxification of xenobiotics and regulation of cellular events such as cell differentiation, DNA replication, repair, and protein synthesis. This exceptional capability of GSH stems from its highly reductive potential, mainly attributed to a central cysteine residue with nucleophilic properties. GSH serves as a potent scavenger of ROS including H_2O_2, $^1O_2$, OH•, and O_2•−. It safeguards various biomolecules by either forming complexes or reducing them. This triggers the production of oxidized glutathione (GSSG) as a by-product. Moreover, GSH is crucial in replenishing the cellular GSH pool by regenerating AsA, ultimately leading to the production of GSSG. This conversion of GSSG back to GSH takes place through either direct synthesis or enzymatic action by GR. Additionally, GSH contributes to the synthesis of phytochelatins via phytochelatin synthase, facilitating the chelation of heavy metal ions and mitigating other potential sources of ROS, thus maintaining the fine equilibrium between GSH and GSSG which is crucial for preserving the cellular redox equilibrium (Das and Roychoudhury, 2014).

Glutathione assists in the regeneration of AsA, another essential antioxidant, converting it from its oxidized to reduced state via the action of DHAR. Additionally, GSH reduces DHA through a nonenzymatic mechanism under specific conditions, particularly in chloroplasts where pH and GSH concentrations are conducive to this process (Do *et al.*, 2016).

Any alterations in GSH levels and the GSH/GSSG ratio serve as indicators of environmental stress, such as drought, salinity (Hefny and Abdel-Kader, 2009), chilling (Radyuk *et al.*, 2009), and metal stress (Mishra *et al.*, 2011). Modulating GSH biosynthesis through enzyme overexpression affects plant tolerance to specific stresses, with varied responses observed (Zhu *et al.*, 1999; Eltayeb *et al.*, 2010).

α-Tocopherols

α-Tocopherols belong to the family of lipophilic antioxidants, known for its ability to effectively neutralize ROS. This characteristic renders it an integral element of cellular membranes (Holländer-Czytko *et al.*, 2005). Of the four isomers (α-, β-, γ-, δ-), α-tocopherol exhibits the highest antioxidant capacity.

Tocopherols are predominantly found in green plant tissues (Hasanuzzaman *et al.*, 2014). α-Tocopherol is derived from γ-tocopherol through the action of γ-tocopherol-methyl-transferase (γ-TMT). Tocopherols safeguard the lipids and other components of chloroplast membranes by engaging with oxygen and neutralizing its excessive energy, thereby shielding PSII. Additionally, tocopherol interacts with lipid radicals (RO•), peroxyl radicals (ROO•), and alkoxyl radicals (RO*) at the interface between membranes and water, reducing them and converting itself into a tocopherol radical (TOH•). The TOH• radical is then recycled to its reduced state through interactions with GSH and AsA.

Carotenoids

Carotenoids are lipophilic antioxidants predominantly localized within the plastids of plant tissues. Carotenoids absorb light within the 450–570 nm range and subsequently relay this energy to chlorophyll molecules. The antioxidative activity of carotenoids manifests in several ways. These include halting the chain reactions by interacting with lipid peroxidation products (LPO), scavenging singlet oxygen ($^1O_2$), and impeding $^1O_2$ formation (Das and Roychoudhury, 2014).

Flavonoids

Flavonoids are phytochemical compounds distributed across the plant world, often located in leaves, floral parts, and pollen grains. They are generally categorized into four structural groups: flavonols, flavones, isoflavones, and anthocyanins. These compounds serve various functions, including imparting color to different plant parts like flowers, fruits, and seeds, contributing to plant fertility and pollen germination, and acting as a defense mechanism against plant pathogens. Flavonoids are also recognized as a secondary system for neutralizing ROS in plants when the photosynthetic machinery is impaired due to excessive excitation energy (Fini *et al.*, 2011). Moreover, they participate in the scavenging of $^1O_2$ and help alleviate damage to the chloroplast membrane (Agati *et al.*, 2012).

Proline

Proline is essential within the antioxidant system, playing a vital role in stress reduction. It also plays crucial roles in flowering, pollen development, embryo formation, and leaf expansion. During stress, proline levels are commonly elevated in the cytosolic fraction to modulate the osmotic milieu (Mushtaq *et al.*, 2023). Increased proline content under stress is often linked to either accelerated synthesis or lower rates of breakdown (Verbruggen and Hermans, 2008). Accumulation of proline increases plant resistance to HM stress (Siddique *et al.*, 2018) and improves resistance to drought or salinity stress (Hmida-Sayari *et al.*, 2005) in plants.

14.4 Plant Responses to Abiotic Oxidative Stress

Under typical growth conditions, plants maintain low levels of ROS. However, when faced with environmental stressors, ROS production escalates significantly, disrupting the equilibrium of O2•−, •OH, and H_2O_2 within the cell. The impact of diverse challenges on ROS generation is discussed in detail below.

14.4.1 Heavy metals

Heavy metals including lead (Pb), cadmium (Cd), mercury (Hg), and arsenic (As) accumulate in soil and water through various human activities including mining, industrial operations, and agricultural practices. These metals can enter plant tissues through root uptake and disrupt essential physiological and biochemical processes. As HMs frequently occur as pollutants in both soil and water, research on the toxicity of HMs in various plant species has garnered significant attention.

Heavy metals disrupt essential homeostatic processes in plants, such as water uptake, transport, and transpiration. These disruptions eventually cause cell death. HMs induce chlorosis, necrosis, wilting, inhibited root growth, stunted overall plant growth, modified stomatal function, reduced water potential, changes in membrane activity, inhibition of photosynthesis, metabolic alterations, and modifications of vital enzyme activities (Sharma and Dubey, 2007).

Prolonged exposure to redox-active HMs leads to serious oxidative injury. HMs induce the generation of ROS, either through their redox activity or by affecting metabolism in a site-targeted fashion. HMs like Fe, Cr, Cu, V, and Co catalyze redox reactions, leading to hydroxyl radical (OH•) and H_2O_2 formation. Exposure to HMs triggers substantial responses from antioxidative machinery, but the nature of these responses varies and mostly depends on factors such as plant species, the specific HM involved, and the severity of the metal stress (Schutzendubel and Polle, 2002).

A decline in AsA content, AsA/DHA ratio, GSH, and GSH/GSSG ratio was observed in mung beans exposed to Cd stress. However, reductions were less pronounced in the tolerant cultivar, indicating the essential role of AsA and GSH reservoirs against Cd stress (Anjum *et al.*, 2011). A study demonstrated that externally administered GSH on rice subjected to Cd stress significantly mitigated growth inhibition induced by Cd stress (Cai *et al.*, 2010). Another study suggested that inherent GSG levels in *Arabidopsis* were sufficient to confer Cd tolerance (Wójcik and Tukiendorf, 2011).

Enzymes like SOD, CAT, and enzymes of the Halliwell-Asada cycle are crucial for clearing ROS and mitigating oxidative damage induced by HMs. Studies indicate that coordination among these antioxidative enzymes enhances tolerance to HMs, facilitating normal plant development even under stress (Gill *et al.*, 2011). Further, antioxidant enzymes like SOD, CAT, and GPX are considerably influenced by HM exposure levels, suggesting a dose-dependent relationship between HM stress and antioxidative defense mechanisms. For instance, plants exposed to lead stress exhibited a dose-dependent rise in the activities of SOD and APX, whereas CAT activity decreased notably (Qureshi *et al.*, 2007). In wheat, APX and SOD activities demonstrated a decrease at low levels of As but increase at higher levels. Conversely, CAT activity showed an initial rise at As concentrations below 1 mg/kg, followed by a subsequent decline.

14.4.2 Salinity

Salinity constrains the productivity of roughly 15% of total arable land, exerting adverse effects on crop yields worldwide. The anticipated escalation in agricultural land salinity is projected to cause detrimental global repercussions in the near future (Mahajan and Tuteja, 2005).

Under salt stress, plants may experience reduced CO_2 availability, hindering carbon fixation. This makes chloroplasts susceptible to surplus excitation energy, potentially increasing ROS generation (Gill and Tuteja, 2010). Elevated ROS prompts the activation of antioxidant defense mechanisms. Exogenous administration of antioxidants like AsA aids in mitigating salt stress-induced damages, enhancing seedling survival rates and inhibiting lipid peroxidation (Shalata and Neumann, 2001; Hamada and Al-Hakimi, 2009). Similarly, GSH maintains the redox state, with a high GSH/GSSG ratio indicating salt tolerance. Sumithra *et al.* (2006) revealed that mung bean leaves exposed to salinity had elevated GSH levels and lower lipid peroxidation and H_2O_2 concentration. Enzyme activities of SOD, CAT, and POD are usually linked with enhanced salt stress tolerance, varying among plant cultivars, duration of stress, and dose applied (El-Bastawisy, 2010).

Additionally, the modulation of ROS detoxification systems through exogenous applications of compounds like selenium (Se) or nitric oxide (NO) donors enhances antioxidant defense and enhances resilience against salinity-induced oxidative damage (Hasanuzzaman *et al.*, 2012).

14.4.3 Drought

Drought is one of the major destructive stressors that negatively affects overall plant health. It is marked by reduced overall cell growth, low water content and water potential, loss of turgor pressure, stomatal closure, etc. (Jaleel *et al.*, 2009; Mingch *et al.*, 2010; Din *et al.*, 2011).

Excessive ROS production during drought arises from compromised electron transport in chloroplasts and mitochondria, leading to a disproportion in electron generation and utilization due to PSII downregulation. This causes dissipation of excess light energy, producing $^1O_2$, H_2O_2, and OH•.

The antioxidants AsA and GSH play a significant role in shielding plants from detrimental impacts of drought (Yazdanpanah *et al.*, 2011). Additionally, the GSH system protects soluble proteins from oxidation during drought conditions. Drought stress alters the GSH/GSSG ratio, with acclimated plants maintaining a higher redox pool compared to nonacclimated ones (Selote and Khanna-Chopra, 2006). Moreover, elevated α-tocopherol levels are correlated with improved stress tolerance (Munné-Bosch *et al.*, 2009). Enzymes like SOD, APX, MDHAR, DHAR, and GR are crucial in mitigating oxidative stress brought on by drought (Sharma and Dubey, 2005). Increased enzyme activity correlates with higher drought tolerance, as observed in various plant species like mung beans (Hossain and Fujita, 2009), rice (Shehab *et al.*, 2010), and maize (Mohammadkhani and Heidari, 2007).

14.4.4 Flooding

Flooding or waterlogging (WL) in plants leads to low oxygen conditions (anoxia or hypoxia), resulting in the formation of toxic compounds that disrupt plant metabolism and result in excess ROS formation (Loreti *et al.*, 2016).

The WL-sensitive genotype of *Sorghum bicolor* accumulated malondialdehyde (MDA) following WL treatment (Zhang *et al.*, 2019). A study by Anee *et al.* (2019) on the WL-sensitive *Sesamum indicum* demonstrated that MDA and H_2O_2 levels increased with longer durations of WL stress. Similarly, an increase in these parameters was seen in *S. lycopersicum* (Rasheed *et al.*, 2018). Additionally, *Deschampsia antarctica*, an Antarctic plant, exhibited higher MDA and H_2O_2 content under WL conditions (Park and Lee, 2019).

14.5 Conclusion

Reactive oxygen species are inevitable by-products generated during routine cellular metabolism, arising from different metabolic pathways occurring in various compartments. While ROS production is typically low under normal circumstances, prolonged exposure to environmental insults such as drought, salinity, and metal toxicity can disturb cellular homeostasis and escalate ROS production. In plants, ROS serve dual roles: they act as signaling molecules in certain contexts but they also induce oxidative damage when present in excess. Elevated ROS levels cause oxidative damage to lipids, proteins, and DNA, subsequently compromising membrane integrity and altering membrane fluidity, ion transport, and enzyme activity. Higher plants have evolved a sophisticated antioxidative defense system to mitigate oxidative damage. Through the action of different antioxidant components, plants are able to scavenge and detoxify excess ROS, demonstrating their adaptability to diverse stress conditions.

Overall, this chapter underscores the importance of understanding ROS dynamics in plants and the critical role of antioxidant defenses in maintaining cellular homeostasis and enhanced plant resilience against environmental challenges.

References

Agati, G., Azzarello, E., Pollastri, S. and Tattini, M. (2012) Flavonoids as antioxidants in plants: Location and functional significance. *Plant Science* 196, 67–76.

Anee, T.I., Nahar, K., Rahman, A., Mahmud, J.A., Bhuiyan, T.F. *et al.* (2019) Oxidative damage and antioxidant defense in *Sesamum indicum* after different waterlogging durations. *Plants* 8(7), 196.

Anjum, N.A., Umar, S., Iqbal, M. and Khan, N.A. (2011) Cadmium causes oxidative stress in mung bean by affecting the antioxidant enzyme system and ascorbate-glutathione cycle metabolism. *Russian Journal of Plant Physiology* 58(1), 92–99.

Anjum, N.A., Sofo, A., Scopa, A., Roychoudhury, A., Gill, S.S. *et al.* (2015) Lipids and proteins – major targets of oxidative modifications in abiotic stressed plants. *Environmental Science and Pollution Research International* 22(6), 4099–4121.

Apel, K. and Hirt, H. (2004) Reactive oxygen species: Metabolism, oxidative stress, and signal transduction. *Annual Review of Plant Biology* 55, 373–399.

Arora, A., Sairam, R.K. and Srivastava, G.C. (2002) Oxidative stress and antioxidative system in plants. *Current Science* 82, 1227–1238.

Ashraf, M.J.B.A. (2009) Biotechnological approach of improving plant salt tolerance using antioxidants as markers. *Biotechnology Advances* 27(1), 84–93.

Blokhina, O. and Fagerstedt, K.V. (2010) Reactive oxygen species and nitric oxide in plant mitochondria: Origin and redundant regulatory systems. *Physiologia Plantarum* 138(4), 447–462.

Cabiscol, E., Piulats, E., Echave, P., Herrero, E. and Ros, J. (2000) Oxidative stress promotes specific protein damage in *Saccharomyces cerevisiae*. *Journal of Biological Chemistry* 275(35), 27393–27398.

Cai, Y., Lin, L., Cheng, W., Zhang, G. and Wu, F. (2010) Genotypic dependent effect of exogenous glutathione on Cd-induced changes in cadmium and mineral uptake and accumulation in rice seedlings (*Oryza sativa*). *Plant, Soil and Environment* 56(11), 516–525.

Chaves, M.M., Pereira, J.S., Maroco, J., Rodrigues, M.L., Ricardo, C.P.P. *et al.* (2002) How plants cope with water stress in the field? Photosynthesis and growth. *Annals of Botany* 89(7), 907.

Chen, S., Yin, C., Strasser, R.J., Yang, C. and Qiang, S. (2012) Reactive oxygen species from chloroplasts contribute to 3-acetyl-5-isopropyltetramic acid-induced leaf necrosis of *Arabidopsis thaliana*. *Plant Physiology and Biochemistry* 52, 38–51.

Das, K. and Roychoudhury, A. (2014) Reactive oxygen species (ROS) and response of antioxidants as ROS-scavengers during environmental stress in plants. *Frontiers in Environmental Science* 2, 53.

Din, J., Khan, S.U., Ali, I. and Gurmani, A.R. (2011) Physiological and agronomic response of canola varieties to drought stress. *Journal of Animal and Plant Sciences* 21, 78–82.

Do, H., Kim, I.-S., Jeon, B.W., Lee, C.W., Park, A.K. *et al.* (2016) Structural understanding of the recycling of oxidized ascorbate by dehydroascorbate reductase (OsDHAR) from *Oryza sativa* L. *japonica*. *Scientific Reports* 6(1), 19498.

El-Bastawisy, Z.M. (2010) Variation in antioxidants among three wheat cultivars varying in tolerance to NaCl. *General and Applied Plant Physiology* 36(3/4), 189–203.

Eltayeb, A.E., Kawano, N., Badawi, G.H., Kaminaka, H., Sanekata, T. *et al.* (2007) Overexpression of monodehydroascorbate reductase in transgenic tobacco confers enhanced tolerance to ozone, salt and polyethylene glycol stresses. *Planta* 225(5), 1255–1264.

Eltayeb, A.E., Yamamoto, S., Habora, M.E.E., Matsukubo, Y., Aono, M. *et al.* (2010) Greater protection against oxidative damages imposed by various environmental stresses in transgenic potato with higher level of reduced glutathione. *Breeding Science* 60(2), 101–109.

Evans, M.D., Dizdaroglu, M. and Cooke, M.S. (2004) Oxidative DNA damage and disease: Induction, repair and significance. *Mutation Research/Reviews in Mutation Research* 567(1), 1–61.

Fini, A., Brunetti, C., Ferdinando, M., Ferrini, F. and Tattini, M. (2011) Stress-induced flavonoid biosynthesis and the antioxidant machinery of plants. *Plant Signaling & Behavior* 6(5), 709–711.

Gharibi, S., Tabatabaei, B.E.S., Saeidi, G. and Goli, S.A.H. (2016) Effect of drought stress on total phenolic, lipid peroxidation, and antioxidant activity of *Achillea* species. *Applied Biochemistry and Biotechnology* 178(4), 796–809.

Gill, S.S. and Tuteja, N. (2010) Reactive oxygen species and antioxidant machinery in abiotic stress tolerance in crop plants. *Plant Physiology and Biochemistry* 48(12), 909–930.

Gill, S.S., Khan, N.A. and Tuteja, N. (2011) Differential cadmium stress tolerance in five Indian mustard (*Brassica juncea* L.) cultivars: An evaluation of the role of antioxidant machinery. *Plant Signaling & Behavior* 6(2), 293–300.

Goraya, G.K. and Asthir, B. (2016) Magnificant role of intracellular reactive oxygen species production and its scavenging encompasses downstream processes. *Journal of Plant Biology* 59(3), 215–222.

Guha, A., Sengupta, D., Rasineni, G.K. and Reddy, A.R. (2012) Non-enzymatic antioxidative defence in drought-stressed mulberry (*Morus indica* L.) genotypes. *Trees* 26(3), 903–918.

Hamada, A. and Al-Hakimi, A. (2009) Exogenous ascorbic acid or thiamine increases the resistance of sunflower and maize plants to salt stress. *Acta Agronomica Hungarica* 57(3), 335–347.

Hasanuzzaman, M., Hossain, M.A., Silva, J.A.T. and Fujita, M. (2012) Plant response and tolerance to abiotic oxidative stress: Antioxidant defense is a key factor. In: Venkateswarlu, B., Shanker, A., Shanker, C. and Maheswari, M. (eds) *Crop Stress and Its Management: Perspectives and Strategies*. Springer, Singapore, pp. 261–315.

Hasanuzzaman, M., Nahar, K. and Fujita, M. (2014) Role of tocopherol (vitamin E) in plants: Abiotic stress tolerance and beyond. In: Ahmad, P. and Rasool, S. (eds) *Emerging Technologies and Management of Crop Stress Tolerance*. Academic Press, New York, pp. 267–289.

Hasanuzzaman, M., Raihan, M.R.H., Masud, A.A.C., Rahman, K., Nowroz, F. *et al.* (2021) Regulation of reactive oxygen species and antioxidant defense in plants under salinity. *International Journal of Molecular Sciences* 22(17), 9326.

Hefny, M. and Abdel-Kader, D.Z. (2009) Antioxidant-enzyme system as selection criteria for salt tolerance in forage sorghum genotypes (*Sorghum bicolor* L *Moench*). *International Journal of Plant Breeding and Genetics* 1, 38–53.

Heyno, E., Mary, V., Schopfer, P. and Krieger-Liszkay, A. (2011) Oxygen activation at the plasma membrane: Relation between superoxide and hydroxyl radical production by isolated membranes. *Planta* 234(1), 35–45.

Hmida-Sayari, A., Gargouri-Bouzid, R., Bidani, A., Jaoua, L., Savouré, A. *et al.* (2005) Overexpression of δ1-pyrroline-5-carboxylate synthetase increases proline production and confers salt tolerance in transgenic potato plants. *Plant Science* 169(4), 746–752.

Holländer-Czytko, H., Grabowski, J., Sandorf, I., Weckermann, K. and Weiler, E.W. (2005) Tocopherol content and activities of tyrosine aminotransferase and cystine lyase in *Arabidopsis* under stress conditions. *Journal of Plant Physiology* 162(7), 767–770.

Hossain, M.A. and Fujita, M. (2009) Exogenous glycine betaine and proline modulate antioxidant defense and methylglyoxal detoxification systems and reduce drought-induced damage in mung bean seedlings (Vigna radiata L). In: *Proceedings of the 3rd International Conference on Integrated Approaches to Improve Crop Production Under Drought-Prone Environments*, pp. 11–16.

Jaleel, C.A., Paramasivam, M., Wahid, A., Farooq, M., Al-Juburi, H.J. *et al.* (2009) Drought stress in plants: A review on morphological characteristics and pigments composition. *International Journal of Agriculture and Biology* 11(1), 100–105.

Janků, M., Luhová, L. and Petřivalský, M. (2019) On the origin and fate of reactive oxygen species in plant cell compartments. *Antioxidants* 8(4), 105.

Jebara, S., Jebara, M., Limam, F. and Aouani, M.E. (2005) Changes in ascorbate peroxidase, catalase, guaiacol peroxidase and superoxide dismutase activities in common bean (*Phaseolus vulgaris*) nodules under salt stress. *Journal of Plant Physiology* 162(8), 929–936.

Liu, J.X., Feng, K., Duan, A.Q., Li, H., Yang, Q.Q. *et al.* (2019) Isolation, purification and characterization of an ascorbate peroxidase from celery and overexpression of the AgaAPX1 gene enhanced ascorbate content and drought tolerance in *Arabidopsis*. *BMC Plant Biology* 19, 1–13.

Loreti, E., Veen, H. and Perata, P. (2016) Plant responses to flooding stress. *Current Opinion in Plant Biology* 33, 64–71.

Lum, M.S., Hanafi, M.M., Rafii, Y.M. and Akmar, A.S.N. (2014) Effect of drought stress on growth, proline and antioxidant enzyme activities of upland rice. *Journal of Animal & Plant Sciences* 24(5), 1487–1493.

Mahajan, S. and Tuteja, N. (2005) Cold, salinity and drought stresses: An overview. *Archives of Biochemistry and Biophysics* 444(2), 139–158.

Maheshwari, R. and Dubey, R.S. (2009) Nickel-induced oxidative stress and the role of antioxidant defence in rice seedlings. *Plant Growth Regulation* 59(1), 37–49.

Mallick, N. and Mohn, F.H. (2000) Reactive oxygen species: Response of algal cells. *Journal of Plant Physiology* 157(2), 183–193.

Mhamdi, A., Queval, G., Chaouch, S., Vanderauwera, S., Van Breusegem, F. *et al.* (2010) Catalase function in plants: A focus on *Arabidopsis* mutants as stress-mimic models. *Journal of Experimental Botany* 61(15), 4197–4220.

Mingch, L., Xiangli, L., Jing, H. and Lihong, G. (2010) Effect of simulated drought stress on plant growth, yield and fruit properties of tomato. *Acta Horticulturae* 856(856), 193–202.

Mishra, S., Jha, A.B. and Dubey, R.S. (2011) Arsenite treatment induces oxidative stress, upregulates antioxidant system, and causes phytochelatin synthesis in rice seedlings. *Protoplasma* 248(3), 565–577.

Mittler, R. (2002) Oxidative stress, antioxidants and stress tolerance. *Trends in Plant Science* 7(9), 405–410.

Mohammadkhani, N. and Heidari, R. (2007) Effects of drought stress on protective enzyme activities and lipid peroxidation in two maize cultivars. *Pakistan Journal of Biological Sciences* 10(21), 3835–3840.

Møller, I.M. and Kristensen, B.K. (2004) Protein oxidation in plant mitochondria as a stress indicator. *Photochemical & Photobiological Sciences* 3(8), 730–735.

Munné-Bosch, S., Falara, V., Pateraki, I., López-Carbonell, M., Cela, J. *et al.* (2009) Physiological and molecular responses of the isoprenoid biosynthetic pathway in a drought-resistant Mediterranean shrub, *Cistus creticus* exposed to water deficit. *Journal of Plant Physiology* 166(2), 136–145.

Murphy, M.P. (2009) How mitochondria produce reactive oxygen species. *Biochemical Journal* 417(1), 1–13.

Mushtaq, N.U., Saleem, S., Rasool, A., Shah, W.H., Tahir, I. *et al.* (2022) Functional characterization of the antioxidant enzymes in plants exposed to environmental stresses. In: Aftab, T. and Hakeem, K.R. (eds) *Antioxidant Defense in Plants: Molecular Basis of Regulation*. Springer Nature, Singapore, pp. 15–30.

Mushtaq, N.U., Alghamdi, K.M., Saleem, S., Tahir, I., Bahieldin, A. *et al.* (2023) Exogenous zinc mitigates salinity stress by stimulating proline metabolism in proso millet (*Panicum miliaceum* L.). *Frontiers in Plant Science* 14, 1053869.

Noctor, G., Veljovic-Jovanovic, S.O.N.J.A., Driscoll, S., Novitskaya, L. and Foyer, C.H. (2002) Drought and oxidative load in the leaves of C3 plants: A predominant role for photorespiration? *Annals of Botany* 89(7), 841–850.

Noctor, G., Paepe, R. and Foyer, C.H. (2007) Mitochondrial redox biology and homeostasis in plants. *Trends in Plant Science* 12(3), 125–134.

Nyathi, Y. and Baker, A. (2006) Plant peroxisomes as a source of signalling molecules. *Biochimica et Biophysica Acta-Molecular Cell Research* 1763(12), 1478–1495.

Pang, C.H. and Wang, B.S. (2008) Oxidative stress and salt tolerance in plants. In: Lüttge, U., Beyschlag, W. and Murata, J. (eds) *Progress in Botany*. Springer, Berlin, pp. 231–245.

Pardo, J.M. (2010) Biotechnology of water and salinity stress tolerance. *Current Opinion in Biotechnology* 21(2), 185–196.

Park, J.S. and Lee, E.J. (2019) Waterlogging induced oxidative stress and the mortality of the Antarctic plant, *Deschampsia antarctica*. *Journal of Ecology and Environment* 43(1), 1–8.

Qureshi, M.I., Abdin, M.Z., Qadir, S. and Iqbal, M. (2007) Lead-induced oxidative stress and metabolic alterations in *Cassia angustifolia* Vahl. *Biologia Plantarum* 51(1), 121–128.

Radyuk, M.S., Domanskaya, I.N., Shcherbakov, R.A. and Shalygo, N.V. (2009) Effect of low above-zero temperature on the content of low-molecular antioxidants and activities of antioxidant enzymes in green barley leaves. *Russian Journal of Plant Physiology* 56(2), 175–180.

Rasheed, R., Iqbal, M., Ashraf, M.A., Hussain, I., Shafiq, F. *et al.* (2018) Glycine betaine counteracts the inhibitory effects of waterlogging on growth, photosynthetic pigments, oxidative defence system, nutrient composition, and fruit quality in tomato. *Journal of Horticultural Science and Biotechnology* 93(4), 385–391.

Romero-Puertas, M.C., Palma, J.M., Gómez, M., Del Río, L.A. and Sandalio, L.M. (2002) Cadmium causes the oxidative modification of proteins in pea plants. *Plant, Cell & Environment* 25(5), 677–686.

Sachdev, S., Ansari, S.A., Ansari, M.I., Fujita, M. and Hasanuzzaman, M. (2021) Abiotic stress and reactive oxygen species: Generation, signaling, and defense mechanisms. *Antioxidants* 10(2), 277.

Schutzendubel, A. and Polle, A. (2002) Plant responses to abiotic stresses: Heavy metal-induced oxidative stress and protection by mycorrhization. *Journal of Experimental Botany* 53(372), 1351–1365.

Selote, D.S. and Khanna-Chopra, R. (2006) Drought acclimation confers oxidative stress tolerance by inducing co-ordinated antioxidant defense at cellular and subcellular level in leaves of wheat seedlings. *Physiologia Plantarum* 127(3), 494–506.

Shah, K., Kumar, R.G., Verma, S. and Dubey, R.S. (2001) Effect of cadmium on lipid peroxidation, superoxide anion generation and activities of antioxidant enzymes in growing rice seedlings. *Plant Science* 161(6), 1135–1144.

Shalata, A. and Neumann, P.M. (2001) Exogenous ascorbic acid (vitamin C) increases resistance to salt stress and reduces lipid peroxidation. *Journal of Experimental Botany* 52(364), 2207–2211.

Sharma, P. and Dubey, R.S. (2005) Drought induces oxidative stress and enhances the activities of antioxidant enzymes in growing rice seedlings. *Plant Growth Regulation* 46(3), 209–221.

Sharma, P., Jha, A.B., Dubey, R.S. and Pessarakli, M. (2012) Reactive oxygen species, oxidative damage, and antioxidative defense mechanism in plants under stressful conditions. *Journal of Botany* 1–26. Available at: https://doi.org/10.1155/2012/217037

Sharma, P. and Dubey, R.S. (2007) Involvement of oxidative stress and role of antioxidative defense system in growing rice seedlings exposed to toxic concentrations of aluminum. *Plant Cell Reports* 26(11), 2027–2038.

Shehab, G.G., Ahmed, O. and El-Beltagi, H.S. (2010) Effects of various chemical agents for alleviation of drought stress in rice plants (*Oryza sativa* L). *Notulae Botanicae Horti Agrobotanici Cluj-Napoca* 38(1), 139–148.

Siddique, A., Kandpal, G. and Kumar, P. (2018) Proline accumulation and its defensive role under diverse stress condition in plants: An overview. *Journal of Pure and Applied Microbiology* 12(3), 1655–1659.

Slade, D. and Radman, M. (2011) Oxidative stress resistance in *Deinococcus radiodurans*. *Microbiology and Molecular Biology Reviews* 75(1), 133–191.

Sumithra, K., Jutur, P.P., Carmel, B.D. and Reddy, A.R. (2006) Salinity-induced changes in two cultivars of *Vigna radiata*: Responses of antioxidative and proline metabolism. *Plant Growth Regulation* 50(1), 11–22.

Sun, C., Du, W., Cheng, X., Xu, X., Zhang, Y. *et al.* (2010) The effects of drought stress on the activity of acid phosphatase and its protective enzymes in pigweed leaves. *African Journal of Biotechnology* 9(6), 825–833.

Tanou, G., Molassiotis, A. and Diamantidis, G. (2009) Induction of reactive oxygen species and necrotic death-like destruction in strawberry leaves by salinity. *Environmental and Experimental Botany* 65(2–3), 270–281.

Taïbi, K., Taïbi, F., Ait Abderrahim, L., Ennajah, A., Belkhodja, M. *et al.* (2016) Effect of salt stress on growth, chlorophyll content, lipid peroxidation and antioxidant defence systems in *Phaseolus vulgaris* L. *South African Journal of Botany* 105, 306–312.

Torres, M.A., Dangl, J.L. and Jones, J.D. (2002) Arabidopsis gp91^{phox} homologues *AtrbohD* and *AtrbohF* are required for accumulation of reactive oxygen intermediates in the plant defense response. *Proceedings of the National Academy of Sciences* 99(1), 517–522.

Tripathy, B.C. and Oelmüller, R. (2012) Reactive oxygen species generation and signaling in plants. *Plant Signaling & Behavior* 7(12), 1621–1633.

Tuna, A.L., Kaya, C., Altunlu, H. and Ashraf, M. (2013) Mitigation effects of non-enzymatic antioxidants in maize ('*Zea mays*' L.) plants under salinity stress. *Australian Journal of Crop Science* 7(8), 1181–1188.

Turrens, J.F. (2003) Mitochondrial formation of reactive oxygen species. *Journal of Physiology* 552(2), 335–344.

Upadhyaya, C.P., Young, K.E., Akula, N., Kim, H. soon, Heung, J.J. *et al.* (2009) Over-expression of strawberry D-galacturonic acid reductase in potato leads to accumulation of vitamin C with enhanced abiotic stress tolerance. *Plant Science* 177(6), 659–667.

Verbruggen, N. and Hermans, C. (2008) Proline accumulation in plants: A review. *Amino Acids* 35(4), 753–759.

Weisany, W., Sohrabi, Y., Heidari, G., Siosemardeh, A. and Ghassemi-Golezani, K. (2012) Changes in antioxidant enzymes activity and plant performance by salinity stress and zinc application in soybean ('*Glycine max*' L.). *Plant Omics* 5(2), 60–67.

Wójcik, M. and Tukiendorf, A. (2011) Glutathione in adaptation of *Arabidopsis thaliana* to cadmium stress. *Biologia Plantarum* 55(1), 125–132.

Xie, X., He, Z., Chen, N., Tang, Z., Wang, Q, *et al.* (2019) The roles of environmental factors in regulation of oxidative stress in plant. *BioMed Research International* 9732325. Available at: https://doi.org/10.1155/2019/9732325

Yadav, D.K., Kumar, S., Choi, E.H., Chaudhary, S. and Kim, M.H. (2019) Molecular dynamic simulations of oxidized skin lipid bilayer and permeability of reactive oxygen species. *Scientific Reports* 9(1), 4496.

Yamauchi, Y., Furutera, A., Seki, K., Toyoda, Y., Tanaka, K. *et al.* (2008) Malondialdehyde generated from peroxidized linolenic acid causes protein modification in heat-stressed plants. *Plant Physiology and Biochemistry* 46(8–9), 786–793.

Yazdanpanah, S., Baghizadeh, A. and Abbassi, F. (2011) The interaction between drought stress and salicylic and ascorbic acids on some biochemical characteristics of *Satureja hortensis*. *African Journal of Agricultural Research* 6(4), 798–807.

Zaefyzadeh, M., Quliyev, R.A., Babayeva, S.M. and Abbasov, M.A. (2009) The effect of the interaction between genotypes and drought stress on the superoxide dismutase and chlorophyll content in durum wheat landraces. *Turkish Journal of Biology* 33(1), 1–7.

Zhang, C., Liu, J., Zhang, Y., Cai, X., Gong, P. *et al.* (2011) Overexpression of *SlGMEs* leads to ascorbate accumulation with enhanced oxidative stress, cold, and salt tolerance in tomato. *Plant Cell Reports* 30(3), 389–398.

Zhang, R.D., Zhou, Y.F., Yue, Z.X., Chen, X.F., Cao, X. *et al.* (2019) Changes in photosynthesis, chloroplast ultrastructure, and antioxidant metabolism in leaves of sorghum under waterlogging stress. *Photosynthetica* 57(4), 1076–1083.

Zhu, Y.L., Pilon-Smits, E.A., Tarun, A.S., Weber, S.U., Jouanin, L. *et al.* (1999) Cadmium tolerance and accumulation in Indian mustard is enhanced by overexpressing gamma-glutamylcysteine synthetase. *Plant Physiology* 121(4), 1169–1178.

15 Plant Defense Mechanisms: Innate and Induced Immunity in Plants

Sucharita Mohapatra[1], Lellapalli Rithesh[2], Rahul Kumar Tiwari[3], Laxmipriya Behera[4], Milan Kumar Lal[4]* and Ravinder Kumar[5]

[1]Department of Plant Pathology, College of Agriculture, Odisha University of Agriculture and Technology, Bhubaneswar, India; [2]Department of Plant Pathology, College of Agriculture, Kerala Agricultural University, Thiruvananthapuram, India; [3]ICAR-Indian Institute of Sugarcane Research, Lucknow, India; [4]ICAR-National Rice Research Institute, Cuttack, India; [5]ICAR-Indian Agricultural Research Institute, New Delhi, India

Abstract

Plants are sessile organisms that face relentless pathogen attacks. However, plants are protected by their efficient immune systems. They rely on various immune systems to ward off the pathogens, which include innate (before infection) and induced (after infection) with various structural and biochemical modes. The resistance mechanism of plants is now well studied at the biochemical and molecular levels. Pathogens are recognized by host cell surface localized pattern recognition receptors (PRRs), which initiate the first line of defense, also known as pattern-triggered immunity (PTI), that restricts the pathogen invasion. Pathogen effectors are sensed by the nucleotide-binding/leucine-rich repeat (NLR) receptors, which allows the plants to activate a second layer of defense also known as effector-triggered immunity (ETI), providing a more potent and durable defense mechanism. Moreover, both PTI and ETI collaborate synergistically to increase disease resistance. This chapter provides a comprehensive review of plant defense responses, which shows the signal transduction mechanisms in PTI and ETI.

Keywords: sustainable agriculture, plant disease resistance, plant immunity, molecular biology, PTI, ETI

15.1 Introduction

Plants and pathogens are in a constant battle due to their ongoing coevolution. Although a large number of plants inherently withstand most diseases, some microbes can cause severe diseases, reducing yield. To counter these threats, plants have evolved a variety of defensive strategies, both innate (preinfection) and induced (postinfection), involving various structural and biochemical strategies.

Recent studies have investigated plant resistance to both abiotic and biotic stresses at the biochemical and molecular levels. Plants utilize diverse defense mechanisms in order to stave off diseases in various host–pathogen systems. These protective systems can be divided into two categories: innate and induced. The term ‘induced

*Corresponding author: milan2925@gmail.com

DOI: 10.1079/9781800627659.0015

defense mechanism' refers to the defense strategies that plants employ against pathogens after being modified. In contrast, innate defense mechanisms are the natural responses that plants can trigger against invading pathogens due to their inherent characteristics (Agrios, 2005). In plants, innate defense mechanisms can be classified into two categories: preinfectional/preexisting and postinfectional/induced defense mechanisms, and they can both be active simultaneously throughout any stage of infection. Due to a combination of structural characteristics and metabolic processes, plants can recover from significant damage, and continue to grow and produce.

15.2 Preexisting/Preinfectional Defense Mechanisms

This mechanism acts as the plant's natural first line of defense against pathogens, existing prior to the colonization of any pathogen. It is classified into two types: preexisting structural defenses and preexisting biochemical defenses.

15.2.1 Preexisting structural defense mechanisms

This type of defense mechanism consists of the surface-level plant structure, which usually acts as a physical barrier against disease invasion. Preexisting defense structures are composed of the following.

Cell wall

These initial layers of specialized and nonspecialized host cells act as the primary barriers against pathogen invasion (Doughari, 2015). Pathogen entry can be directly prevented or made more difficult by epidermal cells' strong and resilient outer wall. For example, the thick epidermal walls help to protect against blast disease in rice, caused by *Magnaporthe oryzae*.

Waxes and cuticle

Wax, composed of long-chain aliphatic chemicals, is highly hydrophobic and creates a coating that repels water from the aerial parts of plants. This wax layer prevents water retention on plant surfaces, which is necessary for spore germination. Dry leaf surfaces reduce the risk of infection by pathogens, including bacteria, nematodes, and fungal zoospores that need water for motility. All land plants have cuticles on the external layer of their aerial epidermis, which are mostly made of waxes and cutin (Yeats and Rose, 2013). For example, when *Melampsora lini* attacks linseed, cutin serves as a physical barrier. Some pathogens can break down the cuticle enzymatically. For instance, *Monilinia fructicola* can penetrate cherry leaves but not *Ginkgo biloba* leaves which have thicker cuticles.

Natural openings

Many pathogens enter via wounds or natural openings, or through vector-borne introduction. Naturally occurring apertures in plants, such as lenticels and stomata, are often the entry points for pathogenic bacteria and fungi. Nevertheless, under some circumstances, the structure of these apertures may inhibit pathogen penetration (Cao *et al.*, 2001). The Szinkum variety of citrus is resistant to the citrus canker bacterium caused by *Xanthomonas campestris* pv. *citri* due to its small, extremely narrow stomatal apertures, which are surrounded by elevated, broad-lipped structures that prevent the passage of water droplets carrying germs (Singh, 2005). Lenticels are openings in the outer walls of plants that facilitate gas exchange. The shape and internal structure of lenticels can significantly influence the plant's susceptibility to pathogen attack. For example, small and suberized lenticels can provide resistance against the potato scab pathogen *Streptomyces scabies*.

15.2.2 Preexisting biochemical defense mechanisms

This defense process, involving various metabolic responses, occurs even before a pathogen invades plant cells or tissues. In order to avoid or lessen infection, plants release a variety of chemicals that reduce and prevent pathogen infection. These chemicals, along with the associated biochemical conditions, may negatively impact the invaders directly or indirectly by promoting the activity of antagonistic plant surface microflora (Doughari, 2015).

Inhibitors present in plant cells

Certain natural compounds called inhibitors are present in plant cells and are mostly protective against particular diseases. For example, chlorogenic acid in potatoes inhibits the potato scab pathogen *Streptomyces scabies*. Other naturally occurring compounds with antifungal membrane-lysis action include tomatine in tomatoes and avenacin in oats (Hammond-Kosack and Jones, 1997).

Inhibitors released by the plant cell

Plants release a wide variety of substances from both their above-ground parts and roots during their growth and development. These include gases and various organic compounds such as organic acids, sugars, amino acids, glycosides, and enzymes. Some of these substances have inhibitory effects, directly influencing microorganisms by either inhibiting their growth or promoting the dominance of beneficial groups that may act as antagonists to pathogens. For example, some red-scale onion cultivars that are resistant to onion smudge disease caused by *Colletotrichum circulinans* secrete fungitoxic exudates, such as catechol and protocatechuic acid. These exudates disrupt the development of fungal spores, preventing germination and thereby protecting the plant from disease (Singh, 2005).

15.2.3 Postinfectional or induced defense mechanisms

While preexisting/preinfectional defense systems aid plants in repelling many pathogens, they often prove insufficient to stop the spread of infection entirely. Consequently, host plants deploy specific structures or processes such as cytoplasmic defenses, cell wall defenses, and histological defenses, which can lead to necrosis or death of the invading cells. These defense mechanisms are activated only after pathogen colonization and are categorized into induced structural defenses and induced biochemical defenses.

Postinfectional/induced structural defense mechanisms

These defense mechanisms may form cytoplasmic, cellular, or histological structures.

HISTOLOGICAL DEFENSE STRUCTURES. Histological defense structures include the cork layer, abscission layer, tyloses, and gum depositions, which are formed when chemicals are deposited in tissues around infections.

FORMATION OF CORK LAYERS. Beyond the infection site, plant pathogens induce the host plant to form multiple layers of cork cells. In addition, to prevent the movement of nutrients and water from healthy to infected areas, these cells function to stop the inward flow of toxic substances produced by the infections (Hutcheson, 1998). For example, common potato scab caused by *Streptomyces scabies* induces the host plant to form cork cells as the mode of defense.

FORMATION OF ABSCISSION LAYERS. An abscission layer develops to facilitate the shedding of ripe fruits and older leaves from plants. It consists of spaces created between healthy and diseased tissues. Over time, infected areas gradually die, shrivel, and slough off, carrying away the pathogens. In stone fruits, abscission layers form on young, actively growing leaves as a response to pathogen infection (Odjakova and Hadjiivanova, 2001).

TYLOSE FORMATION. Tyloses are projections of protoplasts from neighboring parenchyma cells that extend into xylem vessels. Their formation effectively blocks these vessels, preventing the spread of pathogens. For example, in sweet potatoes, wilt disease is caused by the fungus *Fusarium oxysporum* f.sp. *batatas*. However, resistant varieties of sweet potatoes counteract this by quickly producing extensive tyloses before pathogen invasion. These tyloses densely occupy the xylem channels, completely obstructing them and preventing the fungus from spreading further.

GUM DEPOSITION. Many plants produce various types of gums around lesions following pathogen infection. These gum deposits serve to protect and differentiate diseased tissues from healthy ones, effectively starving the pathogen and leading to its death. For example, in stone fruits that are resistant to silver leaf disease caused by *Stereum purpureum*, gum deposition plays an important role to stop the infection from spreading.

15.2.4 Cellular defense structure

Cellular defense mechanisms include hyphal sheathing, in which the inward expansion of the cell wall causes fungal hyphae to encase themselves in a sheath. The protective layer mainly hinders entry and provides a limited restriction to the spread of pathogen. For example, in potatoes affected by late blight caused by *Phytophthora infestans*, hyphal sheathing is a significant defense mechanism. This process creates a barrier that partially checks the pathogen's entry, contributing to the plant's overall defense strategy.

15.2.5 Induced biochemical defense mechanisms

Hypersensitive response

The pathogen infection site in the host plant initiates this particular type of localized cell defense. As a result of several defensive responses by the affected and surrounding cells, toxic substances are released, which causes the hypersensitive response (HR). By killing the pathogen, the invaded cells, and surrounding cells, these substances frequently stop the infection from spreading (Agrios, 2005).

Phytoalexins

Toxic antimicrobial compounds called phytoalexins are generated by plants in reaction to mechanical damage by a variety of phytopathogenic microbes. For example, when pea pods are exposed to *Aschochyta pisi* attack, pisatin is produced. Phaseollin is produced in bean pods and leaves, providing resistance against bean anthracnose disease caused by *Colletotrichum lindemuthianum*. Additionally, when potato blight caused by *Phytophthora infestans* attacks the tubers, rishitin is produced. Phytoalexins prevent infections from growing and spreading and so are essential to plant defense.

Pathogenesis-related proteins

Pathogenesis-related (PR) proteins are the proteins that plants produce in reaction to infection by pathogens (Fig. 15.1). PR proteins were divided into 14 families by Van Loon and Van Strien (1999). PR proteins effectively activate

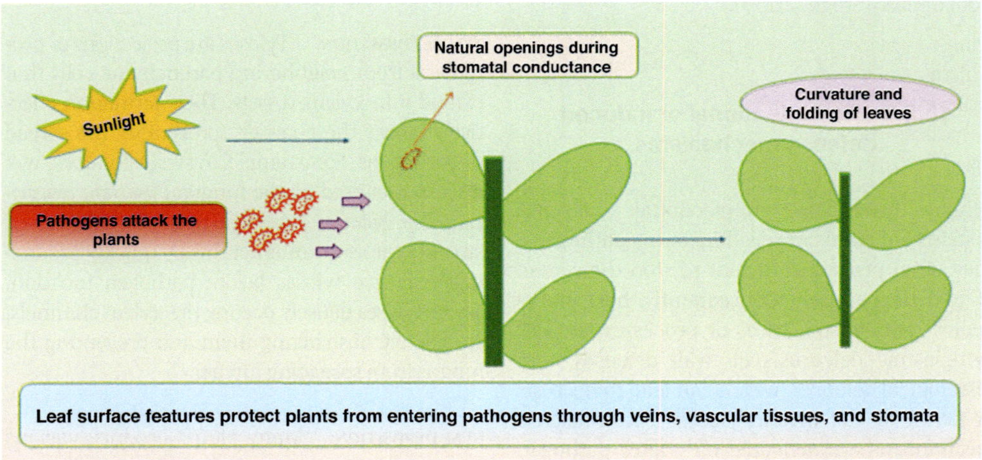

Fig. 15.1. The role of sunlight in driving stomatal conductance, showing how natural openings on leaf surfaces regulate gas exchange and influence pathogen entry points.

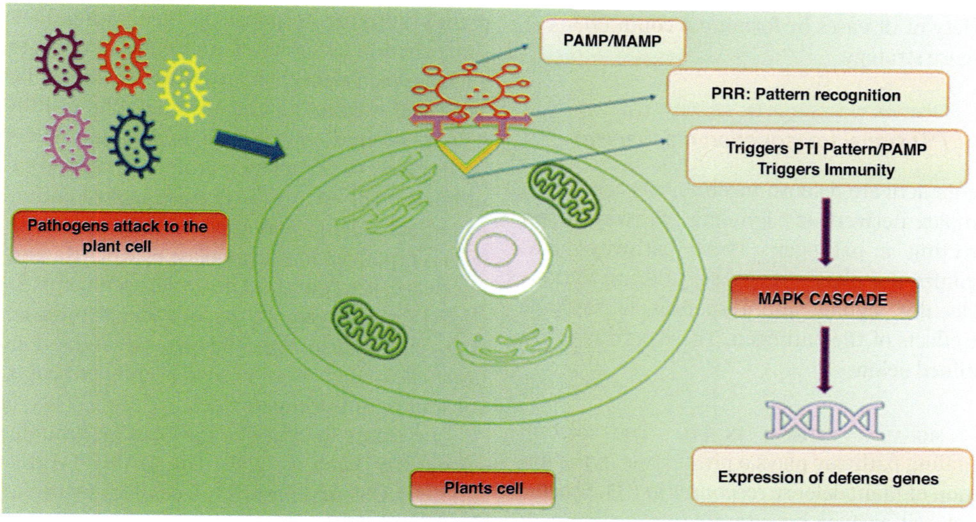

Fig. 15.2. Schematic overview of the expression of defense genes for pathogen attack and subsequent plant defense, highlighting how pathogen effector molecules are detected by NB-LRR proteins, triggering hypersensitive responses and ROS production to confine infection.

host defense mechanisms to restrict infection and prevent pathogen colonization in plant cells (Wang *et al.*, 2021). Some PRs have enzymatic activity, such as β-glucanases (PR-2), chitinases (PR-3, PR-8, PR-11), peroxidases (PR-9), and ribonucleases (PR-10).

15.3 Molecular Basis of Plant Resistance

Plants have evolved sophisticated defense mechanisms to protect themselves from pathogens, primarily through receptor and elicitor-mediated responses (Fig. 15.2). These defenses are categorized into two main systems: pattern-triggered immunity (PTI) and effector-triggered immunity (ETI). Both systems activate a series of induced defenses that typically repel pathogen attacks. Together, these recognition mechanisms enable plants to mount effective responses to infections.

15.3.1 Pattern-triggered immunity

Pattern-triggered immunity (PTI) is the first line of defense in plants. When plant cell surface receptors identify conserved microbial patterns, they start a quick reaction to the possible dangers (Rao *et al.*, 2018). PTI, which is also known as 'pathogen-triggered immunity' or 'PAMP-triggered immunity,' is triggered by both damage-associated molecular patterns (DAMPs), such as peptides and oligosaccharides resulting from plant–pathogen interactions, and pathogen-associated molecular patterns (PAMPs), such as bacterial flagellin, fungal chitin, glucans and glycoproteins from oomycetes, bacterial elongation factor-Tu (EF-Tu), and lipooligosaccharides of Gram-negative bacteria and microbe-associated molecular patterns (MAMPs) derived from nonpathogenic bacteria can also cause PTI.

Conversely, pathogen virulence effector proteins frequently target PTI signaling components, resulting in enhanced pathogen virulence and decreased plant defenses. In response to this, several plant proteins known as resistance proteins have developed the ability to bind with cytoplasmic immune receptors in order to detect effector activity.

Effector-triggered immunity (ETI), a second layer of defense, is activated by this detection. ETI successfully improves the plant's defense response against pathogen attacks by making it more robust and specific. PTI and ETI enable plants to mount a strong defense against a variety of diseases by forming a comprehensive defense strategy.

PTI signal transduction mechanism

Plants activate defense mechanisms through a complex network of signaling pathways upon detecting a pathogen. These pathways synchronize the plant's response, utilizing various molecular signals and responses to mitigate the effects of the pathogen. These pathways are outlined below.

CA^{2+} SIGNALING PATHWAY IN PTI. The calcium signaling pathway plays a pivotal role in the initiation of plant defense responses in PTI. Specific plant cell surface receptors cause an increase in the cytosolic calcium levels in response to detection of a pathogen. This is also known as the Ca^{2+} burst, which is one of the first physiological responses to MAMP/DAMP detection. A primary mechanism in Ca^{2+} signaling is that an increase in intracellular Ca^{2+} content can trigger a biological reaction (Berridge *et al.*, 2000). This response typically occurs within 30 s to 2 min and peaks around 4–6 min after perception of the microbial or damage-associated molecular patterns (MAMPs/DAMPs) (Nomura *et al.*, 2006). In PTI, a key step is a brief increase in cytoplasmic Ca^{2+} levels, regulated by botrytis-induced kinase 1 (BIK1) and related proteins (Li *et al.*, 2014). This Ca^{2+} influx activates membrane transporters, which leads to H^+ influx and K^+, Cl^-, and NO_3^- efflux. These ion changes can cause extracellular alkalinization within a minute and plasma membrane depolarization within 1–3 min.

ROS BURST. Reactive oxygen species (ROS) are essential regulators of intracellular signaling pathways in plants. When pathogens attack, plants detect MAMPs and trigger an extracellular production of ROS, a response known as the 'ROS burst.' This burst takes place within 2–3 min of MAMP recognition and reaches its peak at 10–14 min.

In the model plant *Arabidopsis*, the ROS burst is facilitated by an NADPH oxidase enzyme, known as respiratory burst oxidase homolog D (RBOHD), located in the plasma membrane. RBOHD plays a crucial role in the plant's immune response by generating ROS that function as signaling molecules to activate various defense mechanisms (Ranf *et al.*, 2011). The rapid production and accumulation of ROS help to strengthen cell walls and signal neighboring cells to commence protective measures, ultimately enhancing the defense against pathogen attack. Membrane-impermeable superoxide (O^{2-}) is produced in the apoplast by RBOHD, and afterward, it is converted into hydrogen peroxide (H_2O_2) by the enzyme superoxide dismutase. H_2O_2 can diffuse into the cytosol and other organelles since it is a membrane-permeable substance, unlike superoxide.

Increases in cytosolic Ca^{2+} can be brought on by ROS, such as H_2O_2. The MAMP/DAMP-induced ROS burst positively feeds back cytosolic Ca^{2+} levels, which results in an additional increase in cytosolic Ca^{2+}. This feedback can lead to a second peak of elevated Ca^{2+}. This intricate interplay between ROS and Ca^{2+} signaling amplifies the plant's defense responses, ensuring a vigorous reaction to microbial invasion (Rentel and Knight, 2004).

REACTIVE NITROGEN SPECIES. Reactive nitrogen species, which include nitric oxide (NO) and its derivatives, are essential for several phases of signal transduction. By modifying the cysteine S-nitrosylation of NPR1, which is a master regulator of defense gene expression, they control its oligomeric state. Moreover, reactive nitrogen species can inhibit the activity of RBOHD by cysteine S-nitrosylation. These modifications influence the signaling pathways, which in turn enhance the plant's defense mechanisms. The involvement of reactive nitrogen species in this process highlights their importance in enhancement of immune responses and maintaining the balance between different signaling components during pathogen attack.

Plant pattern recognition receptors

Plant pattern recognition receptors (PRRs) are important components of the immune system, acting as receptor-like kinases (RLKs) or receptor-like proteins (RLPs). These receptors detect MAMPs/ PAMPs that signal the presence of pathogens. PRRs are characterized by their extracellular domains, which usually feature nucleotide-binding/leucine-rich repeat

(NLR) or lysine motif (LysM) structures. These domains are crucial for recognizing and binding to specific MAMPs/PAMPs, initiating a signaling cascade that activates the plant's defense responses. Binding usually causes receptor oligomerization and interaction with coreceptors, which is essential for signaling. Sensitivity and specificity define the PRR response. Recognizing a conserved epitope enables plants that express the appropriate receptor to sense the associated MAMPs/PAMPs even at subnanomolar concentrations.

Using genetic and biochemical methods, PRRs can be introduced into many plant species to improve their defenses against pathogens. Whatever the stage of the pathogen's life cycle, PRRs can be designed to enhance ligand recognition and intracellular immunological responses expressed constitutively in the host.

Phytohormones

Plant responses are facilitated by phytohormones, which function as signaling molecules that coordinate cell-to-cell activities. Usually, salicylic acid (SA) signaling is essential for defense against infections that feed on living host tissue and are classified as either biotrophic or hemibiotrophic. In contrast, defense against necrotrophic pathogens, pathogens that kill host tissue before feeding on the dead matter, usually requires jasmonic acid (JA) and ethylene (ET) signaling (Glazebrook, 2005). Salicylic acid, a precursor of aspirin, is important for plant defense, particularly in systemic acquired resistance. It binds to two specific proteins in plant cell membranes. One of these proteins, which normally breaks down hydrogen peroxide, is inhibited by SA, causing hydrogen peroxide to accumulate. This buildup acts as a signal to enhance the plant's resistance to pathogens. The second protein binds SA with high affinity and directly activates the expression of defense-related genes. Salicylic acid levels increase rapidly around areas of infection and stay elevated in plants that have developed resistance, helping to protect against further attacks.

Soil-borne microbes that colonize roots and benefit plant growth can also trigger systemic resistance mechanisms in leaves. Plant growth-promoting rhizobacteria (PGPR), or beneficial rhizobacteria, induce systemic resistance (ISR) against the pathogen infection. These ISR-inducing microorganisms have the capability to influence the plant's morphological, physiological, and molecular responses, enhancing overall plant health and defense.

Jasmonic acid induces defenses against necrotrophic pathogens. It is a distinct chemical class predominantly produced by the oxidation of fatty acids and contributes to systemic acquired resistance. JA often works in combination with SA and ET to produce defensive components. Ethylene mediates defense against necrotrophic pathogens (Rithesh *et al.*, 2024). The recognition of MAMPs initiates the production of SA, JA, and ET. In particular, SA production starts between 3 and 6 hours after flg22 treatment and peaks at approximately 9 hours. These hormones work together to enhance the plant's resistance to different biotic stresses (Tsuda *et al.*, 2008). According to observations made on 2-week-old *Arabidopsis* Col-0 seedlings treated with 0.2 mM flg22, ET production begins around an hour after treatment and peaks about 4 hours later (Liu and Zhang, 2004). On the other hand, Pep-13, a MAMP, can trigger the formation of JA in potato derived from *Phytophthora* species causing late blight of potato (Halim *et al.*, 2009).

15.3.2 Effector-triggered immunity

Effector-triggered immunity is a crucial mechanism that evolved as part of the intricate and dynamic relationship between plants and their pathogens. The concept of ETI builds on foundational outcomes in the mid-20th century, predominantly H. H. Flor's 'gene-for-gene' hypothesis. Flor's work with flax rust pathogen *Melampsora lini* revealed that specific plant resistance (R) genes correspond to pathogen avirulence (*Avr*) genes, establishing that plants can recognize and respond to particular pathogen strains through these specific interactions (Yuan *et al.*, 2021). This breakthrough laid the groundwork for understanding how plants detect and defend against pathogens at the molecular level.

In the 1980s and 1990s, advances in molecular biology enabled the cloning and characterization of the first plant R genes, such as the tomato '*Cf-9*' gene and the tobacco *N* gene,

Table 15.1. Plant defense mechanisms (innate and induced immunity).

Name of crop	Scientific name	Immunity	References
Rice	*Oryza sativa*	Induced immunity (PTI, ETI, SAR)	Yang *et al.* (2022)
Tomato	*Solanum lycopersicum*	Induced immunity (ETI, RNAi)	Oh and Martin (2011)
Potato	*Solanum tuberosum*	Induced immunity (ETI, PTI)	Martin *et al.* (2020)
Wheat	*Triticum aestivum*	Innate (physical barriers), induced (SAR)	Bertini *et al.* (2003)
Maize	*Zea mays*	Induced immunity (ISR, PTI)	Ziemann *et al.* (2018)
Onion	*Allium cepa*	Innate (physical barriers), induced (PTI)	Abdelrahman *et al.* (2016)
Soybean	*Glycine max*	Induced immunity (ETI, ISR)	Zhang *et al.* (2022)
Barley	*Hordeum vulgare*	Innate (physical barriers), induced (ETI)	Walters *et al.* (2011)
Arabidopsis	*Arabidopsis thaliana*	Induced immunity (PTI, ETI, SAR)	Dong *et al.* (2015)
Grapevine	*Vitis vinifera*	Induced resistance (MAMP, DAMP)	Héloir *et al.* (2019)
Cucumber	*Cucumis sativus*	Induced immunity (ISR, SAR)	Vlot *et al.* (2021)
Tobacco	*Nicotiana tabacum*	Induced tmmunity (ETI, SAR)	Pallas *et al.* (1996); Crabill *et al.* (2010)
Pepper	*Capsicum annuum*	Induced immunity (ETI, PTI)	Priller *et al.* (2016)
Pea	*Pisum sativum*	Induced immunity (PTI, SAR)	Li *et al.* (2021)
Tea	*Camellia sinensis*	Induced immunity (PTI, ISR)	Rui *et al.* (2023)
Cotton	*Gossypium hirsutum*	Induced immunity (ETI, PTI)	Qiu *et al.* (2020)
Pear	*Pyrus communis*	Innate (physical barriers), induced (SAR)	Sparla *et al.* (2004)

DAMP, damage-associated molecular pattern; ETI, effector-triggered immunity; ISR, induced systemic resistance; MAMP, microbe-associated molecular pattern; PTI, pattern-triggered immunity; SAR, systemic acquired resistance.

conferring resistance to *Cladosporium fulvum* and tobacco mosaic virus (TMV), respectively (Kaur *et al.*, 2021). Concurrently, pathogen effectors (encoded by *Avr* genes) were identified, providing crucial insights into how pathogens interact with plant hosts. For instance, the bacterial effector AvrBs2 from *Xanthomonas campestris* was one of the first effectors characterized, emphasizing the diverse strategies pathogens use to influence host processes and promote infection (Timilsina *et al.*, 2020).

Effector-triggered immunity comprises the recognition of pathogen effectors by plant R proteins, typically nucleotide-binding NLR proteins (Table 15.1). These R proteins consist of three main domains: a central nucleotide-binding (NB) domain, a C-terminal leucine-rich repeat (LRR) domain, and an N-terminal domain that can be either a Toll/interleukin-1 receptor (TIR) or a coiled-coil (CC) domain. Upon identifying specific effectors, either directly or indirectly through adjustments in host proteins, R proteins initiate a robust immune response, comprising the HR, the production of ROS, and the expression of PR proteins (Fig. 15.3). The evolutionary arms contest between plants and pathogens drives the diversification of *R* genes and pathogen effectors. As pathogens evolve new effectors to evade detection, plants continuously develop new *R* genes to recognize these evolving threats. This coevolutionary dynamic is evident in the extensive polymorphism observed in *R* gene families, such as the *Mla* (Mildew locus a) gene family in barley and the RPP (resistance to the

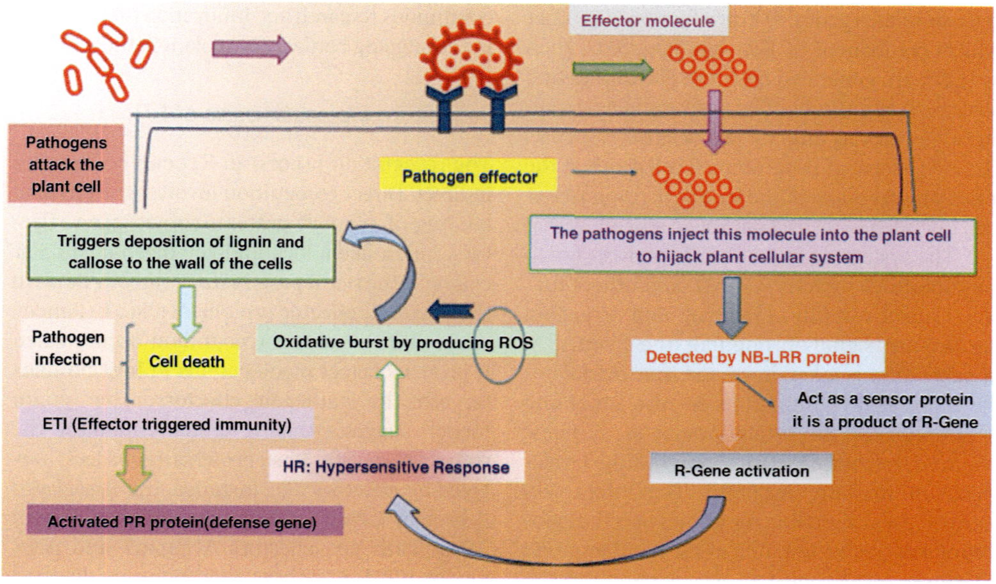

Fig. 15.3. Schematic representation of plant immunity.

oomycete *Hyaloperonospora arabidopsidis*) gene family in *Arabidopsis*, which provide resistance to various strains of powdery mildew and downy mildew pathogens, respectively (Tyler and Rouxel, 2012). This genetic diversity ensures that plant populations retain some level of resistance against rapidly evolving pathogens.

Pathogen effectors, conversely, are diverse molecules produced by pathogens to facilitate infection by manipulating host cell processes. These effectors can be proteins, small RNAs, or other molecules that interfere with host cell function. They are naturally delivered into plant cells by various secretion systems, such as the Type III secretion system in bacteria. Effectors target key components of the plant immune system, aiming to suppress PTI and manipulate host cellular processes to benefit the pathogen. For instance, the bacterial effector AvrPtoB from *Pseudomonas syringae* can ubiquitinate and degrade host kinases involved in PTI, thereby dampening the plant's initial immune response (Liu *et al.*, 2022).

The recognition of pathogen effectors by R proteins can occur through two main mechanisms: direct and indirect recognition. Direct recognition involves the physical binding of the R protein to the effector. An example of this is the interaction between the *Arabidopsis* R protein resistance to *Phakopsora pachyrhizi* 1 (RPP1) and the effector type 1AngII (ATR1) from the oomycete pathogen *H. arabidopsidis* (Tyler and Rouxel, 2012). Indirect recognition, as exemplified by the 'guard model' and the 'decoy model,' involves R proteins monitoring host proteins that are altered by pathogen effectors. In the guard model, R proteins 'guard' these host proteins and trigger an immune response when modifications are detected. The *Arabidopsis* R protein resistance to *P. syringae* pv. *maculicola*1 (RPM1) monitors the host protein (RPM1)-interacting protein 4 (RIN4), which is modified by the pathogen *P. syringae* effector AvrRpm1 and avirulence effector protein (AvrB) (Zhao *et al.*, 2021). The decoy model suggests that some host proteins act as decoys, mimicking genuine effector targets to lure effectors and facilitate their recognition by R proteins.

Signal transduction following effector recognition involves a cascade of molecular events leading to the activation of defense responses. Key players in this process include mitogen-activated protein kinases (MAPKs) and calcium-dependent protein kinases (CDPKs), which are rapidly activated upon R protein activation. These kinases phosphorylate various downstream targets, leading to the activation of transcription factors that regulate the expression of defense-related genes. ROS production is another critical component of ETI signaling. ROS, such as H_2O_2, are generated by NADPH oxidases and serve both as antimicrobial agents and as signaling molecules to amplify the defense response. The HR, a form of programmed cell death at the infection site, is one of the hallmark responses of ETI, intended to restrict pathogen spread.

The systemic nature of ETI involves signaling molecules such as SA, NO, and others that play crucial roles in amplifying and spreading the defense signal throughout the plant. SA, in particular, is a key hormone that modulates the expression of numerous defense genes and is essential for the establishment of systemic acquired resistance (SAR), a long-lasting, broad-spectrum immune response. The interaction between these signaling molecules and the transcriptional reprogramming of the plant immune system ensures a robust and coordinated defense against a wide range of pathogens.

Considering the molecular components of ETI provides crucial insights into plant immunity and informs strategies for developing disease-resistant crops. Advances in genomics and biotechnology, including the use of CRISPR-Cas9 for precise genome editing, offer new avenues for enhancing ETI by introducing or modifying *R* genes. By leveraging these molecular insights, researchers can design crops with improved resistance profiles, contributing to sustainable agriculture and food security. Modern advances in genomics and bioinformatics have revolutionized the study of ETI, enabling the rapid identification of *R* genes and pathogen effectors. Techniques like genome-wide association studies (GWAS) and high-throughput sequencing have facilitated the discovery of resistance loci and effector repertoires, providing deeper insights into the coevolutionary history of plant–pathogen interactions (Xiao *et al.*, 2022). For instance, CRISPR-Cas9 has been used to enhance disease resistance in rice by targeting susceptibility genes and stacking multiple *R* genes (Zafar *et al.*, 2020).

Understanding ETI is crucial for developing sustainable agricultural practices and ensuring food security, as scientists can engineer crops with durable resistance, reducing reliance on chemical pesticides and promoting environmental sustainability. The discovery and evolution of ETI highlight the complex and adaptive nature of plant immune systems, and the importance of continuous research and innovation to stay ahead in the ongoing battle against plant pathogens.

Molecular basis of ETI

The recognition process in ETI can be direct or indirect. Direct recognition involves the physical binding of plant R proteins to pathogen effectors, much like a lock-and-key mechanism. For example, the rice R protein Xa21 directly binds to the bacterial effector protein AvrXa21 (Pandey *et al.*, 2023). Indirect recognition occurs when R proteins detect modifications in host proteins targeted by pathogen effectors. The 'guard model' suggests that R proteins monitor the status of specific host proteins (guardees) targeted by effectors. For instance, the *Arabidopsis* R protein RPM1 monitors the guardee protein RIN4. Pathogen effectors AvrRpm1 and AvrB modify RIN4, and these alterations are detected by RPM1, triggering ETI (Zhao *et al.*, 2021). Once R proteins distinguish pathogen effectors, they undergo conformational changes that lead to their activation. Most R proteins involved in ETI are NLR proteins. The NB domain binds and hydrolyzes ATP or GTP, essential for R protein activation, while the LRR domain is involved in recognizing pathogen effectors.

Upon activation, these proteins initiate a cascade of downstream signaling events that amplify the defense response. The activation of R proteins triggers immediate early signaling events, including the rapid activation of MAPKs and CDPKs. These kinases phosphorylate various downstream targets, leading to the activation of transcription factors that regulate the expression of defense-related genes. One of the early responses in ETI is the production of ROS, such as H_2O_2 (see Table 15.1). ROS have antimicrobial properties and also function as signaling molecules to further propagate the defense response. The production of ROS is mediated by enzymes such as NADPH oxidases and peroxidases, which are rapidly activated upon ETI initiation (see Table 15.1). Additionally, ETI induces the expression of PR proteins, which have various antimicrobial activities, including chitinases, glucanases, and protease inhibitors that degrade pathogen cell walls and inhibit their enzymes.

Effector-triggered immunity can also trigger SAR, a long-lasting, broad-spectrum immune response throughout the plant.

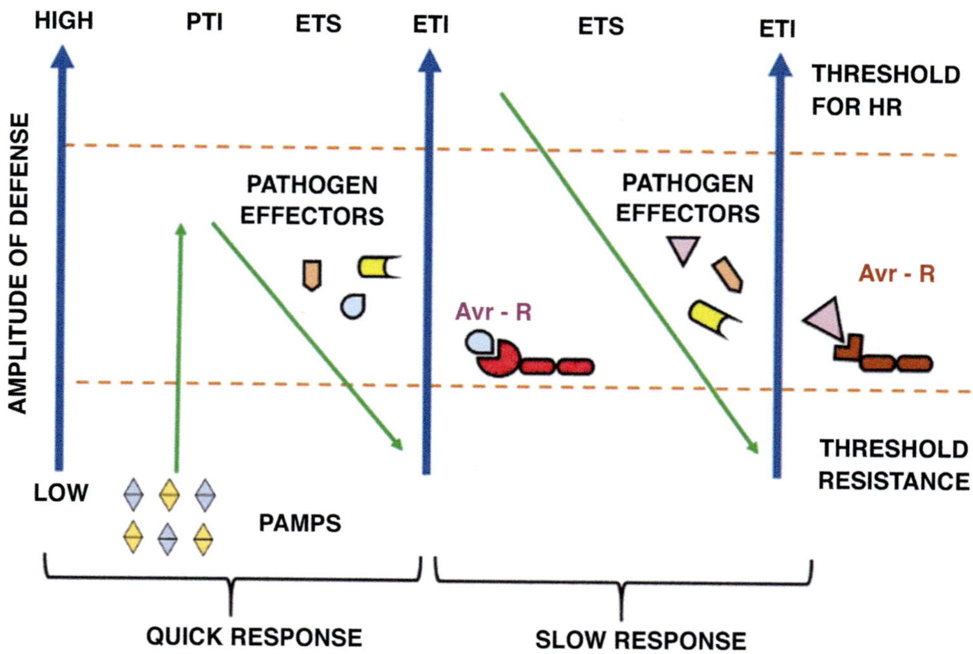

Fig. 15.4. The integration of pattern-triggered immunity (PTI) and effector-triggered immunity (ETI) in plant immunity.

Signaling molecules such as SA, MeSA, and azelaic acid, which travel from the infection site to distant tissues, mediate SAR, priming them for enhanced defense. SAR involves the upregulation of a distinct set of PR genes and other defense-related genes in systemic tissues. The NPR1 plays a central role in SAR (Fig. 15.4). In response to SA, NPR1 translocates to the nucleus and activates SAR-related gene expression, thereby enhancing the plant's overall defense capacity. The mechanisms of ETI involve a highly coordinated and dynamic interaction between pathogen recognition, signal transduction, and activation of robust defense responses. This ongoing evolutionary arms race between plants and pathogens drives the continuous adaptation and diversification of *R* genes and pathogen effectors.

Effector-triggered immunity and systemic acquired resistance

R proteins, often NLR proteins, identify effectors delivered into plant cells by pathogens and undergo conformational modifications that trigger a signaling cascade. This cascade consists of early signaling events, such as the initiation of MAPKs and CDPKs, which phosphorylate downstream targets, leading to the activation of transcription factors and the expression of defense-related genes. A key reaction of ETI is the HR, a localized programmed cell death that limits pathogen spread by sacrificing infected and neighboring cells. SAR is a long-lasting, broad-spectrum immune response mediating ETI's defense signals throughout the plant. Elicited by localized infections, SAR confers enhanced resistance to subsequent infections by various pathogens in distant tissues. Signaling molecules, including SA, MeSA, and azelaic acid, mediate SAR. SA accumulates at the infection site and in distal tissues, priming the plant's immune system for rapid future responses. SA's transformation to MeSA facilitates long-distance signaling through the vascular system. SAR establishment involves complex molecular signaling networks, with the NPR1 being essential. Elevated SA levels induce a redox-mediated conformational change in NPR1, allowing it to enter the nucleus and activate SAR-related genes, including antimicrobial PR proteins (Niu *et al.*, 2016). Advances in molecular biology and genomics have revealed deeper perceptions into ETI and SAR mechanisms, identifying numerous genes and regulatory networks. Biotechnological tools like CRISPR-Cas9 genome editing enable precise control of these genes, offering strategies for engineering disease-resistant crops. Combining traditional breeding with modern biotechnological approaches promises to improve next-generation crops with enhanced disease resistance, promoting sustainable agricultural practices and food security.

15.4 Conclusion

Both PTI and ETI represent a critical defense mechanism in plants, involving the specific recognition of pathogen PAMPs, such as fungal chitin or bacterial flagellin and effectors by R proteins respectively, leading to robust immune responses like the HR and SAR. Despite challenges such as rapid pathogen evolution and the complexity of *R* gene networks, future prospects in both PTI and ETI research are promising. Advances in effector and *R* gene discovery through high-throughput sequencing and biotechnological tools like CRISPR-Cas9 offer new avenues for developing durable disease resistance in crops. Integrating genomic insights with traditional breeding strategies, marker-assisted selection (MAS), and innovative pest management approaches will enhance crop resilience and contribute to sustainable agriculture worldwide. Global collaboration, data sharing, and education efforts will be crucial in translating PTI/ETI research into practical solutions, ensuring food security in the face of emerging pathogen threats and changing environmental conditions.

References

Abdelrahman, M., Abdel-Motaal, F., El-Sayed, M., Jogaiah, S., Shigyo, M. *et al.* (2016) Dissection of *Trichoderma longibrachiatum*-induced defense in onion (*Allium cepa* L.) against *Fusarium oxysporum* f. sp. cepa by target metabolite profiling. *Plant Science* 246, 128–138.

Agrios, G.N. (2005) *Plant Pathology*. Academic Press, London.

Berridge, M.J., Lipp, P. and Bootman, MD. (2000) The versatility and universality of calcium signalling. *Nature Reviews Molecular Cell Biology* 1(1), 11–21.

Bertini, L., Leonardi, L., Caporale, C., Tucci, M., Cascone, N. *et al.* (2003) Pathogen-responsive wheat PR4 genes are induced by activators of systemic acquired resistance and wounding. *Plant Science* 164(6), 1067–1078.

Cao, H., Baldini, R.L. and Rahme, L.G. (2001) Common mechanisms for pathogens of plants and animals. *Annual Review of Phytopathology* 39, 259–284.

Crabill, E., Joe, A., Block, A., Rooyen, J.M. and Alfano, J.R. (2010) Plant immunity directly or indirectly restricts the injection of type III effectors by the *pseudomonas syringae* type III secretion system. *Plant Physiology* 154(1), 233–244.

Dong, X., Jiang, Z., Peng, Y.L. and Zhang, Z. (2015) Revealing shared and distinct gene network organization in *Arabidopsis* immune responses by integrative analysis. *Plant Physiology* 167(3), 1186–1203.

Doughari, D. (2015) An overview of plant immunity. *Journal of Plant Pathology & Microbiology* 6(2), 1–11.

Glazebrook, J. (2005) Contrasting mechanisms of defense against biotrophic and necrotrophic pathogens. *Annual Review of Phytopathology* 43, 205–227.

Halim, V.A., Altmann, S., Ellinger, D., Eschen-Lippold, L., Miersch, O. *et al.* (2009) PAMP-induced defense responses in potato require both salicylic acid and jasmonic acid. *Plant Journal* 57(2), 230–242.

Hammond-Kosack, K.E. and Jones, J.D.G. (1997) Plant disease resistance genes. *Annual Review of Plant Physiology and Plant Molecular Biology* 48(1), 575–607.

Héloir, M.-C., Adrian, M., Brulé, D., Claverie, J., Cordelier, S. *et al.* (2019) Recognition of elicitors in grapevine: From MAMP and DAMP perception to induced resistance. *Frontiers in Plant Science* 10, 1117.

Hutcheson, S.W. (1998) Current concepts of active defense in plants. *Annual Review of Phytopathology* 36, 59–90.

Kaur, B., Bhatia, D. and Mavi, G.S. (2021) Eighty years of gene-for-gene relationship and its applications in identification and utilization of *R* genes. *Journal of Genetics* 100, 50.

Li, L., Li, M., Yu, L., Zhou, Z., Liang, X. *et al.* (2014) The FLS2-associated kinase BIK1 directly phosphorylates the NADPH oxidase RBOHD to control plant immunity. *Cell Host & Microbe* 15(3), 329–338.

Li, P., Zhao, L., Qi, F., Htwe, N.M.P.S., Li, Q. *et al.* (2021) The receptor-like cytoplasmic kinase RIPK regulates broad-spectrum ROS signaling in multiple layers of plant immune system. *Molecular Plant* 14(10), 1652–1667.

Liu, Y. and Zhang, S. (2004) Phosphorylation of 1-aminocyclopropane-1-carboxylic acid synthase by MPK6, a stress-responsive mitogen-activated protein kinase, induces ethylene biosynthesis in *Arabidopsis*. *Plant Cell* 16(12), 3386–3399.

Liu, Y., Mahmud, M.R., Xu, N. and Liu, J. (2022) The *Pseudomonas syringae* effector AvrPtoB targets abscisic acid signaling pathway to promote its virulence in *Arabidopsis*. *Phytopathology Research* 4(1), 5.

Martin, R.L., Le Boulch, P., Clin, P., Schwarzenberg, A., Yvin, J.-C. *et al.* (2020) A comparison of PTI defense profiles induced in *solanum tuberosum* by PAMP and non-PAMP elicitors shows distinct, elicitor-specific responses. *PLoS One* 15(8), e0236633.

Niu, D., Wang, X., Wang, Y., Song, X., Wang, J. *et al.* (2016) Bacillus cereus AR156 activates PAMP-triggered immunity and induces a systemic acquired resistance through a NPR1-and SA-dependent signaling pathway. *Biochemical and Biophysical Research Communications* 469(1), 120–125.

Nomura, K., Debroy, S., Lee, Y.H., Pumplin, N., Jones, J. *et al.* (2006) A bacterial virulence protein suppresses host innate immunity to cause plant disease. *Science* 313(5784), 220–223.

Odjakova, M. and Hadjiivanova, C. (2001) The complexity of pathogen defence in plants. *Bulgarian Journal of Plant Physiology* 27(1/02), 101–109.

Oh, C.S. and Martin, G.B. (2011) Effector-triggered immunity mediated by the Pto kinase. *Trends in Plant Science* 16(3), 132–140.

Pallas, J.A., Paiva, N.L., Lamb, C. and Dixon, R.A. (1996) Tobacco plants epigenetically suppressed in phenylalanine ammonia-lyase expression do not develop systemic acquired resistance in response to infection by tobacco mosaic virus. *Plant Journal* 10(2), 281–293.

Pandey, S., Mishra, P., Monya, T. and Rai, R. (2023) Virulence determinants and host defence factors in *Xanthomonas*-rice interaction. In: Srivastava, D., Shamim, M., Ahmad, M. and Singh, K. (eds) *Bacterial Diseases of Rice and Their Management*. CRC Press, Boca Raton, pp. 73–90.

Priller, J.P.R., Reid, S., Konein, P., Dietrich, P. and Sonnewald, S. (2016) The *Xanthomonas campestris* pv. vesicatoria type-3 effector XopB inhibits plant defence responses by interfering with ROS production. *PLoS One* 11(7), e0159107.

Qiu, T., Wang, Y., Jiang, J., Zhao, J., Wang, Y. *et al.* (2020) *GbAt11* gene cloned from *Gossypium barbadense* mediates resistance to Verticillium wilt in *Gossypium hirsutum*. *Journal of Cotton Research* 3(1), 9.

Ranf, S., Eschen-Lippold, L., Pecher, P., Lee, J. and Scheel, D. (2011) Interplay between calcium signalling and early signalling elements during defence responses to microbe- or damage-associated molecular patterns. *Plant Journal* 68(1), 100–113.

Rao, S., Zhou, Z., Miao, P., Bi, G., Hu, M. *et al.* (2018) Roles of receptor-like cytoplasmic kinase VII members in pattern-triggered immune signaling. *Plant Physiology* 177(4), 1679–1690.

Rentel, M.C. and Knight, M.R. (2004) Oxidative stress-induced calcium signaling in *Arabidopsis*. *Plant Physiology* 135(3), 1471–1479.

Rithesh, L., Krishna, G.V., Rao, S.S. and Shiva, B. (2024) Emerging role of melatonin in integrated management of crop pathogens. In: Kumar, R., Altaf, M., Lal, M. and Tiwari, R. (eds) *Melatonin in Plants: A Pleiotropic Molecule for Abiotic Stresses and Pathogen Infection*. Springer Nature, Singapore, pp. 195–221.

Rui, L., Su, J.-Y., Li, T., Sun, J.-M. and Wu, G.-H. (2023) Molecular regulation of immunity in tea plants. *Molecular Biology Reports* 50(3), 2883–2892.

Singh, R.P. (2005) *Plant Pathology*. Kaylani Publishers, New Delhi.

Sparla, F., Rotino, L., Valgimigli, M.C., Pupillo, P. and Trost, P. (2004) Systemic resistance induced by benzothiadiazole in pear inoculated with the agent of fire blight (*Erwinia amylovora*). *Scientia Horticulturae* 101(3), 269–279.

Timilsina, S., Potnis, N., Newberry, E.A., Liyanapathiranage, P., Iruegas-Bocardo, F. *et al.* (2020) *Xanthomonas* diversity, virulence and plant-pathogen interactions. *Nature Reviews Microbiology* 18(8), 415–427.

Tsuda, K., Sato, M., Glazebrook, J., Cohen, J.D. and Katagiri, F. (2008) Interplay between MAMP-triggered and SA-mediated defense responses. *Plant Journal* 53(5), 763–775.

Tyler, B.M. and Rouxel, T. (2012) Effectors of fungi and oomycetes: Their virulence and avirulence functions and translocation from pathogen to host cells. *Molecular Plant Immunity* 123–167.

Van Loon, L.C. and Van Strien, E.A. (1999) The families of pathogenesis-related proteins, their activities, and comparative analysis of PR-1 type proteins. *Physiological and Molecular Plant Pathology* 55(2), 85–97.

Vlot, A.C., Sales, J.H., Lenk, M., Bauer, K., Brambilla, A. *et al.* (2021) Systemic propagation of immunity in plants. *New Phytologist* 229(3), 1234–1250.

Walters, D.R., Havis, N.D., Paterson, L., Taylor, J. and Walsh, D.J. (2011) Cultivar effects on the expression of induced resistance in spring barley. *Plant Disease* 95(5), 595–600.

Wang, Y., Liu, M., Wang, X., Zhong, L., Shi, G. *et al.* (2021) A novel β-1,3-glucanase gns6 from rice possesses antifungal activity against *Magnaporthe oryzae*. *Journal of Plant Physiology* 265, 153493.

Xiao, Q., Bai, X., Zhang, C. and He, Y. (2022) Advanced high-throughput plant phenotyping techniques for genome-wide association studies: A review. *Journal of Advanced Research* 35, 215–230.

Yang, Y., Kim, N.H., Cevik, V., Jacob, P., Wan, L. *et al.* (2022) Allelic variation in the *Arabidopsis* TNL CHS3/CSA1 immune receptor pair reveals two functional cell-death regulatory modes. *Cell Host & Microbe* 30(12), 1701–1716.

Yeats, T.H. and Rose, J.K.C. (2013) The formation and function of plant cuticles. *Plant Physiology* 163(1), 5–20.

Yuan, M., Ngou, B.P.M., Ding, P. and Xin, X.-F. (2021) PTI-ETI crosstalk: An integrative view of plant immunity. *Current Opinion in Plant Biology* 62, 102030.

Zafar, K., Khan, M.Z., Amin, I., Mukhtar, Z., Yasmin, S. *et al.* (2020) Precise CRISPR-Cas9 mediated genome editing in super basmati rice for resistance against bacterial blight by targeting the major susceptibility gene. *Frontiers in Plant Science* 11, 575.

Zhang, Q., Chen, S., Bao, Y., Wang, D., Wang, W. *et al.* (2022) Functional diversification analysis of soybean malectin/malectin-like domain-containing receptor-like kinases in immunity by transient expression assays. *Frontiers in Plant Science* 13.

Zhao, G., Guo, D., Wang, L., Li, H., Wang, C. *et al.* (2021) Functions of RPM1-interacting protein 4 in plant immunity. *Planta* 253(1), 1–11.

Ziemann, S., Linde, K., Lahrmann, U., Acar, B., Kaschani, F. *et al.* (2018) An apoplastic peptide activates salicylic acid signalling in maize. *Nature Plants* 4(3), 172–180.

16 Alley Cropping: A Sustainable Agroforestry System for Enhancing Productivity, Soil Health and Biodiversity

Sagar Maitra[1]*, Masina Sairam[1], Lalichetti Sagar[1], Sumit Ray[1], D.T Santosh[2], Dinkar Jagannath Gaikwad[2], Anjana J. Atapattu[3] and Harun I. Gitari[4]

[1]Department of Agronomy and Agroforestry, M. S. Swaminathan School of Agriculture, Centurion University of Technology and Management, Paralakhemundi, India; [2]Research Centre for Smart Agriculture, Centurion University of Technology and Management, Paralakhemundi, India; [3]Agronomy Division, Coconut Research Institute of Sri Lanka, Lunuwila, Sri Lanka; [4]Department of Agricultural Science and Technology, School of Agriculture and Environmental Sciences, Kenyatta University, Nairobi, Kenya

Abstract

To ensure food and environmental security by 2050, it is imperative to adopt strategies that bring less fertile and marginal lands into crop production as fertile and productive lands are already engaged. In this context, alley cropping plays a crucial role for crop diversification and agricultural sustainability as besides improving farm output per unit area, it also helps to mitigate global warming by cultivating food, forage, and specialty crops in the space available between rows of perennial crops and contributing towards carbon sequestration. In this chapter it is observed that crop choice usually depends mainly on regional conditions and farmers' preferences. In general, perennial tree species with fewer roots are selected for hedgerows while short-duration, hardy, and partially shade-tolerant annual crops are likely to be chosen for cultivation in alleys. The total production of biomass will be greater in closed spacing of alleys. In areas with low rainfall, wider alleys are planted to avoid competition for soil moisture. Pruning the tree crops is vital to reduce the shade impact when the annual crops are produced. In order to prevent shadowing the annual crop with a low canopy height and to allow both crops to access sunshine, a low pruning height of 15–30 cm would be ideal. Further, the addition of pruned biomass was observed to improve the nutrient yield, and legume hedgerows contributed more nutrients (especially N) than nonlegumes. In addition, alley cropping systems also helped to conserve and improve soil properties by both their above- and below-ground interactions. In the light of the above, this chapter examines alley cropping's potential for achieving sustainability.

Keywords: alley cropping, crop diversification, sustainability, global warming, carbon sequestration

*Corresponding author: sagar.maitra@cutm.ac.in

DOI: 10.1079/9781800627659.0016

16.1 Introduction

In the current situation, crop intensification is of the utmost importance to increase agricultural production in relation to ever-diminishing and degrading resources. The FAO estimated that by 2050, we will need to increase food production by 60% to feed a world population of 9.3 billion (UN, 2012). To reach this goal, a basket of technologies is to be adopted which will connect agroecosystems and improve ecosystem services by reducing the carbon footprint of food production (Maitra *et al.*, 2023; Santosh *et al.*, 2024).

In the current context of climate change, it is imperative to reduce anthropogenic greenhouse gas emissions to restrict further damage caused by global warming. In this regard, agroforestry could offer ample opportunity to enhance farm output with multifaceted benefits (Xu *et al.*, 2019; Maitra *et al.*, 2024; Chappa *et al.*, 2024; Gitari *et al.*, 2020, 2024). Forests are known to maintain ecological balance in many ways so the advantages of forestry can be included in agroecosystems in the form of agroforestry which enables forest products and crops to be grown together (Nungula *et al.*, 2024).

Among various agroforestry systems, alley cropping (Fig. 16.1) is an option where crop growing can harness the benefits of forests (Solaimalai *et al.*, 2005). In erosion-prone, low fertile and marginal lands, successful raising of crops is a difficult task but alley cropping, which is the combination of selected crops and forests, can ensure economically viable farm productivity by offering short-, medium-, and long-term benefits. As more farm output is required to meet future demand, fertile and productive lands are already engaged. There is further scope to bring less fertile and marginal lands into crop production by adoption of agroforestry as well as alley cropping for crop diversification and agricultural sustainability.

Alley cropping is a form of intercropping system, also termed hedgerow intercropping, where food, forage, and specialty crops are cultivated in the space available between rows of perennial crops. Similar to other intercropping

Fig. 16.1. Alley cropping of trees and annual crops.

systems, two or more different types of crops coexist in alley cropping and appropriate agronomic management is provided to all component crops. In general, annual food crops or forages are grown in the alleys (free space) between two rows of perennial plants (hedgerows) which may be trees or shrubs (Korwar, 1990).

Alley cropping is a common agroforestry system in which perennial trees or shrubs are established in distant rows as per recommended spacings, thus creating a space for raising annual crops in the initial years of perennial crops prior to formation of their full canopy (Gold and Garrett, 2009; Maitra *et al.*, 2021). Such mixed stands are economic as they provide a regular cashflow from annual crops and offer medium- to long- term products from trees or shrubs (Xu *et al.*, 2019). In general, mixed stands result in interspecies competition for available resources such as water, light, and nutrients from the common pool under resource-poor conditions (Jose *et al.*, 2006; Maitra, 2023). Earlier research evidenced yield reduction in alley cropping for both perennial and annual crops; however, the combined output was higher than sole forestry systems as the land equivalent ratio (LER) was greater than unity (Nerlich *et al.*, 2013; Singh *et al.*, 2014; Razouk *et al.*, 2016). Miah *et al.* (2018) recorded that in a jackfruit-based agroforestry system in Bangladesh, jackfruit yield was enhanced by 32.7%; however, the lower storey papaya and brinjal yields declined by 22.8 and 17.4% respectively compared to their pure stand due to insufficient light penetration at the lower level. Moreover, by raising of alley crops, agronomic management supported yield enhancement of jackfruit with an increase in the LER.

In this chapter, we explore the potential of alley cropping as a viable agroforestry practice that can contribute to sustainable agricultural intensification.

16.2 Choice of Crops

The crops considered in alley cropping are tree, shrubs, and annuals. Perennial trees and shrubs are planted in rows and annuals (generally food or forage crops) are grown in the alleys between two rows. The woody tree and shrub species commonly chosen for alley cropping include *Acacia auriculiformis*, *Acacia barterii*, *Albizia lebbeck* (saras or frywood), *Alchomea cordifolia*, *Azadirachta indica* (neem), *Cajanus cajan*, *Desmanthus virgatus*, *Dalbergia sissoo* (sisu or north Indian rose wood), *Dactyladenia barteri* (syn. *Acioa barteri*), *Calliandra calothyrsus*, *Erythrina poeppigiana*, *Flemingia congesta*, *Flemingia macrophylla*, *Gliricidia sepium*, *Gmelina arborea*, *Leucaena leucocephala* (subabul), *Paraserianthes falcataria* (syn. *Albizia falcataria*), etc. In tropical north and south America, *Inga edulis* and *I. oerstediana* are selected for hedgerows. On basic soils, *Leucaena leucocephala* and *Glificidia sepium* perform well, but on acidic Ultisols and Oxisols, growth of these plants is reduced because of low phosphorus and calcium and higher aluminum. *Acacia angustissima* and *Calliandra callothyrsus* are chosen in the western highlands of Cameroon (Akume *et al.*, 2015).

The common annual crops grown in alley cropping are cereals (maize, sorghum, wheat, upland rice), leguminous crops (beans, cowpea, soybean), root crops (cassava, yam), and different types of vegetables. Companion crops are sown in the alleys between the hedgerows. The choice of companion crop is highly region specific and depends greatly on the farmer's choice. Generally, food and forage crops are grown in alleys but fruits and specialty crops are also considered and sometime biomass-producing crops are chosen.

During the initial period of hedgerow establishment, there will be less shade and food crops can be easily grown. But gradually the trees and shrubs will grow and although these will be pruned periodically, they will influence the growing environment and alley crops may get reduced resources. At that time, growing of shade-tolerant and specialty crops will be wise. Integration of specialty crops is also a practice in alley cropping. The common specialty crops are fruits and vegetables, tree nuts, horticulture and nursery crops. When the hedgerows are fully grown, it is better to grow shade-loving or shade-tolerant specialty crops in the alleys.

16.3 Crop Establishment

In erosion-prone lands, perennial crops are grown by direct seeding or transplanting of seedlings along the contour or across the slope. In arable areas when wind erosion control is one of the objectives of alley cropping, it is better to close up plant spacing within the rows of tree species. Sometimes closer intra-hedgerow planting is done in trenches rather than in holes or pits. Later on, closely growing trees may be thinned as required. After establishment of tree species, regular agronomic management is essential to achieve proper growth and desired yield from hedgerows. The tree species used for hedgerows should have faster growth, less mortality, faster regrowth after repeated prunings, ability to fix N biologically (legumes), tolerance of abiotic stress and suitable for multipurpose use (Solaimalai *et al.*, 2005). Further, the tree species should have fewer roots at the soil surface to reduce competition with crops grown in the alleyway (Hodge *et al.*, 1999).

For annual food crops, direct sowing is commonly observed and dense seeding is advised. At the early seedling stage, extra plants are thinned. But for direct sowing, a good seed bed is required to facilitate germination and uniform stand. In humid and high rainfall areas, direct seeding is preferred but in dry zones, transplantation of seedlings is advised for proper stand establishment. The crops should be of short duration, ecologically hardy, with wide adaptability to different climatic conditions, tolerant to partial shade and abiotic stress, and shallow rooted.

16.4 Alley Width

In alley cropping, narrow alleys will lead to maximum competition between hedgerow crops and annual crops. The tree and/or shrub rows are planted on the basis of the purpose. The planting geometry depends on factors including growth potential of hedgerows, whether single or multiple species are to be planted, whether single or multiple rows are to be planted, length of the slope, breadth of the field, requirement of light for the crops grown in between alleys, and farmer choice. In semiarid regions where moderately good amounts of rainfall are received, wider alleys are recommended and a generally good return is obtained from tree and annual crops (Singh *et al.*, 1989a). Hedgerows of *Leucaena leucocephala* yielded more with narrow alley width (4.0 m) over wider alley width of 6.0m as mentioned by Solaimalai *et al.* (2005).

When food production is the major goal for smallholders, wider hedgerow width is preferred to accommodate more annual food crops in the alleys. The total production of biomass will be higher with close spacing of alleys. In areas with low rainfall, wider alleys are planted to avoid completion for soil moisture and to reduce shading to annual agricultural crops. In one study, the wheat yield remained on par in both *Gliricidia sepium* (3.63 t/ha) and *Leucaena leucocephala* (3.31 t/ha) planted in 6m alley widths in Bangladesh (Ferdush *et al.*, 2018). Madhu *et al.* (2019) suggested planting of tree crops *Gliricidia maculata* and *Leucaena leucocephala* with 5–10m width in eastern regions of India. For *Azadirachta indica* (neem), 8m spacing between two rows can be used to minimize competition with annual crops.

In temperate areas, there is limited scope for adoption of alley cropping and in general a food crop is grown with timber species (Wolz and DeLucia, 2018). However, Wolz *et al.* (2018) suggested diversification of woody species and adoption of alley cropping with components like food crops and fodders in temperate regions.

16.5 Management of Crops in Alley Cropping

Tree species are cut frequently and should be kept pruned throughout the growing period of annual crops to minimize shading and reduce competition with food crops. However, they are allowed to grow freely to cover the land when there is no food crop. Alley management mainly includes choice of crops (trees, shrubs, and annual crops), maintenance of alley width, pruning, and nutrient management.

16.5.1 Pruning height and frequency of pruning

Pruning height is an important aspect of hedgerow management. If annual crops are to be cultivated, it is essential to prune the tree crops to minimize the shading effect. A low pruning height of 15–30 cm for *Leucaena* would be desirable to avoid shading the annual crop with low canopy height. A low pruning height of 20–40 cm for *Leucaena* caused its wood to split, making it prone to termite attack. In one study it was noted that the trees survived with stumps as short as 7.5 cm, with greater herbage yield than taller stumps (Korwar, 1990). In general, pruning height of 30–90 cm is satisfactory. The biomass production of *Leucaena leucocephala* was greater when pruned at a height of 90 cm compared to 60 cm and 30 cm (Maiti and Chatterjee, 1986). Actually, pruning height depends on the component crops in the mixed stand. In general, the lower the hedgerow and taller the crop, the less frequently pruning is needed. The main objective of pruning is to facilitate sunlight to both the crops and maximum use of it.

The time of pruning the hedgerows is important in low fertile lands. In Hawaii, *Leucaena* harvested when 4 months old gave maximum leaf yield (at height of 120–150 cm) and there was a gradual decrease in yield with increase of plant height and delayed cutting (Korwar, 1990). But in the Philippines, the first cutting of *Leucaena* is deferred until plants are 6 months old. In semiarid regions *Leucaena* needs more time to produce a sizeable yield which generally takes about a year. Fast-growing crops like *Gliricidia* and *Leucaena* require pruning at every 5–6 weeks during the cropping season. Further, too low or too frequent pruning should be avoided as there will be a chance of dieback disease infection.

16.5.2 Root pruning

Interaction between tree species and annual crops is very important to obtain benefit from the mixed cropping. There will always be competition for available resources, even during the early stages of establishment of tree species, and the competition may decrease yield of annual crops because both species will absorb soil moisture and nutrients from a common pool (Noordwijk *et al.*, 2015).

To reduce the competition, root pruning of tree crops is practiced in alley cropping. Root pruning is done by tilling the soil up to 30 cm depth on both sides of hedgerows. In India, root pruning is done during the monsoon season. To restrict the horizontal growth of roots, a polythene sheet may be placed as a root barrier at 0.3m from the hedgerow trunk and up to 0.5–1.2m depth (Singh *et al.*, 1989b; Gillespie *et al.*, 2000; Solaimalai *et al.*, 2005). If root pruning is not adopted, tree species will spread their roots into the alleys and annual crop yield will be reduced.

16.5.3 Nutrient balance and management

Legume hedgerows contribute more nutrients (especially N) than nonlegumes and annual crops benefit from this. On average, the biomass production of *Leucaena* varies from 1.1 to 3.5 t/ha/year and the quantities of N and P recycled will be 33–150 and 2.2–7.0 kg respectively. Not all the recycled nutrients are used by annual crops and it has been noted that annual crops get only 10–20% of N in a given season (Solaimalai *et al.*, 2005).

In Kenya, *Leucaena* prunings added 52 kg N and 3 kg P ha/season to the soil, but maize + *Leucaena* alley cropping caused a positive balance of 15 kg for N and a negative balance of 4 kg for P ha/season. In this situation, negative P balance could be corrected by application of cattle manure (Lupwayi and Haque, 1999). In the dry tropical savannah of northern Kenya, a low nutrient input region, enhancement of N and K balance in alley cropping systems of *Acacia* + sorghum was observed due to mulching crop residues and *Acacia* leaves which resulted in a positive nutrient balance (Lehmann *et al.*, 1999). In Malawi, in a *Gliricidia sepium* + maize alley cropping system, maize yielded 3.3 t/ha compared to 1.0 t/ha in pure stands of maize without application of chemical nutrients (Akinnifesi *et al.*, 2006) which indicated the benefit of tree species in providing nutrients to annual crops. Several studies have clearly indicated that pruning and leaf fall can add moderately good quantities of N, but for other nutrients, exogenous application is necessary to achieve a good harvest from annual crops in alley cropping. Beule *et al.* (2019) observed that in Germany, cereals and oilseeds raised in 12m wide poplar hedgerows in temperate conditions enhanced soil microorganisms and soil N cycling genes, facilitating soil fertility.

16.5.4 Nutrient yield in alley cropping

In alley cropping, woody perennials accumulate biomass primarily through enhanced carbon sequestration. The biomass is supplemented with essential nutrients needed for crop growth either through deep nutrient capture from the soil or biological nitrogen fixation (BNF). Nutrient capture from the soil involves the roots of woody perennials exploring deeper soil layers and harvesting nutrients that shallow-rooted annuals are unable to reach. In BNF, specific microorganisms convert atmospheric nitrogen into ammonia, making it available to plants (Nair *et al.*, 2021; Sánchez-Navarro *et al.*, 2022). As trees shed their leaves and twigs over time or during pruning, these materials gradually decompose, releasing stored nutrients back into the soil (Lehmann *et al.*, 1995). This process contributes to nutrient enrichment in the top soil which provides the nutrient needs of short-duration annuals cultivated in the alleys, making the system sustainable in the long term. The nutrient uptake of grains under different alley cropping systems is presented in Table 16.1.

In a study conducted at Mbalmayo, southern Cameroon by Hauser (2019), the early phase prunings of *Flemingia macrophylla* contributed higher nutrient yield compared to *Senna spectabilis* and *Dactyladenia barteri*. The mean nutrient contributions by *Flemingia macrophylla* over 3 years were 36.05 kg/ha of nitrogen, 3.99 kg/ha of phosphorus, 13.56 kg/ha of potassium, 7.85 kg/ha of magnesium, and 8.47 kg/ha of calcium. In another study by Mafongoya and Jiri (2016), it was found that the *Sesbania* fallow system contributed 13.4N kg/ha over 3 years, which was significantly higher than the *Tephrosia* fallow system, which added 10.9N kg/ha in the same period. Both *Sesbania* and *Tephrosia* fallows outperformed grass fallow and unfertilized maize in terms of adding inorganic nitrogen to the topsoil.

16.6 Impact of Alley Cropping on Soil Properties and Soil Conservation

The role of alley cropping in slowing down runoff water, acting as a shelterbelt to withstand heavy winds and exploring deeper soil layers to

Table 16.1. Nutrient uptake in grains under different alley cropping systems.

Alley cropping systems	Region	Nutrient uptake (kg/ha) N	P	K	References
Sesbania sesban + *Cenchrus ciliaris* + pearl millet	Rajasthan, India	93.9	14.6	71.9	Choudhary *et al.* (2020)
Sesbania sesban + *Cenchrus ciliaris* + cluster bean		90.8	8.9	38.2	
Populus deltoides + wheat	Harayana, India	108.9	16.6	22.4	Dahiya *et al.* (2022)
Eucalyptus tereticornis + barley		81.2	14.5	21.5	
Dactyladenia barteri + groundnut	Southern Cameroon	54.4	4.8	42.6	Hauser (2020)
Flemingia macrophylla + groundnut		44.7	3.2	30.7	
Senna spectabilis + groundnut		49.7	3.9	37.3	
Leucaena + pearl millet	Maharastra, India	25.2	5.60	–	Durgude and Patil (1997)
Poplar + wheat	Uttar Pradesh, India	47.2	7.6	11.2	Sharma *et al.* (2012)

secure fragile soil contribute to conserving soil and its fertility (Kinama *et al.*, 2007). Besides enhancing soil properties, this system often adds nutrient-rich litter, which contributes to an increase in soil organic matter and reduces soil erosion (Ivezić *et al.*, 2022).

In a study in Laguna, Philippines, Paningbatan *et al.* (1995) found that by adopting alley cropping with *Desmanthus virgatus* as the hedgerow along the contour with maize in the alley during *kharif* and mung bean during *rabi* season, the average annual soil loss was reduced from 150 t/ha to 5 t/ha mainly by reducing total runoff volume and sediment concentration. In another study, better light penetration and firm anchorage of root to soil by adoption of pineapple as a low hedge plant resulted in minimum soil loss of 8.89 t/ha compared to *Indigofera*, *Bogamedula*, and Napier grass in hilly areas of Bangladesh (Islam *et al.*, 2020). Furthermore, a metaanalysis revealed that alley cropping improved soil organic carbon. The response was larger in regions with low initial soil carbon than in those with high initial soil carbon, and the alley cropping system's soil organic carbon stocks increased only after 10 years of alley cropping (Ivezić *et al.*, 2022).

According to Adhikary *et al.* (2017), increased nitrogen cycling and decreased nutrient leaching were two ways by which alley cropping contributes to the improvement of soil properties. During pruning, the nitrogen-fixing trees shed their nodules which degrade faster than leaf litter, aiding in enrichment of soil biochemical properties (Jose *et al.*, 2000).

Although soil physical properties are influenced by alley cropping systems, during extreme weather conditions this influence will be relatively low due to low biomass contribution. Gangwar *et al.* (2004) reported *Leucaena leucocephala* as one of the species cultivated in hedgerows of alley cropping systems which lowered the soil bulk density of sandy loam from 1.7 to 1.5 g/cm^3. This influence on bulk density might be the reason for the role of alley cropping in determining soil moisture content. The potential of a *Gliricidia* hedgerow in conserving soil moisture was comparatively greater than a *Leucaena* hedgerow (Adhikary *et al.*, 2017). Moreover, cultivating leguminous trees like *Gliricidia* and *Leucaena* helped to add nitrogen to the soil nitrogen pool and more soil organic carbon due to their high biomass-producing ability. Additionally, this contribution of biomass over the long term produces organic acids upon decomposition which results in higher cation exchange capacity. Gangwar *et al.* (2004) reported 23% more potassium with addition of pruning material from a *L. leucocephala* alley cropping system compared with no alley cropping system.

16.6.1 Above- and below-ground interaction in alley cropping

The overall productivity of an alley cropping system is mainly determined by the interaction between above- and below-ground interactions (Ong *et al.*, 1991). The annual crops grown in the alleyways may face competition for sunlight from the tree canopies, which could negatively impact their productivity by casting shade (Kang *et al.*, 2008). However, by altering the microclimate, serving as a shelterbelt, and adding organic matter through leaf litter, which improves soil fertility, these tree canopies can also safeguard the annual crops (Quinkenstein *et al.*, 2017). Similarly, the tree roots compete with crops in the alleys for nutrients and water besides aiding in nutrient cycling, improving soil structure, preventing erosion, and promoting the activity of beneficial soil microorganisms (Fahad *et al.*, 2022).

In alley cropping systems, as trees are grown alongside annual crops, the sunlight is mostly blocked from reaching the crops by the tree canopies, which shades the crops and causes competition for light, which eventually lowers the output of the annual crops (Swieter *et al.*, 2022). The results of Mantino *et al.* (2021), which demonstrated that the presence of trees had a detrimental impact on alfalfa biomass output mostly because of a reduction in light availability at the herbaceous layer, substantially corroborated this discussion. Similarly, Varella *et al.* (2011) observed a roughly 70% production decline in alfalfa over a 2-year field study, which they attributed to a 49% drop in mean daily photosynthetic photon flux density transmittance relative to full sunlight.

However, trees in hedgerows can serve as a natural barrier and lessen the negative effects of unfavorable weather conditions, such as strong winds, unusually high temperatures, etc., on annual crops. This discussion was substantially supported by the findings of Veldkamp *et al.* (2023), who showed that the addition of tree rows to temperate hedgerows decreased wind speed below the risk level. Moreover, air temperature in neighboring crop alleys was found to be lower near the tree rows than it was further away because of the trees' increased evapotranspiration rates or trees shading the area (Bohm and Tsonkova, 2018).

The below-ground interactions between trees and crops in an alley cropping system play a crucial role in understanding their competition for growth factors – nutrients and moisture (Swieter *et al.*, 2022). In general, tree roots often penetrate deep into the soil and extract moisture and nutrients from soil layers that are inaccessible to shallow-rooted annuals which are mostly concentrated in the top soil. However, competition for space also exists between trees and crops in an alley cropping system, due to which maize yields were reduced by 50% in a span of 10 years when cultivated in fallows of black walnut and red oak tree (Gillespie *et al.*, 2000). In another study, compared to maple tree fallows with root barriers, maize productivity in fallows without barriers was lower (Majaura *et al.*, 2024).

16.7 Multifaceted Benefits of Alley Cropping

Alley cropping provides an ideal diversification with the fullest scope for utilization of available resources and increase of short-, medium-, and long-term returns including regular cashflow from marginal lands. Like other intercropping systems, it offers other benefits which are discussed below.

16.7.1 Greater use of resources

In alley cropping, two or more crops are cultivated together. Maximum utilization of space (land area) is ensured as more plants are included in a unit area and other natural resources such as soil moisture, atmospheric carbon dioxide, and sunlight are better utilized than pure stands of forestry. The poor and marginal areas are chosen for alley cropping in general where commercial production of annual crops is not preferred. In this way, it adds value in marginal lands.

16.7.2 Performance of crops and yield advantage

Earlier research carried out worldwide evidenced that alley cropping was able to enhance crop productivity from marginal lands in most cases. However, crop yield depends on several factors, including growing situation of crops, impact of agroclimatic conditions, choice of crops, and agronomic management provided. For example, in an experiment conducted at Jhansi, sesame yield was increased in *Leucaena* alleys but peanut and red gram did not perform well (Singh, 1983), which might be due to a combination of legume (*Leucaena*) and nonlegume (sesame). In contrast, when alley cropped with peanut, perennial red gram yield benefit was recorded compared to perennial red gram with sorghum alley cropping (Solaimalai *et al.*, 2005). Alley cropping of cowpea yielded more than sole cropping when grown in alleys of *Gliricidia* and *Leucaena* in Guyana (Chesney *et al.*, 2010).

16.7.3 Regular cashflow

In degraded, poorly fertile, and marginal lands, forestry can be adopted but there is also scope for adoption of different agroforestry systems inclusive of alley cropping. Alley cropping has the potential to produce considerable farm output even from marginal and degraded lands. A combination of annual and specialty crops can be grown in alleys composed of shrubs and perennial trees. In such cases, specialty crops and annual crops will ensure regular cashflow (which is important for small farmers) and other perennial crops will yield fodder, fuel and timber in the medium and long term. Degraded and marginal lands are mostly managed by poor farmers where investment in agriculture is less and the regular cashflow helps the smallholders to operate the farm efficiently and manage their livelihood.

16.7.4 Soil conservation and prevention of nutrient loss

Marginal lands are highly prone to soil erosion (more than 8% slope) even when annual crops are cultivated. In alley cropping, hedgerows act as barriers in restricting soil erosion. Further, mulching with pruned leaves and twigs will also check soil erosion. In sloping lands (8–10% slope) of eastern India, planting of perennial *Gliricidia maculata* and *Leucaena leucocephala* with 5–10m alley width is suggested to prevent soil loss. The space between alleys is used for growing of annual food crops or forages. Experimental results evidenced that alley cropping with *Gliricidia* decreased runoff by 27.9–28.2% and soil loss by 49.3–51.1% over no alley cropping while a *Leucaena* alley with miniature trench reduced runoff by 18.3–18.7% and soil loss by 37.2–43.0% (Madhu *et al.*, 2019). Moreover, alley cropping has the potential to reduce nutrient loss from top soil. Both *Gliricidia* and *Leucaena* alleys are known to conserve nutrient loss by checking loss of topsoil. *Gliricidia* alleys conserved organic carbon, nitrogen, phosphorus, and potassium by 63.4, 5.0, 0.3 and 2.4 kg/ha, respectively, over no alley cropping. Similarly, *Leucaena* alleys conserved organic carbon, nitrogen, phosphorus, and potassium by 57.7, 4.6, 0.3 and 2.2 kg/ha, respectively, over no alley cropping (Madhu *et al.*, 2019).

16.7.5 Carbon sequestration

Carbon sequestration is of great importance in the context of global warming and carbon footprint in agriculture (Table 16.2). Alley cropping potentially increases carbon sequestration and adds woody biomass to the soil and thus can be considered as an ecofriendly cropping system. Research findings evidenced that agroforestry increased carbon stock in dry regions of Kenya (Reppin *et al.*, 2020). In the Philippines, jatropha and corn alley cropping markedly increased organic carbon in the soil (Marin, 2016). Alley cropping with *Gliricidia maculata* and *Leucaena leucocephala* stored 4.26 and 4.52% more soil organic carbon, respectively, over no alley cropping (Madhu *et al.*, 2019). In the temperate Mediterranean region also, alley cropping produced more carbon sequestration (Cardinael *et al.*, 2015).

16.7.6 Enhancement of biodiversity

Biodiversity enhancement of alley cropping is observed by many ways (Xu *et al.*, 2019; Staton *et al.*, 2022). First, plant species grown together enrich crop diversity more than pure stands of crops. Second, alley cropping supports the habitat of birds and bats (Harvey and Villalobos, 2007). Third, increased below-ground biodiversity is noted in the form of microorganisms,

Table 16.2. Rate of carbon sequestration in different alley cropping systems.

Agroforestry systems	Region	Rate of carbon sequestration (t/ha/year)	References
Bamboo-based alley cropping system	Assam	21.4	Nath and Das (2012)
Poplar + wheat	Central Himalayas	7.7	Adhikari *et al.* (2020)
Mango + wheat		6.8	
Populus-based alley cropping system	Uttarakhand	6.2	Arora *et al.* (2014)
Poplar-based alley cropping system	Indo-Gangetic plains	10.3	Chavan *et al.* (2023)
Eucalyptus-based alley cropping system		12.7	
Eucalyptus-based alley cropping system	Haryana	12.9	Ahlawat *et al.* (2019)

arthropods, and other fauna. Ashraf *et al.* (2018) recorded the abundance of arthropods and species diversity in the soil where alley cropping adopted.

16.7.7 Climate change mitigation and adaptation

Alley cropping has huge potential to sequester carbon as trees sequester more than three-quarters of the global carbon pool (Zomer *et al.*, 2016). Reppin *et al.* (2020) reported that in Kenya, adoption of agroforestry increased carbon storage in soil and upgraded livelihoods of smallholder farmers. A sizeable amount of carbon dioxide emissions in agriculture is caused by anthropogenic interventions and carbon sequestration is a top priority of sustainable agriculture. Alley cropping sequesters carbon in two ways, namely above-ground biomass (leaves, twigs, etc.) and below-ground biomass (roots). Thus, alley cropping as well as agroforestry is potentially a climate-smart cropping system (Yirefu *et al.*, 2019). Presently, it is considered as one of the important adaptation strategies to minimize the adverse effects of climate change. Alley cropping can play a buffering role by creating microclimates under climatic extremes for yield stability and thus, it can be considered as an adaptation strategy against climate change (Reyes *et al.*, 2021).

16.8 Alley Cropping: Limitations and Remarks

Although it has several benefits, alley cropping also has a few limitations as multiple crops are grown at the same time on the same land.

First, for intensive management of hedgerows, specialized equipment and skills are required to manage multiple crops. However, once alley cropping is adopted by farmers in a region, demand is created and suppliers are able to provide this need as the market for these specialized tools and equipment grows simultaneously. Second, in the beginning alley cropping may not result in lucrative returns (as income comes only from annual crops). In general, alley cropping is adopted for the long term on marginal and problematic lands where lucrative return is not always expected. Third, alley cropping requires proper marketing of yield output from perennial crops. To overcome this issue, the necessary marketing facilities must be created and crops should be chosen based on local demand. Fourth, tree species may create competition for annual crops which can be solved by choosing crops with greater compatibility. Fifth, herbicide application is limited as it increases the cost of cultivation; this could be minimized by adoption of integrated weed management. Lastly, it offers employment opportunities to underutilized family laborers for smallholders.

16.9 Future Perspectives

Alley cropping has the potential to contribute significantly to climate-smart agriculture in the present era of climate change. Therefore, further research should be focused on optimizing species selection, alley width, and management practices for different agroecological regions. To increase adoption of this system, it should always be supported by policy frameworks and incentives. Moreover, integration of modern technologies, such as remote sensing and precision agriculture, may enhance the scalability and efficiency of alley cropping systems.

16.10 Conclusion

In resource-poor marginal lands, investment in agriculture is less and farmers do not use costly inputs. In this situation, alley cropping may be a suitable option for efficient land use and sustainable agricultural production. In some cases, alley cropping may yield less from annual crops than pure stands, but its multifaceted benefits should be considered. Proper planting geometry, choice of suitable crops, and appropriate management can make alley cropping more profitable than sole crops as evidenced in different countries around the world.

References

Adhikary, P.P., Hombegowda, H.C., Barman, D., Jakhar, P. and Madhu, M. (2017) Soil erosion control and carbon sequestration in shifting cultivated degraded highlands of Eastern India: Performance of two contour hedgerow systems. *Agroforestry Systems* 91(4), 757–771.

Adhikari, B., Lodhiyal, N. and Lodhiyal, L. (2020) Assessment of crop yield, productivity and carbon sequestration in agroforestry systems in central Himalaya, India. *Agroforestry Systems* 94(1), 281–296.

Ahlawat, K.S., Daneva, V., Sirohi, C. and Dalal, V. (2019) Production potential of agricultural crops under *eucalyptus tereticornis* based agrisilviculture system in semi-arid region of haryana. *International Journal of Current Microbiology and Applied Sciences* 8(6), 2725–2731.

Akinnifesi, F.K., Makumba, W. and Kwesiga, F.R. (2006) Sustainable maize production using gliricidia/maize intercropping in Southern Malawi. *Experimental Agriculture* 42(4), 441–457.

Akume, N.D., Christopher, S., Manga, M.A., Francis, N.A.S., Yaya, F.V. *et al.* (2015) Effect of tree hedgerow pruning on maize yield in Santa, Cameroon. *International Journal of Agriculture Innovation and Research* 3(6), 1750–1756.

Arora, G., Chaturvedi, S., Kaushal, R., Nain, A., Tewari, S. *et al.* (2014) Growth, biomass, carbon stocks, and sequestration in an age series of *Populus deltoides* plantations in Tarai region of central himalaya. *Turkish Journal of Agriculture and Forestr* 38, 550–560.

Ashraf, M., Zulkifli, R., Sanusi, R., Tohiran, K.A., Terhem, R. *et al.* (2018) Alley-cropping system can boost arthropod biodiversity and ecosystem functions in oil palm plantations. *Agriculture, Ecosystems & Environment* 260, 19–26.

Beule, L., Corre, M.D., Schmidt, M., Göbel, L., Veldkamp, E. *et al.* (2019) Conversion of monoculture cropland and open grassland to agroforestry alters the abundance of soil bacteria, fungi and soil-N-cycling genes. *PLoS One* 14(6), e0218779.

Bohm, C. and Tsonkova, P. (2018) Effekte des agrarholzanbaus auf mikroklimatische kenngrößen. In: Veste, M. and Böhm, C. (eds) *Grarholz—Schnellwachsende Bäume in Der Landwirtschaft*. Springer, Berlin, pp. 335–389.

Cardinael, R., Chevallier, T., Barthès, B.G., Saby, N.P.A., Parent, T. *et al.* (2015) Impact of alley cropping agroforestry on stocks, forms and spatial distribution of soil organic carbon – A case study in a Mediterranean context. *Geoderma* 259–260, 288–299.

Chappa, L.R., Nungula, E.Z., Makwinja, Y.H., Ranjan, S., Sow, S. *et al.* (2024) Outlooks on major agroforestry systems. In: Raj, A., Jhariya, M.K., Banerjee, A., Jha, R.K. and Singh, K.P. (eds) *Agroforestry*. Wiley-Scrivener, Hoboken, pp. 21–48.

Chavan, S.B., Dhillon, R.S., Sirohi, C., Uthappa, A.R., Jinger, D. *et al.* (2023) Carbon sequestration potential of commercial agroforestry systems in Indo-Gangetic plains of India: Poplar and eucalyptus-based agroforestry systems. *Forests* 14(3), 559.

Chesney, P.E.K., Simpson, L.A., Cumberbatch, R.N., Homenauth, O. and Benjamin, F. (2010) Cowpea yield performance in an alley cropping practice on an acid infertile soil at Ebini, Guyana. *Open Agriculture Journal* 4(1), 80–84.

Choudhary, M., Kantwa, S.R., Dixit, A.K. and Nagar, R.P. (2020) Alley cropping system improves forage productivity and profitability in rainfed semi-arid Rajasthan. *Indian Journal of Agroforestry* 22(2), 21–25.

Dahiya, G., Bhardwaj, K.K., Devi, S., Ahlawat, I., Singh, C. *et al.* (2022) Influence of different agroforestry systems on yield and nutrient uptake in semi-arid region of Haryana. *Indian Journal of Agroforestry* 24(2), 19–23.

Durgude, A.G. and Patil, J.D. (1997) Recycling of *Leucaena leucocephala* to pearl millet. *Indian Journal of Dryland and Agricultural Research Development* 12(1), 1–7.

Fahad, S., Chavan, S.B., Chichaghare, A.R., Uthappa, A.R., Kumar, M. *et al.* (2022) Agroforestry systems for soil health improvement and maintenance. Sustainability 14(22), 14877.

Ferdush, J., Karim, M.M., Himel, R.M., Saha, S.R. and Ahamed, T. (2018) Impact of alley cropping on wheat productivity. *IOSR Journal of Agriculture and Veterinary Science* 11(2), 17–25.

Gangwar, K.S., Sharma, S.K. and Tomar, O.K. (2004) Alley cropping of Subabul (*Leucaena leucocephala*) for sustaining higher crop productivity and soil fertility of rice (*Oryza sativa*)-wheat (*Triticum aestivum*) system in semi-arid conditions. *Indian Journal of Agronomy* 49(2), 84–88.

Gillespie, A.R., Jose, S., Mengel, D.B., Hoover, W.L., Pope, P.E. *et al.* (2000) Defining competition vectors in a temperate alley cropping system in the mid-Western USA. *Agroforestry Systems* 48(1), 25–40.

Gitari, H.I., Nyawade, S.O., Kamau, S., Karanja, N.N., Gachene, C.K.K. *et al.* (2020) Revisiting intercropping indices with respect to potato-legume intercropping systems. *Field Crops Research* 258, 107957.

Gitari, H.I., Nungula, E.Z., Chappa, L.R., Raza, M.A., Ranjan, S. *et al.* (2024) Agroforestry for climate security. In: Raj, A., Jhariya, M.K., Banerjee, A., Jha, R.K. and Singh, K.P. (eds) *Agroforestry*. Wiley-Scrivener, Hoboken, pp. 317–342.

Gold, M.A. and Garrett, H.E. (2009) *Agroforestry Nomenclature, Concepts, and Practices*, 2nd. American Society of Agronomy, Madison, pp. 45–56.

Harvey, C.A. and Villalobos, J.A.G. (2007) Agroforestry systems conserve species-rich but modified assemblages of tropical birds and bats. *Biodiversity and Conservation* 16(8), 2257–2292.

Hauser, S. (2019) Impact of planted hedgerow fallows on nutrient balances in a groundnut/maize/cassava intercrop. *Archives of Agronomy and Soil Science* 66(3), 386–397.

Hauser, S. (2020) Impact of planted hedgerow fallows on nutrient balances in a groundnut/maize/cassava intercrop. *Archives of Agronomy and Soil Science* 66(3), 386–397.

Hodge, S., Garrett, H.E. and Bratton, J. (1999) Alley Cropping: An Agroforestry Practice, Agroforestry Notes (USDA-NAC). Available at: https://digitalcommons.unl.edu/agroforestnotes

Islam, M., Hossain, M.B., Islam, K.N., Za, N.M. and Alam, M.K. (2020) Alley cropping and hedge species effects on soil erosion and crop yield hilly areas of Bangladesh. *Business, Social and Scientific Research* 8(3), 9–14.

Ivezić, V., Lorenz, K. and Lal, R. (2022) Soil organic carbon in alley cropping systems: A meta-analysis. *Sustainability* 14(3), 1296.

Jose, S., Williams, R. and Zamora, D. (2006) Belowground ecological interactions in mixed-species forest plantations. *Forest Ecology and Management* 233(2–3), 231–239.

Jose, S., Gillespie, A.R., Seifert, J.R. and Biehle, D.J. (2000) Defining competition vectors in a temperate alley cropping system in the midwestern USA: 2 Competition for Water. *Agroforestry Systems* 48(1), 41–59.

Kang, H., Shannon, D.A., Prior, S.A. and Arriaga, F.J. (2008) Hedgerow pruning effects on light interception, water relations and yield in alley-cropped maize. *Journal of Sustainable Agriculture* 31(4), 115–137.

Kinama, J.M., Stigter, C.J., Ong, C.K., Ng'ang'a, J.K. and Gichuki, F.N. (2007) Contour hedgerows and grass strips in erosion and runoff control on sloping land in semi-arid Kenya. *Arid Land Research and Management* 21(1), 1–19.

Korwar, G.R. (1990) Alley cropping. In: Hosmani, M.M., Chittapur, B.M. and Hiremath, S.M. (eds) *Intercropping Principles and Practices. Department of Agronomy, College of Agriculture, Dharwad*. pp. 155–160.

Lehmann, J., Schroth, G. and Zech, W. (1995) Decomposition and nutrient release from leaves, twigs and roots of three alley-cropped tree legumes in central Togo. *Agroforestry Systems* 29(1), 21–36.

Lehmann, J., Weigl, D., Droppelmann, K., Huwe, B. and Zech, W. (1999) Nutrient cycling in an agroforestry system with runoff irrigation in Northern Kenya. *Agroforestry System* 43, 49–70.

Lupwayi, N.Z. and Haque, I. (1999) Leucaena hedgerow intercropping and cattle manure application in the ethiopian highlands.III Nutrient balances. *Biology and Fertility of Soils* 28(2), 204–211.

Madhu, M., Hombegowda, H.C., Beer, K., Adhikary, P.P., Jakhar, P. *et al.* (2019) Status of natural resources and resource conservation technologies in Eastern region of India. In: Karma, B. (ed.) *Resource Conservation in Eastern Region of India: Lead Papers of FFCSWR, 2019. Indian Association of Soil and Water Conservationists, Dehradun*. pp. 60–82.

Mafongoya, P. and Jiri, O. (2016) Nutrient availability following planted tree fallows and benefits to subsequent maize crops. *International Journal of Agriculture Innovations and Research* 4(5), 818–823.

Maiti, S. and Chatterjee, B.N. (1986) Proceedings of national seminar on recent trends in forage production. In: IGFRI, pp. 88–89.

Maitra, S. (2023) NIPA genx electronic resources and solutions. In: *Intercropping System: Theory and Practices*. New Delhi.

Maitra, S., Hossain, A., Brestic, M., Skalicky, M., Ondrisik, P. *et al.* (2021) Intercropping – a low input agricultural strategy for food and environmental security. *Agronomy* 11(2), 343.

Maitra, S., Sahoo, U., Sairam, M., Gitari, H.I., Rezaei-chiyaneh, E. *et al.* (2023) Cultivating sustainability: A comprehensive review on intercropping in a changing climate. *Research on Crops* 24, 702–715.

Maitra, S., Bhattacharya, U., Pramanick, B., Sagar, L., Gaikwad, D.J. *et al.* (2024) Agroforestry: A resource conserving technology for efficient utilization of agricultural inputs, leads to food and environmental security. In: Jatav, H.S., Rajput, V.D., Minkina, T., Hullebusch, E.D. and Dutta, A. (eds) *Agroforestry to Combat Global Challenges*, Vol. 36. Sustainable Development and Biodiversity. Springer, Singapore, pp. 15–52.

Majaura, M., Böhm, C. and Freese, D. (2024) The influence of trees on crop yields in temperate zone alley cropping systems: A review. *Sustainability* 16(8), 3301.

Mantino, A., Tozzini, C., Bonari, E., Mele, M. and Ragaglini, G. (2021) Competition for light affects alfalfa biomass production more than its nutritive value in an olive-based alley-cropping system. *Forests* 12(2), 233.

Marin, R.A. (2016) Jatropha-based alley cropping system's contribution to carbon sequestration. *International Journal of Agronomy & Agricultural Research* 8(1), 1–9.

Miah, Md.G., Islam, M.M., Rahman, Md., Ahamed, T., Islam, Md.R. *et al*. (2018) Transformation of jackfruit (*Artocarpus heterophyllus* lam.) orchard into multistory agroforestry increases system productivity. *Agroforestry Systems* 92(6), 1687–1697.

Nair, P.K.R., Kumar, B.M. and Nair, V.D. (2021) Biological nitrogen fixation and nitrogen fixing trees. In: Nair, P.R., Kumar, B.M. and Nair, V.D. (eds) *An Introduction to Agroforestry: Four Decades of Scientific Developments*. Springer, Singapore, pp. 413–443.

Nath, A.J. and Das, A.K. (2012) Carbon pool and sequestration potential of village bamboos in the agroforestry system of North-East India. *Tropical Ecology* 53, 287–293.

Nerlich, K., Graeff-Hönninger, S. and Claupein, W. (2013) Erratum to: Agroforestry in europe: A review of the disappearance of traditional systems and development of modern agroforestry practices, with emphasis on experiences in Germany. *Agroforestry Systems* 87(5), 1211–1211.

Noordwijk, M., Lawson, G., Hairiah, K. and Wilson, J. (2015) Root distribution of trees and crops: Competition and/or complementarity. In: Black, C.K. and Wilson, J. (eds) *Tree-Crop Interactions: Agroforestry in a Changing Climate*, 2nd edn. CABI, Wallingford, pp. 221–257.

Nungula, E.Z., Chappa, L.R., Ranjan, S., Sow, S. and Alnemari, A.M. (2024) Ecosystem services through agroforestry systems and its sustainability. In: Raj, A., Jhariya, M.K., Banerjee, A., Jha, R.K. and Singh, K.P. (eds) *Agroforestry*. Wiley-Scrivener, Hoboken, pp. 221–252.

Ong, C.K., Corlett, J.E., Singh, R.P. and Black, C.R. (1991) Above and below ground interactions in agroforestry systems. *Forest Ecology and Management* 45(1–4), 45–57.

Paningbatan, E.P., Ciesiolka, C.A., Coughlan, K.J. and Rose, C.W. (1995) Alley cropping for managing soil erosion of hilly lands in the Philippines. *Soil Technology* 8(3), 193–204.

Quinkenstein, A., Tsonkova, P., Freese, D., Dagar, J. and Tewari, V. (2017) *Alley cropping with short rotation coppices in the temperate region: A land-use strategy for optimizing microclimate, soil organic carbon and ecosystem service provision of agricultural landscapes*. Anecdotal to Modern Science. Springer, Singapore, Agroforestry, pp. 263–297.

Razouk, R., Daou, K., Ramdani, A. and Chergaoui, A. (2016) Optimal distance between olive trees and annual crops in rainfed intercropping system in Northern Morocco. *Journal of Crop Science Research* 1, 23–32.

Reppin, S., Kuyah, S., Neergaard, A., Oelofse, M. and Rosenstock, T.S. (2020) Contribution of agroforestry to climate change mitigation and livelihoods in Western Kenya. *Agroforestry Systems* 94(1), 203–220.

Reyes, F., Gosme, M., Wolz, K.J., Lecomte, I. and Dupraz, C. (2021) Alley cropping mitigates the impacts of climate change on a wheat crop in a Mediterranean environment: A biophysical model-based assessment. *Agriculture* 11(4), 356.

Sánchez-Navarro, V., Shahrokh, V., Martínez-Martínez, S., Acosta, J.A., Almagro, M. *et al*. (2022) Perennial alley cropping contributes to decrease soil CO_2 and N_2O emissions and increase soil carbon sequestration in a Mediterranean almond orchard. *Science of the Total Environment* 845, 157225.

Santosh, D.T., Debnath, S., Maitra, S., Sairam, M., Sagar, L. *et al*. (2024) Alleviation of climate catastrophe in agriculture through adoption of climate-smart technologies. In: Chatterjee, U., Shaw, R., Kumar, S., Raj, A. and Das, S. (eds) *Climate Crisis: Adaptive Approaches and Sustainability*. Springer Nature, Cham, pp. 307–332.

Sharma, N.K., Singh, R.J. and Kumar, K. (2012) Dry matter accumulation and nutrient uptake by wheat (*Triticum aestivum* l.) under poplar (*Populus deltoides*). *International Scholarly Research Network* 359673.

Singh, G.B. (1983) Role of agroforestry in improving the environment. *Indian Farming* 33, 15–19.

Singh, R.P., Saharan, N. and Ong, C.K. (1989a) Above and below ground interactions in alley-cropping in semi-arid India. *Agroforestry Systems* 9(3), 259–274.

Singh, R.P., Van den Beldt, R.J., Hocking, D. and Korwar, G.R. (1989b) Alley farming in the semi-arid regions of India. In: Kang, B.T. and Reynolds, L. (eds) *Proceedings of an international workshop, Ibadan*, pp. 108–122.

Singh, B., Bishnoi, M., Baloch, M.R. and Singh, G. (2014) Tree biomass, resource use and crop productivity in agri-horti-silvicultural systems in the dry region of Rajasthan. *Archives of Agronomy and Soil Science* 60(8), 1031–1049.

Solaimalai, A., Muralidaran, C. and Subburamu, K. (2005) Alley cropping in rainfed agro-ecosystem – A review. *Agricultural Reviews* 26(3), 157–172.

Staton, T., Breeze, T.D., Walters, R.J., Smith, J. and Girling, R.D. (2022) Productivity, biodiversity trade-offs, and farm income in an agroforestry versus an arable system. *Ecological Economics* 191, 107214.

Swieter, A., Langhof, M. and Lamerre, J. (2022) Competition, stress and benefits: trees and crops in the transition zone of a temperate short rotation alley cropping agroforestry system. *Journal of Agronomy and Crop Science* 208(2), 209–224.

UN (2012) Feeding the world sustainably. Available at: www.un.org/en/chronicle/article/feeding-world-sustainably#:~:text=According%20to%20estimates%20compiled%20by,world%20population%20of%209.3%20billion (accessed 16 March 2024).

Varella, A.C., Moot, D.J., Pollock, K.M., Peri, P.L. and Lucas, R.J. (2011) Do light and alfalfa responses to cloth and slatted shade represent those measured under an agroforestry system? *Agroforestry Systems* 81(2), 157–173.

Veldkamp, E., Schmidt, M., Markwitz, C., Beule, L., Beuschel, R. *et al.* (2023) Multifunctionality of temperate alley-cropping agroforestry outperforms open cropland and grassland. *Communications Earth & Environment* 4(1), 20.

Wolz, K. J. and DeLucia, E.H. (2018) Alley cropping: Global patterns of species composition and function. *Agriculture, Ecosystems & Environment* 252, 61–68.

Wolz, K.J., Lovell, S.T., Branham, B.E., Eddy, W.C., Keeley, K. *et al.* (2018) Frontiers in alley cropping: Transformative solutions for temperate agriculture. *Global Change Biology* 24(3), 883–894.

Xu, H., Bi, H., Gao, L. and Yun, L. (2019) Alley cropping increases land use efficiency and economic profitability across the combination cultivation period. *Agronomy* 9(1), 34.

Yirefu, T., Yenenesh, H. and Zemedkun, S. (2019) Potential of agroforestry for climate change mitigation through carbon sequestration: Review paper. *Agricultural Research & Technology: Open Access Journal* 22(3), 556196.

Zomer, R.J., Neufeldt, H., Xu, J., Ahrends, A., Bossio, D. *et al.* (2016) Global tree cover and biomass carbon on agricultural land: The contribution of agroforestry to global and national carbon budgets. *Scientific Reports* 6(1), 29987.

17 Regulatory and Ethical Issues for Plant Biotechnology and Genetic Engineering

Ankur Singh[1] and Aryadeep Roychoudhury[2*]
[1]KIIT-Technology Business Incubator (KIIT-TBI), KIIT University, Bhubaneswar, India; [2]Discipline of Life Sciences, School of Sciences, Indira Gandhi National Open University, New Delhi, India

Abstract
Plant biotechnology and genetic engineering is a growing branch of plant research which promises to offer remarkable social and economic advantage in the future. The global agricultural area used for the cultivation of various genetically modified crops has significantly increased in the last two decades. However, this wide cultivation of transgenic crops has evoked both praise and adverse reaction, leading to a global debate over its potential for good or ill. Large-scale cultivation of transgenic crops promises better yields together with better quality of grain. However, horizontal gene transfer from transgenic to nontransgenic crops is considered to reduce the diversity of species and alter the balance of the surrounding ecosystem. To control this adverse effect of transgenic crops on human health and environment, rules, regulations, and laws governing the development and use of genetically engineered crops are strictly implemented worldwide. Ethics in plant biotechnology is one of the most controversial issues. Plant bioethics codes and policies include protecting plants from possible unfair manipulation, integrity, responsible publication, intellectual rights recognition, social responsibility, honesty, objectivity, openness, confidentiality, legality, and competency. Ethics of experimental research in agriculture involves all activities done before, during, and after the study, consisting of personal, research, and social ethics. In this chapter, we briefly discuss the ethical aspects associated with research in plant biotechnology and genetic engineering.

Keywords: plant biotechnology, bioethics, transgenic crops, horizontal gene transfer, environmental hazard

17.1 Introduction

Ethics may be considered as the deciding factor between good and bad conditions that are acceptable to a definite social group. Ethics is a kind of scientific integrity, a thinking principle based on honesty, applying conceptual discipline in judging what is bad or good as a moral obligation through determined behavioral rules and standards. According to Feynman (1974), ethics reflects the communal morality of a society.

Experimental research discovers the effect of certain variables on other variables in controlled conditions. In agricultural disciplines, for instance, experimental research may be found in several branches, such as planting culture,

*Corresponding author: aryadeep.rc@gmail.com

 Plant Biotechnology and Sustainable Agriculture (ed. Tariq Aftab)
DOI: 10.1079/9781800627659.0017

plant pest and disease management, agricultural technology, and agricultural socio-economics. Ethics in experimental research consists of all the activities that are performed before, during, and after conducting the experiments (Blakstad, 2008). Planning and designing of an experiment are part of preactivity ethics. Ethics during work cover issues such as use of organisms, collection of data and their analysis, and ethics after work deals with writing and publishing the correct data. McLaren (2007) categorized bioethics in three different classes: social, personal, and research. Social ethics deals with the effect of consequences on the environment and human society due to the execution of research. Personal ethics deals with morality, responsibility, and openness during and after conducting any research, and research ethics guide the work of the research itself.

Science and research in agriculture have developed enormously, enabling researchers to provide solutions to various problems faced by the agricultural sector. Plant biotechnology is one of the most developed branches that results in genetically engineered plants. According to Bryant and Hughes (2017) and von Braun (2010), such genetically modified plants were developed to counter problems such as low crop yield and production, nutrient deficiency, and low resistance to environmental stresses. However, in tandem, the agricultural products of genetic engineering and biotechnological manufacturing created economic, social, health, environmental, political, legal, and ethical issues concerning the products themselves. The risk that arises due to engineered plants concerns society and thus their use has become controversial. Therefore, it is imperative to study the rules and regulations and follow the guidelines before conducting any agricultural research.

In this chapter, we cover the ethical aspects that need to be considered during and after conducting any research in plant biotechnology and engineering.

17.2 Role of Ethics in Agricultural Research

As stated above, bioethics in experimental science should be applied before, during, and after the work. McLaren (2007) categorized ethics in three different classes and pointed out that personal ethics cover issues such as avoiding misconduct, maintaining responsibility and high morality. Social ethics must also be considered when performing any experimental work on plants, since the consequences of the work may turn out to be unacceptable to society.

Before carrying any research in the agricultural sector, careful planning is necessary. The hypothesis needs to be accurately constructed, with proper relevance and limiting the biases which might occur. The theory must be carefully and thoroughly examined so that duplication of the data can be avoided (George, 2016). During execution of the research, ethics can be correlated with methodology, data collection, and data analysis. Data collection must be done with high accuracy to avoid manual error and with care and honesty (Burkhardt *et al.*, 2005). Examples of questionable data analysis include inaccurate statistical procedures, neglecting biases/nonrandom errors, negative conclusion due to incorrect statistical calculations, or practices such as cutting, falsifying, or 'delineating' data (Baumans, 2004; Punia, 2006; McLaren, 2007; Ridwan, 2013).

After completion of the research, it is very important to present the correct data and appreciate the work of all the researchers. The authors conducting the experiments and generating data are the main intellectual contributors of the research; therefore, they are directly responsible for the content of the report or publication, and all these contributors should be able to explain the how and why of the research (Grimm, 2008).

According to Resnik (2015), the code of bioethics in conducting any experimental work includes the following: integrity, honesty, vigilance, objectivity, openness, mentoring responsibility, honoring intellectual property, confidentiality, social responsibility, responsible publication, protection in utilizing plants as the object/subject of the research, competency, and legality. Honesty must be exercised in all aspect of bioethics, while conducting the research and communicating its results along with data and methodology. Protection of plants must be exercised by avoiding unnecessary trials or carrying out experiments with poor design. Ethical issues in agricultural research should consider the fact that agricultural sciences are applied sciences, closely related to social and ecological systems. Thus, ethics in agriculture may be interpreted as norms related to resource management, agricultural systems, distribution, food processing, marketing, and consumption (Shelley-Egan and Rodrigues, 2015).

17.3 Ethics in Transgenic Crop Production

The discussion about ethical aspects of genetically modified organism (GMO) creation is based on two critical aspects, extrinsic and intrinsic. The intrinsic aspect states that creation of genetically modified (GM) crops is an unnatural process. According to Weale (2010) and Bhumiratana and Kongsawat (2008), creation of GM crops might interfere with the tolerance ability of the plants and also disturb the environmental balance. Based on various opinions, the use of GM crops in the agricultural field is widely rejected by society. According to Amin *et al.* (2011), changing the fate of plants by genetic alteration is similar to playing God. Therefore, those who hold this opinion state that genetic engineering is interfering with the natural process and thus is theologically unethical. Another opinion that rejects the use of GM crops states that newer technology and genetics may change the balance of the world which is absolutely under the control of God. Both these statements are countered by researchers who state that these opinions are not rational and use of GM crops will reduce the use of fertilizers and increase the productivity of plants, thereby making them more resistant to environmental stresses, and will also increase the biodiversity of the land (Zadoks and Waibel, 2000). The third opinion that rejects the use of GM crops states that genetic engineering blurs the concept of species and removes the boundary between species by transferring genes from one species of plant to another through horizontal gene transfer (Ankeny, 2017). This group also states that the plant species which undergo genetic transfer will suffer. According to Cowan (2015), intrinsic opinions mainly arise from personal fanaticism, born out of religious and spiritual beliefs and egocentrism.

The second, extrinsic aspect is mainly based on the damage that may arise due to GM crops in agricultural land. The negative aspect of GM crops that most worries society is their negative impact on human health and environment (Kelly *et al.*, 2010). According to Hoffman (2016) and Arcieri (2016), transgenic crops can cross-pollinate with pure nontransgenic plants which may disturb the genetic diversity of species and contaminate the environment. In addition, GM crops that are consumed by humans or animals can also initiate allergenic reactions in their bodies (Batista *et al.*, 2005; Holst-Jensen, 2009), as happened with transgenic soybean incorporating the gene encoding methionine-rich 2S storage albumin protein from Brazil nut or maize containing the *cry9C* gene encoding the insecticidal crystal protein endotoxin (Star Link corn).

Another negative impact of GMO crops relates to commercial marketing, where small-scale farmers might be pushed out by the large-scale agricultural enterprises that grow GM crops, eventually enabling them to achieve a monopoly and fully control the market price and policy without considering their effects on poor farmers (Hake *et al.*, 2016). Thus, GM crops might appear beneficial only to large-scale enterprises.

Terminator technology, once used by multinational companies to produce first-generation viable and fertile seeds but infertile second-generation seeds, by incorporation of a lethal gene in the GM crop, created controversy and opposition regarding the acceptability of this technology. Poor farmers were forced to purchase new seeds from the seed companies every season at a high price, thereby questioning the claim of genetic engineering being for the benefit of poor people.

The best way to deal with this extrinsic aspect is by careful formulation of legally binding rules and regulations on using and developing GMOs, so as to contain damages to environment and human health or economic disruption.

Genetically modified crops also pose a serious threat to the environment. Mahrus (2014) reported that in Indonesia, generally crops are cultivated in small areas surrounded by various weeds, where cross-pollination of weeds with transgenic crops can enhance their tolerance to herbicides. Losey *et al.* (1999)

reported higher mortality rates in monarch butterfly larvae when fed with milkweed previously covered with pollen from transgenic corn plants. Another similar report was published by Hansen Jesse and Obrycki (2000) in which they showed that higher levels of Bt corn pollen in the field is harmful to monarch butterfly larvae.

Major risks associated with transgenic plants are competition with natural species, risk of horizontal gene transfer, and higher selection pressure on non-GM crops (Prakash *et al.*, 2011). Thus, it can be said that GMO development is a two-edged sword: it creates hopes for higher yield and better quality, but also exerts adverse effects on the environment and human health.

17.4 Global Status of Transgenic Crops

Genetically engineered crops might provide the solution to the scarcity of food and other human needs. According to Chaturvedi *et al.* (2016), genetic transfer can create new transgenic crops that will have higher production and better grain quality with greater nutritional value and tolerance against various environmental stresses.

According to a report published by the International Service for the Acquisition of Agri-biotech Applications (ISAAA, 2020), out of the 29 countries that permitted the plantation of transgenic crops in agricultural fields, 24 are developing countries and five are industrial countries. This shows that developing countries have planted more transgenic crops than industrial countries. The major countries involved in planation of transgenic crops are India, Pakistan, China, Argentina, Brazil, Canada, and the USA, with an area of 190.4 million hectares (Si and Zhao, 2016; ISAAA, 2020). Important transgenic crops that are cultivated in most parts of the world are rice, maize, canola, and cotton. Apart from food, transgenic crops are also important for other commodities such as industrial raw materials and animal feed.

The role of transgenic crops in solving human food requirements is beyond doubt. However, in practice, crops which are being developed are those with the possibility of bringing great profit to biotechnology companies. According to Uzogara (2000), transgenic crop development should be directed toward improving the tolerance of plants to various environmental cues. Nevertheless, the major effort in transgenic research has been to design crops that are suited to large-scale agricultural industries rather than small-scale farmers, thus questioning its claim to be of overall societal benefit. The GMO controversy is widely linked with various aspects of life such as health, safety, religious and cultural ethics, and environmental diversity. In agriculture, transgenic crops have the potential to disrupt ecosystem balance. Another negative impact of transgenic crops is that they pose a threat to other living organisms such as microorganisms responsible for soil quality (Wasilah *et al.*, 2019).

17.5 Ethics and Regulatory Policies for GM Crops

By placing rules and regulations on the use and cultivation of GMO crops in agricultural land, the negative impact of these crops on the environment can be minimized. The major concern about the use of transgenic crops on a large scale is horizontal gene transfer which leads to the disappearance of environmental biodiversity, which has provoked international discussion about the use of GM crops (Howlett and Mukherjee, 2016).

The international rule which regulates GMO implementation for ensuring biosafety is known as the Cartagena Protocol on Biosafety. More than 120 countries participated in this biosafety protocol (Fontes, 2003; Watanabe *et al.*, 2004). It aims to regulate GMO transfer mechanisms, management and usage, labeling of GMO products, listing of products generated through biotechnological approaches which may adversely affect conservation and biodiversity functioning in the long term (Dunwell, 2003), and highlight the probable risk to human health (Pertry *et al.*, 2014). Denmark was one of the early countries that adopted this regulation. The Danish administration enacted a law which ensures a safe-distance isolation procedure that has been scientifically analyzed and amended, regulates a permit system for growing transgenic crops and encourages the prevention of damage and disturbance which may arise from mixing transgenic and nontransgenic organisms (Watanabe *et al.*, 2004; Varzakas and Tzanidis, 2016). It also implemented clear and descriptive labeling of genetically engineered products (Syaifuddin and Handayani, 2014; Preston and Wickson, 2016).

Guidelines on the food security of GM products in Indonesia have been drafted by the Ministry of Forestry and Estate Crops, Biological and Food Securities, based on the Joint Ministerial Decree of the Minister of Agriculture, State Minister of Food and Horticulture and Minister of Health on Biological Security and Genetical Engineering Agricultural Food Products Security. These guidelines consist of procedures that are important for genetic engineering to ensure food safety. The rules and regulations guiding the security of genetically engineered food crops can be found in Governmental Decree No. 21/2005 on Genetic Engineering Products Biological Security.

17.6 Conclusion

Rapid growth in the field of agricultural biotechnology has led to the development of newer technologies that greatly affect human life. Technologies such as gene editing, nanobiotechnology, gene therapy, and synthetic biology are being used to address a variety of genetic disorders and induce tolerance capability in plants. However, these technologies present unique regulatory and bioethical conundrums.

Bioethics in experimental research includes all the activities that are performed before, during, and after the investigation, along with personal moral and social ethics. Ethical policies in conducting research include objectivity, honoring intellectual property, honesty, responsible publication, mentoring responsibility, integrity, vigilance, openness, social responsibility, confidentiality, competency, legality, and protection in utilizing plants as the object/subject of research. Many regulations and rules have been formulated regarding the cultivation and use of genetically modified crops. However, implementation and precise monitoring of these rules are not yet possible. It is important for farmers to strictly adhere to these rules during cultivation of GM crops, maintaining a safe isolation zone so that no horizontal gene transfer takes place and ecosystem diversity can be maintained.

References

Amin, L., Sujak, S.F., Ramlee, S.N.S., Samian, A.L., Haron, M.S. *et al.* (2011) Educating the ummah by introducing Islamic bioethics in genetics and modern biotechnology. *Procedia - Social and Behavioral Sciences* 15, 3399–3403.

Ankeny, R.A. (2017) New technologies: Ethics of stem cell research. In: Quah, S.R. and Cockerham, W.C. (eds) *International Encyclopedia of Public Health*, 2nd edn. Elsevier, New York, pp. 248–250.

Arcieri, M. (2016) Spread and potential risks of genetically modified organisms. *Agriculture and Agricultural Science Procedia* 8, 552–559.

Batista, R., Nunes, B., Carmo, M., Cardoso, C., José, H.S. *et al.* (2005) Lack of detectable allergenicity of transgenic maize and soya samples. *Journal of Allergy and Clinical Immunology* 116(2), 403–410.

Baumans, V. (2004) Use of animals in experimental research: An ethical dilemma. *Gene Therapy* 11, S64–6.

Bhumiratana, S. and Kongsawat, C. (2008) Food production use of genetically modified organisms. *Asian Journal of Food and Agro-Industry* 1, 24–28.

Blakstad, O. (2008) *Ethics in Research*. Available at: https://explorable.com/ethicsin-research?gid=1585 (accessed 16 April 2025).

Bryant, J.A. and Hughes, S.G. (2017) Food safety issues. In: Thomas, B., Murray, B.G. and Murphy, D.J. (eds) *Encyclopedia of Applied Plant Sciences*, 2nd edn., Vol. 2. Breeding Genetics and Biotechnology, New York, pp. 13–18.

Burkhardt, J., Comstock, G., Hartel, P.G. and Thompson, P.B. (2005) Agricultural ethics. *Council for Agricultural Science and Technology* 29.

Chaturvedi, S., Srinivas, K.R. and Kumar, A. (2016) Agriculture technology choices and the Responsible Research and Innovation (RRI) framework: Emerging experiences from China and India. *Asian Biotechnology and Development Review* 18, 93–111.

Cowan, T. (2015) *Agricultural Biotechnology: Background, Regulation, and Policy Issues*. Congressional Research Service, Washington DC.

Dunwell, J.M. (2003) Ethics and biosafety: Plant genetic engineering, ecological issues. In: Thomas, B., Murphy, D.J. and Murray, B.G. (eds) *Encyclopedia of Applied Plant Sciences*. Elsevier, New York, pp. 295–301.

Feynman, R.P. (1974) Cargo cult science: Some remarks on science, pseudoscience, and learning how to not fool yourself. *Engineering and Sciences* 37, 10–13.

Fontes, E.M.G. (2003) Legal and regulatory concerns about transgenic plants in Brazil. *Journal of Invertebrate Pathology* 83(2), 100–103.

George, A.J.T. (2016) Research ethics. *Medicine* 44(10), 615–618.

Grimm, H. (2008) Ethical issues in agriculture. UNESCO-EOLSS sample chapter. *Interdisciplinary and Sustainability in Food and Agriculture* 1.

Hake, J.F., Schlör, H., Schürmann, K. and Venghaus, S. (2016) Ethics, sustainability and the water, energy, food nexus approach – A new integrated assessment of urban systems. *Energy Procedia* 88, 236–242.

Hansen Jesse, L.C. and Obrycki, J.J. (2000) Field deposition of Bt transgenic corn pollen: Lethal effects on the monarch butterfly. *Oecologia* 125(2), 241–248.

Hoffman, W. (2016) Ecosystems, food crops, and bioscience: A symbiosis for the Anthropocene. *Asian Biotechnology and Development Review* 18, 39–68.

Holst-Jensen, A. (2009) Testing for Genetically Modified Organisms (GMOs): Past, present and future perspectives. *Biotechnology Advances* 27(6), 1071–1082.

Howlett, M. and Mukherjee, I. (2016) Achieving regulatory excellence in the agri-food biotechnology sector: Building policy capacity. *Asian Biotechnology and Development Review* 18, 5–16.

International Service for the Acquisition of Agribiotech (2020) ISAAA in 2019: Accomplishment Report. Available at: www.isaaa.org/resources/publications/annualreport/2019/default.asp (accessed 16 April 2025).

Kelly, A.H., Ameh, D., Majambere, S., Lindsay, S. and Pinder, M. (2010) 'Like sugar and honey': The embedded ethics of a larval control project in the Gambia. *Social Science & Medicine* 70(12), 1912–1919.

Losey, J.E., Rayor, L.S. and Carter, M.E. (1999) Transgenic pollen harms monarch larvae. *Nature* 399(6733), 214.

Mahrus, M. (2014) Controversy over genetically engineered products consumed by the public. *Jurnal Biologi Tropis* 14, 108–119.

McLaren, A. (2007) A scientist's view of the ethics of human embryonic stem cell research. *Cell Stem Cell* 1(1), 23–26.

Pertry, I., Sabbadini, S., Goormachtig, S., Lokko, Y., Gheysen, G. *et al*. (2014) Biosafety capacity building: Experiences and challenges from a distance learning approach. *New Biotechnology* 31(1), 64–68.

Prakash, D., Verma, S., Bhatia, R. and Tiwary, B.N. (2011) Risks and precautions of genetically modified organisms. *ISRN Ecology* 1–13.

Preston, C.J. and Wickson, F. (2016) Broadening the lens for the governance of emerging technologies: Care ethics and agricultural biotechnology. *Technology in Society* 45, 48–57.

Punia, M.S. (2006) *International Agriculture Research–Initiatives and Ethics*. Department of Genetics and Plant Breeding, Chaudhary Charan Singh Haryana Agricultural University, Hisar, India.

Resnik, D.B. (2015) What is ethics in research and why is it important? National Institutes of Health. Available at: www.niehs.nih.gov/research/resources/bioethics/whatis (accessed 16 April 2025).

Ridwan, E. (2013) Ethics of the use of experimental animals in health research. *Journal of the Indonesian Medical Association: Majalah Kedokteran Indonesia* 63, 112–116.

Shelley-Egan, C. and Rodrigues, R. (2015) Ethics assessment and guidance at the European Union level. Annex 5.a Ethical assessment of research and innovation: A comparative analysis of practices and institutions in the EU and selected other countries. Available at: https://satoriproject.eu/media/5.a-EA-and-Guidance-at-the-EU-level.pdf (accessed 16 April 2025).

Si, T. and Zhao, H. (2016) A brief overview of synthetic biology research programs and roadmap studies in the United States. *Synthetic and Systems Biotechnology* 1(4), 258–264.

Syaifuddin, M. and Handayani, S. (2014) The relationship between law, morals and intellectual property rights (analysis of the legal and moral controversy over genetic engineering of living things in Indonesia). *Jurnal Dinamika Hukum* 14, 97–109.
Uzogara, S.G. (2000) The impact of genetic modification of human foods in the 21st century: A review. *Biotechnology Advances* 18(3), 179–206.
Varzakas, T. and Tzanidis, T. (2016) Genetically modified foods: Risk assessment, legislation, consumer behavior, and ethics. In: Caballero, B., Finglas, P.M. and Toldrá, F. (eds) *Encyclopedia of Food and Health*. Elsevier, New York, pp. 204–210.
von Braun, J. (2010) Food insecurity, hunger and malnutrition: Necessary policy and technology changes. *New Biotechnology* 27(5), 449–452.
Wasilah, U., Rohimah, S. and Su'udi, M. (2019) The development of biotechnology in Indonesia. *Rekayasa* 12(2), 85–90.
Watanabe, K.N., Taeb, M. and Okusu, H. (2004) Putting Cartagena into practice. *Nature Biotechnology* 22(10), 1207–1208.
Weale, A. (2010) Ethical arguments relevant to the use of GM crops. *New Biotechnology* 27(5), 582–587.
Zadoks, J.C. and Waibel, H. (2000) From pesticides to genetically modified plants: History, economics and politics. *NJAS: Wageningen Journal of Life Sciences* 48(2), 125–149.

Index

Abiotic stress, 4
  management, PGPR in, 46–47
  plant responses, 214–215
  tolerance, 16–19
Agrobacterium-mediated transformation, 195
Agrobacterium tumefaciens, 14
Agroecology, 151–154
  advantages, 153, 155–156
  circular economy, 161–162
  cocreation, 158
  culture and food traditions, 160–161
  diversification, 157–158
  efficiency, 159
  elements, 157–162
  human and social values, 160
  knowledge sharing, 158
  recycling, 159
  resilience, 159–160
  responsible governance, 161
  solidarity economy, 161–162
  and sustainability, 155–157
  synergies, 158–159
  technical element, 156
Agroforestry practices
  alley cropping (see Alley cropping)
  in vitro propagation, 89–92
Agronomic biofortification, 129–131
Alley cropping, 235–237
  above- and below-ground interaction, 241–242
  alley width, 238
  benefits, 242–244
  crop
    choice, 237
    establishment, 237–238
    management, 238–240
  future perspectives, 244
  impact on soil properties/conservation, 240–242
  limitations, 244
  nutrient balance, 239–240
  nutrient yield, 240
  pruning height, 238–239
  root pruning, 239
Ammonia borane-loaded hollow mesoporous silica nanoparticle (AB@hMSN), 69
Anther culture, 99–100
Antioxidant defense, in plants
  enzymatic antioxidants, 210–212
  nonenzymatic antioxidants, 212–214
Arabidopsis, 72, 73
Arbuscular mycorrhizal fungi (AMF), 166, 173–176
Argentine North Patagonia
  agricultural perspect, 114–117
  bioeconomy, 119–122
  biotechnology, 120, 121
  food security, 117–119
Ascorbate peroxidase, 211–212
Ascorbic acid, 212–213
Auxins production, 176

Bacillus thuringiensis, 2, 13
Bacteria, 170–172, 175
BIOALI network, 85–86
Bioethics, 250, 253
Biofertilizers, 131–132
Biofortification, 5, 20, 129
  agronomic, 129–131
  genetic approaches, 133–135
  microbe-mediated, 131–132
  wheat
    biotechnology, 135–139

Biofortification (*Continued*)
genome editing using CRISPR, 141–142
genomic selection (GS), 141
marker-assisted selection (MAS), 139–140
quantitative trait loci (QTL) mapping, 139–140
transgenic approaches, 141
Biolistic method, 193, 194
Biopolymers, 35–36
Biotechnology, 34–37, 92
BIOALI network, 85–86
sustainable food production, 84–85
carob trees, 88–89
chañar fruits, 89
in vitro propagation, 89–92
viroid-free plants, 86–88
virus-free plants, 86–88
for wheat biofortification, 135–139
genome editing using CRISPR, 141–142
genomic selection (GS), 141
marker-assisted selection (MAS), 139–140
quantitative trait loci (QTL) mapping, 139–140
transgenic approaches, 141
Biotic stress, 4, 19–20, 45–46
Bowdichia virgiloides, 91

Caffeine, 58
Camellia sinensis, 54–55
Carbon monoxide (CO), 72
Carbon nanotube carriers (CNT), 194–195
Carbon sequestration, 240, 243, 244
Carotenoids, 213
Cas13, 14
Catalase, 72, 211
Catechins, 55, 56, 59–60
Chitin, 35
Chitosan, 35
Chloroplast genome transformation, 192–196
Chloroplasts, 206
Chromosome doubling, 104
Chromosome elimination, in tobacco, 98, 99
Climate change, 3–5, 17, 29, 62, 84, 86, 113, 116, 117, 121, 153, 157, 158, 166, 236, 244
Climate-smart agriculture practices, 62
Cocreation process, 158
Colchicines, 104
Cold stress, 18–19
Common Agricultural Policy (CAP), 153
Conventional breeding, 133
Copper, 32–33
Cost/benefit analysis, 77
CRISPR-Cas9, 1, 3, 4, 11, 12–13
metal/metalloid smart plants engineering, 197–198
plant–microbe interactions, 44, 47–48
stress-related gene mutations, in tea plants, 55, 61
wheat biofortification, 142
Crop intensification, 235
Crop rotation, 30, 35, 36, 154, 155, 157
Crop yield, 8, 20, 73–74
Cry proteins, 13–14
CsMYB5 transcription factor, 58
Cysteine, 189

Dehydroascorbate reductase (DHAR), 212
Disease-suppressive soils, 31–32, 34
Diversification, 157–158
DNA, 210
Doubled haploid (DH), 97, 106–107
application, 104–106
generation, 98–99
origin, 98–99
in wheat, 99–100
Wheat × Maize System (*see* Wheat × Maize System)
Drought, 46
defined, 166
plant growth-promoting microorganisms (PGPM) (*see* Plant growth-promoting microorganisms (PGPM))
plant responses, 215
on plants, 167–170
resistance mechanism, 16–17, 168–169
water deficit, 166, 167

Effector-triggered immunity (ETI), 225, 227–232
Electrolysis, 69
Endophytes, 131
Environmental sustainability, 3
Ethics, 5, 249
in agricultural research, 250–251
for GM crops, 252–253
in transgenic crop production, 251–252
Ethylene, 171
Eukaryotic cells, 30, 31
Exopolysaccharides (EPS), 171, 176

Flavonoids, 213–214
Flooding, 215
Food culture, 160–161
Food insecurity, 8, 67
Food security, 5, 117–119
Food self-sufficiency, 117
Food tradition, 160–161
Fortification, 128–129
Free-living fungi, 172–173

Gene editing, 3
advancements, 1
CRISPR-Cas9, 12–13
plant–microbe interactions, 44
transcription activator-like effector nuclease (TALEN), 11–12
zinc finger nucleases, 11, 12
Genetically modified organisms (GMOs), 2, 55–56, 60
Genetic diversity, 129
Genetic engineering, 9
future prospect, 20–21
gene editing technologies
CRISPR-Cas9, 12–13
transcription activator-like effector nuclease (TALEN), 11–12
zinc finger nucleases, 11, 12
historical perspective, 9–10
for stress tolerance, 14–16
abiotic stress, 16–19
biofortification, 20
biotic stress, 19–20
cold stress, 18–19
drought resistance, 16–17
pest and disease resistance, 20–21
salinity tolerance, 17–18
temperature, 18–19
toxic metal tolerance, 19
techniques, 9–11
transgenic technology, 13–14
for wheat biofortification, 141
Genetic modification (GM), 2, 5
crops ethics and regulatory policies, 252–253
Genetic variation, 9
Genome-wide association studies (GWAS), 139
Genomic selection (GS), 141
Geoffroea decorticans, 89, 90
Glomalin production, 174
Glutaredoxins (Grxs), 192
Glutathione reductase, 212
Glutathione, 213
Grazing, 159
Green Revolution, 8, 9, 155
Guaiacol peroxidase, 211

Haploids, 97–98
Heat stress, 18–19
Heavy metals (HMs), 185, 186, 188, 214–215
Heavy metal toxicity, 19, 47
Heme oxygenase-1 (HO-1), 72
Herbicide-resistant crops, 14
High-affinity potassium transporters (HTKs), 18–19
High-throughput sequencing techniques, 61
Homology-directed repair (HDR), 12
Hydrogen nanobubble water (HNW), 69, 73
Hydrogen-rich water (HRW), 68–69, 68–70, 74, 76, 77
Hypsizygus marmoreus, 72

In vitro cultivation, 97
Indole-3-acetic acid (IAA), 170
Induced doubling, 104
Iron, 32
agronomic biofortification, 129–131
bioavailability, 142–143
deficiency, 127
phytic acid and phytase ratio, 142–143

Jasmonic acid, 227

Knowledge sharing, 158

Land equivalent ratio (LER), 159
Lipids, 208
Livestock husbandry, 159
Lysin motif receptor-like kinases (LysM-RLKs), 48

Manganese, 33
Marker-assisted breeding, 2–3
Marker-assisted selection (MAS), 139–140
Metabolic engineering, tea. *See* Tea plants
Metabolic flux modeling, 58–59
Metagenomics, 36–37
Metalloid metabolism, 187
Metallothioneins (MT), 187, 190–192
Metals/metalloids transporter genes
antioxidant defense machinery, 190–192
cadmium, 186–188
chloroplast genome transformation, 192–196
Cr, 188
metallothioneins (MTs), 190, 191
phytochelatins (PCs), 189, 191
phytoremediation, 186–187, 192–196
remediation, transgenic plants in, 196–197
selenium uptake, 187
silicon (Si) uptake, 187
smart plants engineering, 197–198
sulfur-containing metabolites, 189
trace metals, 188
uptake system, 187
Microbe-mediated biofortification, 131–132
Microbiome, 29, 30, 32–33, 37
assemblage, 47
in nutrient cycling, 32–33
Micronutrients, 126–127
microbe-mediated biofortification, 131–132
solubilization, 132
Mitochondria, 207
Molecular breeding, 133
Molecular hydrogen (H_2), 67, 68, 77, 78
crop yields, 73–74
of fresh material, 74, 75
fumigation, 68, 74
hydrogen nanobubble water (HNW), 69, 73

Manganese (*Continued*)
- hydrogen-rich water (HRW), 68–70, 74, 76, 77
- oxy-hydrogen gas, 70
- plant growth enhancement, 70–73
- post harvest treatment, 74–78
- of processed food products, 76–77
- treatment methods, 68–69

Monodehydroascorbate reductase, 212
MYB transcription factor, 58
Mycorrhizal fungi, 131

Nanomaterials, 13, 33, 34
Nanotechnology, 33–34, 37, 69, 186
Neltuma alpataco, 88–90
Neltuma caldenia, 89, 90
Neltuma flexuosa, 88–90
Next-generation sequencing (NGS), 36–37
Nitrogen, 32, 171–172
Nonhomologous end-joining (NHEJ) repair, 12–13
Nutrient cycling, 32–33
Nutrient recycling, 155
Nutritional enhancement, 3–4
Nutritional value, 5

Omics, 36–37, 61
Organic farming practice
- environmental benefits, 154

Organic matter, 34–35, 155
Osmolytes, 171
Oxy-hydrogen gas, 70

Pathogenesis-related (PR) proteins, 224–225
Pattern recognition receptors (PRRs), 226–227
Pattern-triggered immunity (PTI), 225–227
Peroxidase, 72
Peroxisomes, 207–208
Pest regulation, 155
Pest resistance, 20–21
Phenolic compounds, 119
Phosphorus, 171, 174
Phytic Acid, 142–143
Phytoalexins, 224
Phytochelatins (PCs), 189, 191
Phytohormone, 132, 175, 227
Phytohormone therapy, 103
Phytoremediation, 186–187, 192–196
Plant defense mechanisms, 228
- cellular defense mechanisms, 224
- effector-triggered immunity (ETI), 225, 227–232
- induced biochemical defense mechanisms, 224–225
- pattern-triggered immunity (PTI), 225–227
- postinfectional/induced, 223–224
- preexisting biochemical defense, 222–223
- preexisting structural defense, 222

Plant growth-promoting fungi (PGPF), 172–173
Plant growth-promoting microorganisms (PGPM), 166–167, 177, 178
- benefits, 176–177
- drought stress mitigation, 170
  - arbuscular mycorrhizal fungi (AMF), 173–175
  - bacteria, 170–172
  - free-living fungi, 172–173
- plant growth promotion, 175–177

Plant growth-promoting rhizobacteria (PGPR), 32, 131
- in abiotic stress management, 46–47
- in biotic stress management, 45–46
- heavy metal toxicity, 47

Plant–microbe interactions, 44
- biotechnologically improved, 44
- CRISPR-Cas9, 47–48
- genome editing techniques, 44
- plant-associated microbiome
  - assemblage, 47
  - commensalistic interactions, 45
  - mutualistic relationship, 45
  - parasitic interactions, 45
  - plant growth-promoting rhizobacteria (PGPR), 45–47
  - probiotics, 48–50
- plant growth-promoting rhizobacteria (PGPR)
  - in abiotic stress management, 46–47
  - in biotic stress management, 45–46

Plant resistance
- effector-triggered immunity (ETI), 225, 227–232
  - molecular basis, 230–231
  - pathogen effectors, 228–230
  - and systemic acquired resistance, 231–232
- molecular basis, 225
- pattern-triggered immunity (PTI), 225–226
  - phytohormones, 226–227
  - plant pattern recognition receptors (PRRs), 226–227
  - signal transduction mechanism, 226

Plant tissue culture (PTC), 120
Plasma membrane, 208
Pollination, 103
Polyethylene glycol-mediated transformation, 193, 194
Probiotics, 48–50
Prokaryotic cells, 30, 31
Proline, 171, 214
Proteins, 209–210
Pruning height, 238–239
Prunus sp., 86–87
Pseudomonas fluorescens, 46
Public engagement, 5
Public perception, 5

Quantitative trait loci (QTL) mapping, 139–140

Reactive oxygen species (ROS), 17, 166, 205–206, 216
- biomolecules targeted by, 208, 209
- chloroplasts, 206
- DNA, 210
- generation, 206, 207
- intracellular signaling, 226
- lipids, 208
- mitochondria, 207
- peroxisomes, 207–208
- plasma membrane, 208
- proteins, 209–210

Recycling, 159
Regulatory challenges, 5
Resilience, 159–160
Respiratory burst oxidase homolog D (RBOHD), 226
Rhizobacteria, 131
Rhizophagus intraradices, 174
RNA interference (RNAi), 14, 142–143
Root pruning, 239

Salinity, 17–18, 46, 215
Selenium uptake, 187
Sexual crosses, 97
Siderophores, 132
Silicon (Si) uptake, 187
Soil health, 29–31
- biotechnology, 34–37
- defined, 29
- energy, 29
- eukaryotic cells, 30, 31
- with functional microbial biodiversity, 34
- microbiome, 29, 30, 32–33, 37
- microorganisms, 30, 31
- nanotechnology, 33–34
- prokaryotic cells, 30, 31
- soil management practices, 30

Soil management practices, 30
Soil microbiome, 29, 30
Spermidine (Spd), 192
Superoxide dismutase (SOD), 210
Sustainable agriculture, 151, 154
- components, 155
- environmental benefits, 154
- plant biotechnology, 3–4
- transition to, 155

Sustained crop production, 155
Symbiotic microbes, 48, 49
Synergies, 158–159

Tea plants
- caffeine, 58
- catechins, 55, 56
- challenges, 55–56
- CRISPR-Cas9, 55
- flavor, 55, 60
- metabolic engineering, 55, 57
  - advancement, 61–62
  - biosynthetic genes, overexpression of, 58
  - case studies, 59–60
  - challenges, 60–61
  - metabolic flux modeling, 58–59
  - process, 59
  - in silico approaches, 58–59
  - transcription factors, 58
- terpene biosynthesis, 55
- theaflavins, 57
- theanine, 58

Terpene biosynthesis, 55
Tetrastigma hemsleyanum, 72
Theaflavins, 57
Theanine, 58
α-Tocopherols, 213
Trace metals, 188
Transcription activator-like effector nuclease (TALEN), 11–12
Transcription factors, 48
Transgenic crops, 251–252
Transgenic technology, 13
Transparent communication, 5
Trehalose, 171
Trichoderma, 35
Triticum, 98

UV laser microbeam, 193, 195

Vegetative cover, 155
Vegetative insecticidal proteins (Vips), 13
Volatile organic compounds (VOCs), 172

Water deficit, 166, 167, 175, 177
- drought (*see* Drought)

Wheat × Maize System, 97
- anther culture, 99–100
- benefits, 100
- doubled haploids, 98–99
- environmental factors, 102–103
- genotypes impact, 101–102
- hormone treatment, 103–104
- origin, 100
- phytohormone therapy, 103
- pollination, 103
- principle, 100
- research, advances in, 100–101

Yeasts, 172

Zinc
agronomic biofortification, 129–131
bioavailability, 142–143
deficiency, 127
phytic acid and phytase ratio, 142–143
Zinc finger nucleases (ZFNs), 11, 12

Printed and bound by CPI Group (UK) Ltd, Croydon, CR0 4YY

13/09/2026

14961117-0003